Lecture Notes
in Business Information Processing   **331**

Series Editors

Wil van der Aalst
RWTH Aachen University, Aachen, Germany
John Mylopoulos
University of Trento, Trento, Italy
Michael Rosemann
Queensland University of Technology, Brisbane, QLD, Australia
Michael J. Shaw
University of Illinois, Urbana-Champaign, IL, USA
Clemens Szyperski
Microsoft Research, Redmond, WA, USA

More information about this series at http://www.springer.com/series/7911

Gerhard Satzger · Lia Patrício
Mohamed Zaki · Niklas Kühl
Peter Hottum (Eds.)

Exploring
Service Science

9th International Conference, IESS 2018
Karlsruhe, Germany, September 19–21, 2018
Proceedings

 Springer

Editors
Gerhard Satzger
Karlsruhe Service Research Institute (KSRI)
Karlsruhe Institute of Technology
Karlsruhe
Germany

Niklas Kühl
Karlsruhe Service Research Institute (KSRI)
Karlsruhe Institute of Technology
Karlsruhe
Germany

Lia Patrício
INESC TEC
Universidade do Porto
Porto
Portugal

Peter Hottum
Karlsruhe Service Research Institute (KSRI)
Karlsruhe Institute of Technology
Karlsruhe
Germany

Mohamed Zaki
Institute for Manufacturing
University of Cambridge
Cambridge
UK

ISSN 1865-1348   ISSN 1865-1356   (electronic)
Lecture Notes in Business Information Processing
ISBN 978-3-030-00712-6   ISBN 978-3-030-00713-3   (eBook)
https://doi.org/10.1007/978-3-030-00713-3

Library of Congress Control Number: 2018954926

This Springer imprint is published by the registered company Springer Nature Switzerland AG
The registered company address is: Gewerbestrasse 11, 6330 Cham, Switzerland

Preface

Services comprise over 60% of the gross world product, and represent the fastest-growing sector in emerging economies. On the academic side, the field of Service Science unites researchers in the search for new knowledge and solutions across academic disciplines and across individual institutions ("Service Systems"). Starting with the conference 1.0 in 2010, the series of International Conferences on Exploring Service Science, IESS, has provided a forum for sharing interesting and noteworthy achievements and developments in this field.

In this volume of "Exploring Service Science", we have collected the peer-reviewed papers of IESS 1.8, organized September 19–21, 2018, by the Karlsruhe Service Research Institute (KSRI) at the Karlsruhe Institute of Technology (KIT) in Germany. The event gathered academic scientists and practitioners from the service industry and related disciplines in a collegial and inspiring environment. According to its tradition, IESS 1.8 covered major research and development areas related to Service Science foundations under the theme of "Services in the Digital Era". For the first time, the event was organized in multiple tracks: Service Design and Innovation, Smart Service Processes, Service Business Models, Big Data in Services, Service Exploration, as well as Design Science Research for Services.

A total of 67 submissions were received from authors from 14 countries, out of which 30 quality full papers were selected in a double-blind review process. All submissions were reviewed by at least two members of the International Conference Program Committee, composed of service science experts from over 20 countries.

The book is structured in six parts, based on the six main tracks of the conference. Each track features contributions describing current research in a particular domain of service science. Part 1, Service Design and Innovation, deals with research on the design of new service offerings and customer experience. In part 2, Smart Service Processes, the necessary environments for smart services in the field of IoT are discussed, focusing on the technologies, data models, and optimization tasks. In the third part, Service Business Models, it is discussed how the increasing importance of digital technologies has given rise to new services and, consequently, new kinds of service business models. Part 4, Big Data in Services, covers insights from the application of machine learning and artificial intelligence in different use cases of digital services. Part 5, Service Topics Open Exploration, deals with explorative research in the field of service science, introducing new frameworks, methods, and IT tools. Finally, part 6, Design Science Research in Services, focuses on the development of artifacts for value co-creation by applying methodologies from the field of design science research. Thus, the book offers an extended, ICT-focused vision on services and addresses multiple relevant aspects, including underlying business models, the necessary processes, as well technological capabilities like big data and machine learning. The academic work showcased at the conference should help to advance service science and its application in practice.

We would like to thank the IESS Steering Committee for their support and the opportunity to host this conference. Furthermore, we owe special thanks to the local organizing team at the Karlsruhe Institute of Technology (KIT). Finally, we thank the keynote speakers, Werner Kunz and Gerhard Pfau, for their important contributions to this conference.

We wish you pleasant reading and hope you find the insights from the multifaceted domain of service science both relevant and useful for your future research endeavors.

August 2018

Gerhard Satzger
Lia Patrício
Mohamed Zaki
Niklas Kühl
Peter Hottum

Organization

Program Chairs

Gerhard Satzger	Karlsruhe Institute of Technology (KIT), Germany
Lia Patrício	University of Porto, Portugal
Mohamed Zaki	University of Cambridge, UK

Local Chairs

Niklas Kühl	Karlsruhe Institute of Technology (KIT), Germany
Peter Hottum	Karlsruhe Institute of Technology (KIT), Germany

Track Chairs

Service Design and Innovation

Lia Patrício	University of Porto, Portugal
Niels Feldmann	Karlsruhe Institute of Technology (KIT), Germany

Smart Service Processes

Leonard Walletzky	Masaryk University, Czech Republic
Maria Maleshkova	University of Bonn, Germany

Service Business Models

Tilo Böhmann	University of Hamburg, Germany
Ronny Schüritz	Karlsruhe Institute of Technology (KIT), Germany

Big Data in Services

Mohamed Zaki	University of Cambridge, UK
Niklas Kühl	Karlsruhe Institute of Technology (KIT), Germany

Service Topics Open Exploration

Monica Drâgoicea	University Politehnica of Bucharest, Romania
Henriqueta Nóvoa	University of Porto, Portugal
Melanie Reuter-Oppermann	Karlsruhe Institute of Technology (KIT), Germany

Design Science Research in Services

Tuure Tuunanen	University of Jyväskylä, Finland
Stefan Morana	Karlsruhe Institute of Technology (KIT), Germany

IESS Steering Committee

Theodir Borangiu	University Politehnica of Bucharest, Romania
Fabrizio D'Ascenzo	Sapienza, University of Rome, Italy
Monica Drâgoicea	University Politehnica of Bucharest, Romania
Eric Dubois	Luxembourg Institute of Science and Technology (LIST), Luxembourg
João Falcão e Cunha	University of Porto, Portugal
Michel Léonard	University of Geneva, Switzerland
Marco De Marco	Uninettuno University, Italy
Henriqueta Nóvoa	University of Porto, Portugal
Mehdi Snene	University of Geneva, Switzerland

IESS General Chair

Michel Léonard	University of Geneva, Switzerland

IESS International Program Committee

Sabrina Bonomi	eCampus University, Italy
Theodir Borangiu	University Politehnica of Bucharest, Romania
Antonio Brito	University of Porto, Portugal
Jorge Cardoso	University of Coimbra, Portugal
Fabrizio D'Ascenzo	Sapienza, University of Rome, Italy
María Valeria De Castro	Universidad Rey Juan Carlos, Spain
Sergio Cavalieri	University of Bergamo, Italy
Valentin Cristea	University Politehnica of Bucharest, Romania
Monica Drâgoicea	University Politehnica of Bucharest, Romania
Jose Faria	University of Porto, Portugal
Teresa Fernandes	University of Porto, Portugal
Isabel Horta	University of Porto, Portugal
Peter Hottum	Karlsruhe Institute of Technology (KIT), Germany
Manuele Kirsch-Pinheiro	Université Paris 1 Panthéon Sorbonne, France
Natalia Kryvinska	University of Vienna, Austria
Weiping Li	Peking University, China
Paul Lillrank	Aalto University, Finland
Paul Maglio	University of California, Merced, USA
Marco De Marco	Uninettuno University, Italy
Vera Migueis	University of Porto, Portugal
Jean-Henry Morin	University of Geneva, Switzerland
Henriqueta Nóvoa	University of Porto, Portugal
Lia Patrício	University of Porto, Portugal
Tomas Pitner	Masaryk University, Czech Republic
Geert Poels	Ghent University, Belgium
Jolita Ralyté	University of Geneva, Switzerland

Shai Rozenes   Afeka Tel Aviv Academic College of Engineering, Israel
Gerhard Satzger   Karlsruhe Institute of Technology (KIT), Germany
Miguel Mira Da Silva   Instituto Superior Técnico, Portugal
Mehdi Snene   University of Geneva, Switzerland
Maddalena Sorrentino   University of Milan, Italy
Zhongjie Wang   Harbin Institute of Technology, China
Adi Wolfson   Shamoon College of Engineering, Israel
Stefano Za   eCampus University, Novedrate (CO), Italy

Conference Program Committee

Sabrina Bonomi   eCampus University, Italy
Theodor Borangiu   University Politehnica of Bucharest, Romania
Antonio Brito   University of Porto, Portugal
Jorge Cardoso   University of Coimbra, Portugal
Fabrizio D'Ascenzo   Sapienza, University of Rome, Italy
María Valeria De Castro Universidad Rey Juan Carlos, Spain
Sergio Cavalieri   University of Bergamo, Italy
Valentin Cristea   University Politehnica of Bucharest, Romania
Monica Drâgoicea   University Politehnica of Bucharest, Romania
David Diaz   Universidad de Chile, Chile
Philipp Ebel   University of St. Gallen, Switzerland
Jose Faria   University of Porto, Portugal
Teresa Fernandes   University of Porto, Portugal
Isabel Horta   University of Porto, Portugal
Peter Hottum   Karlsruhe Institute of Technology (KIT), Germany
Patrick Jochem   Karlsruhe Institute of Technology (KIT), Germany
Julia Jonas   University of Erlangen-Nuremberg, Germany
Manuele   Université Paris 1 Panthéon Sorbonne, France
  Kirsch-Pinheiro
Natalia Kryvinska   University of Vienna, Austria
Thang Le Dinh   Université du Québec à Trois-Rivières, Canada
Gréanne Leeftink   University of Twente, The Netherlands
Jan Marco Leimeister   Universität Kassel, Germany
Vera Migueis   University of Porto, Portugal
Benjamin Mueller   University of Groningen, The Netherlands
Christoph Peters   University of St. Gallen, Switzerland
Geert Poels   Ghent University, Belgium
Jens Poeppelbuss   Ruhr-Universität Bochum, Germany
Sebastian Rachuba   University Wuppertal, Germany
Laleh Rafati   Ghent University, Belgium
Jolita Ralyté   University of Geneva, Switzerland
Shai Rozenes   Afeka Tel Aviv Academic College of Engineering, Israel
Gerhard Satzger   Karlsruhe Institute of Technology (KIT), Germany
Susan Sherer   Lehigh University, USA
Maddalena Sorrentino   University of Milan, Italy

Jost Steinhäuser	Universitätsklinikum Schleswig-Holstein, Germany
Maartje van de Vrugt	University Hospital Leiden and University Twente, The Netherlands
Adi Wolfson	Shamoon College of Engineering, Israel
Stefano Za	eCampus University, Novedrate (CO), Italy
Anne Zander	Karlsruhe Institute of Technology (KIT), Germany
Andreas Zolnowski	University of Hamburg, Germany

Contents

Service Design and Innovation

The Effect of Service Modularity on Flexibility in the Digital Age – An
Investigation in the B2B Context 3
  Torben Stoffer, Thomas Widjaja, and Nicolas Zacharias

Modular Sales – Using Concepts of Modularity to Improve the Quotation
Process for B2B Service Providers 16
  Aleksander Lubarski

Omni-Channel Service Architectures in a Technology-Based Business
Network: An Empirical Insight 31
  João Reis, Marlene Amorim, and Nuno Melão

An Approach for Customer-Centered Smart Service Innovation Based
on Customer Data Management 45
  Katharina Blöcher and Rainer Alt

Closing the Gap Between Research and Practice – A Study on the Usage
of Service Engineering Development Methods in German Enterprises 59
  Simon Hagen, Sven Jannaber, and Oliver Thomas

Customers Input via Social Media for New Service Development 72
  Intekhab (Ian) Alam

Employee-Centric Service Innovation: A Viable Proxy
for Customer-Intimacy for Product-Focused Enterprises 88
  *Michael Vössing, Jörg Siegel, Niels Feldmann, Thorsten Wuest,
  and Carina Benz*

Towards Managing Smart Service Innovation: A Literature Review........ 101
  Caroline Götz, Sophie Hohler, and Carina Benz

Open Innovation in Ecosystems – A Service Science Perspective
on Open Innovation. 112
  Carina Benz and Stefan Seebacher

Smart Service Processes

Crowdsensing-Based Road Condition Monitoring Service: An Assessment
of Its Managerial Implications to Road Authorities 127
  Kevin Laubis, Florian Knöll, Verena Zeidler, and Viliam Simko

Digitalization of Field Service Planning: The Role of Organizational
Knowledge and Decision Support Systems........................ 138
  Michael Vössing, Clemens Wolff, and Volkmar Reinerth

Co-creation in Action: An Acid Test of Smart Service Systems Viability.... 151
  Francesco Polese, Luca Carrubbo, Francesco Caputo,
  and Antonietta Megaro

Towards Enabling Cyber-Physical Systems in Brownfield Environments:
Leveraging Environmental Information to Derive Virtual Representations
of Unconnected Assets...................................... 165
  Sebastian R. Bader, Clemens Wolff, Michael Vössing,
  and Jan-Peter Schmidt

Market Launch Process of Data-Driven Services for Manufacturers:
A Qualitative Guideline..................................... 177
  Achim Kampker, Marco Husmann, Philipp Jussen, and Laura Schwerdt

Service Business Models

Success Factors of SaaS Providers' Business Models – An Exploratory
Multiple-Case Study 193
  Sebastian Floerecke

End-to-End Methodological Approach for the Data-Driven Design
of Customer-Centered Digital Services......................... 208
  Jürg Meierhofer and Anne Herrmann

Utilizing Data and Analytics to Advance Service: Towards Enabling
Organizations to Successfully Ride the Next Wave of Servitization........ 219
  Fabian Hunke and Christian Engel

Big Data in Services

Forecast Correction Using Organizational Debiasing in Corporate Cash
Flow Revisioning 235
  Florian Knöll and Katerina Shapoval

A Framework for the Simulation-Based Estimation of Downtime Costs 247
  Clemens Wolff and Michael Voessing

Combining Machine Learning and Domain Experience: A Hybrid-Learning
Monitor Approach for Industrial Machines...................... 261
  Daniel Olivotti, Jens Passlick, Alexander Axjonow, Dennis Eilers,
  and Michael H. Breitner

Exploring the Value of Data – A Research Agenda.   274
  Tobias Enders

Service Topics Open Exploration

Investigating the Alignment Between Web and Social Media Efforts
and Effectiveness: The Case of Science Centres   289
  *Marlene Amorim, Fatemeh Bashashi Saghezchi, Maria João Rosa,
  and Pedro Pombo*

Exploring Customers' Internal Response to the Service Experience:
An Empirical Study in Healthcare.   303
  Gabriela Beirão and Humberto Costa

Health Information Technology and Caregiver Interaction: Building
Healthy Ecosystems.   316
  Nabil Georges Badr, Maddalena Sorrentino, and Marco De Marco

Service Science Research and Service Standards Development   330
  Reinhard Weissinger and Stephen K. Kwan

From Data to Service Intelligence: Exploring Public Safety as a Service   344
  *Monica Drăgoicea, Nabil Georges Badr, João Falcão e Cunha,
  and Virginia Ecaterina Oltean*

Managing Patient Observation Sheets in Hospitals Using Cloud Services. . . .   358
  *Florin Anton, Theodor Borangiu, Silviu Raileanu, Iulia Iacob,
  and Silvia Anton*

Design Science Research in Services

Bringing Design Science Research to Service Design   373
  Jorge Grenha Teixeira, Lia Patrício, and Tuure Tuunanen

Scaling Consultative Selling with Virtual Reality: Design and Evaluation
of Digitally Enhanced Services.   385
  Osmo Mattila, Tuure Tuunanen, Jani Holopainen, and Petri Parvinen

Designing Value Co-creation with the Value Management Platform   399
  Geert Poels, Ben Roelens, Henk de Man, and Theodoor van Donge

Author Index   415

Service Design and Innovation

The Effect of Service Modularity on Flexibility in the Digital Age – An Investigation in the B2B Context

Torben Stoffer[1](✉), Thomas Widjaja[1], and Nicolas Zacharias[2]

[1] University of Passau, Innstraße 41, 94032 Passau, Germany
{torben.stoffer, thomas.widjaja}@uni-passau.de
[2] TU Darmstadt, Karolinenplatz 5, 64289 Darmstadt, Germany
zacharias@bwl.tu-darmstadt.de

Abstract. The goal of this study is to investigate the moderating role of digitalization on the well-known positive effect of service modularity on service flexibility. This is important since research findings on the role of service digitalization in this context are scarce and still equivocal. Following research on digital business strategy, we propose and provide empirical evidence that service digitalization positively moderates the effect of service modularization on service flexibility. By doing this, we furthermore enhance this research by considering service digitalization as a continuum ranging from low (i.e., services mainly provided by personnel) to high (i.e., services mainly provided by IT). In addition, we show that service flexibility has a positive effect on service value which is an important factor for firms' market success. Hereby we aim to contribute to research on service modularization and technology management. Our research is based on survey-data of 147 companies offering IT services in the B2B context and is analyzed using the partial least square method.

Keywords: Service modularity · Service flexibility · Service digitalization B2B · PLS

1 Introduction

In recent years, suppliers of business-to-business (B2B) services have continuously increased the number of digital services in their portfolios by creating new digital services or altering the degree of digitalization of existing services. An example for the alteration of existing services are the McKinsey Solutions which comprise services (e.g., assessment of firms' competitive position) that have traditionally been performed by consultants (i.e., low degree of digitalization) and are now offered completely digitally. Due to this development, it is of high practical and theoretical relevance to understand which parts of our knowledge on service design can be transferred to or have to be adjusted in the context of digital services. Our study aims to contribute to this endeavor by focusing on the moderating role of digitalization on the well-known effect of service modularity on service flexibility (cf., Fig. 1 for our research model) [1].

In this study, we propose to conceptualize service digitalization a continuum ranging from no digitalization to completely digital services. This is in line with, for

© Springer Nature Switzerland AG 2018
G. Satzger et al. (Eds.): IESS 2018, LNBIP 331, pp. 3–15, 2018.
https://doi.org/10.1007/978-3-030-00713-3_1

example, Froehle and Roth [2] as well as Böhmann et al. [3] who suggest that services are provided by making use of personnel and/or IT resources. Following this way of thinking, we define service digitalization as the degree to which the service is provided by IT instead of personnel. Surprisingly, extant literature often has treated service digitalization as a binary concept and focused only on general changes by a high degree of service digitalization (e.g., customer self-service) in comparison to traditional service provisions (e.g., as a personal service) [4, 5]. Therefore, this more nuanced view on service digitalization offers possibilities to add to the understanding of the effects of different degrees of service digitalization.

As stated above, we are focusing on the moderating role of service digitalization on the effect of service modularity on service flexibility. In line with Baldwin and Clark [6] and Vickery et al. [7], we define *service modularity* as the degree to which services consist of service modules that are designed independently to offer a specific functionality. An example for the modularization of a digital service is the analysis of big data by Amazon Web Services. The service modules comprise different analytic frameworks and databases which are used in combination to provide the complete service (i.e., big data analysis). Beside other effects of service modularity (e.g., reduction of cost [8], complexity [9], and risks [1]), extant literature especially highlights the positive effect of service modularization on service flexibility [10]. *Service flexibility* is defined as the suppliers' extent of possibilities to provide different services [11–14].

However, the influence of service digitalization on the relationship between service modularization and service flexibility remains unclear. On the one hand, extant literature highlights the positive effect of service modularization on service flexibility both for services with a low degree of digitalization [15] and services with a high degree of digitalization [16]. On the other hand, literature in the context of digital business strategy posits that modularization offers unprecedented magnitudes of flexibility in combination with digitalization [17, 18]. Therefore, it remains unclear if and how service digitalization influences service modularity. This leads to our research question:

RQ: Does service digitalization moderate the effect between service modularity and service flexibility?

We are aware that service flexibility itself cannot be a primary objective for service suppliers. Therefore, to underscore the practical relevance of our research, we include service value as an important effect of service flexibility in our research model. Following Stock and Zacharias [19], we define s*ervice value* as the superiority of a service in terms of its quality and benefits for the customer. As customer needs are heterogeneous and change over time especially in the B2B context [20], suppliers have to offer services of high value to succeed in the market.

With our research we aim to contribute in three ways. First, we contribute to technology and innovation management research [4, 17, 18, 21], by elaborating on the interplay of service modularization and service digitalization. In particular, we provide insights following the conceptual thoughts of Yoo [17] and Bharadwaj et al. [18]. As services can take various degrees of digitalization, we provide generalized insights on the effect of digitalization on service modularity, which is in line with Nambisan et al. [4] and Iman [21]. Second, we add to the growing research of service modularization [22] by enhancing the understanding of the effect of service modularization on service flexibility in the B2B context. Third, our results help practitioners offering services in the B2B context as well. Suppliers that want to maximize the success of their services can benefit from the results by reconsidering the modularization of their services against the background of the services' degree of digitalization.

The remainder of this paper is structured as follows: The next section introduces the extant research on service modularity and the conceptualization of service digitalization. Then, the research model is developed. The fourth section describes the survey-based sample, comprising 147 companies offering B2B IT services, and the constructs' conceptualization. Afterwards, the research model is assessed. The paper concludes with the discussion, limitations, and recommendations for future research.

2   Conceptual Background

The conceptual background of this paper is divided in two sub-sections. First, we discuss the extant literature on service modularity and, second, we introduce the concept of service digitalization.

2.1   Service Modularity

We build upon early work of Sundbo [23], where he introduces the concept of service modularity and proposes that service modularity could ease the trade-off between standardization and customization. In line with Baldwin and Clark [6] as well as Vickery et al. [7], we define *service modularity* as the degree to which services consist of service modules that are designed independently to offer a specific functionality. Hence, a higher degree of service modularity can be achieved by breaking down services into self-contained service modules [21]. Then, these service modules can be flexibly recombined to provide the respective service which is also known as mixing-and-matching [22, 24, 25].

Extant literature has highlighted various effects of service modularity (e.g., fostering innovation, effective division of labor, mitigating the risks of service adoption, and enhancing customization). Due to the flexible recombination of service modules, suppliers have various options to compose innovative services and can avoid the re-invention of already existing service modules. Hence, service modularity fosters innovation [22] and enables suppliers to effectively divide labor among different actors [17]. For example, a consulting service offering a specific strategic planning for a customer could be divided into service modules regarding the consultants' technical skills which are necessary for the consecutive phases of the service provision (i.e., fact gathering, data analysis, and strategy definition). Thus, this improvement, achieved by service modularity, reduces costs in operations as well as functionality [9]. Xue et al. [1] argue that service modularity mitigates the risk of adopting digital supply chain services by reducing the risks perceived by organizational decision makers regarding the desirable outcomes of the services. One of the most important effects of service modularity highlighted in literature is service flexibility [10]. *Service flexibility* is defined as the suppliers' extent of possibilities to provide different services [11–14] and is achieved by the flexible recombination of service modules [26, 27].

2.2 Service Digitalization

Froehle and Roth [2] classify services based on the role of technology used during the service provision. Analogously Böhmann et al. [3] state that personnel and/or IT resources are used during service provisions. Based on these conceptualizations, we define *service digitalization* as the degree to which the service is provided by IT instead of personnel. Hence, the degree of service digitalization can range from *low*, where the service is mainly provided by personnel (e.g., a consulting service, where the service provision consists of the consultants' work in the first place), to *high*, where the service is mainly provided by IT (e.g., software as a service, which is offered as a self-service with the result that, on the supplier's side, mainly soft- and hardware is involved in the service provision). Additionally, as our conceptualization of service digitalization is a continuum, it can take all intermediate forms between the two anchors (e.g., a project management service, which consists of a consultant's work and a complementary software which is operated by the supplier and used by the customer).

Literature has identified different effects of service digitalization which are related to our research model. Conceptual literature has emphasized the possibilities of digital services for service flexibility [17]. This flexibility of digital services can be achieved by a rapid recombination of service modules without sacrificing cost or quality [25]. The same idea has been pursued by Sambamurthy et al. [28] who state in the domain of organizational IT that suppliers can succeed in competition through agility which is inherent in IT.

3 Research Model and Hypotheses Development

The research model, as illustrated in Fig. 1, contains three hypotheses which are explained in the following. In addition, we include three control variables (i.e., investment cost, firm age, revenue) for our focal construct service flexibility.

Service modularity reflects the degree to which services consist of service modules that are designed independently to offer a specific functionality [6, 7]. For specific service provisions, these distinct service modules are recombined to provide the respective services that are offered to customers [22, 24]. This recombination, also known as mixing-and-matching, comprises the selection of different service modules and/or service modules' sequences [8, 25]. Hence, by making use of service modularity, suppliers increase the flexibility of their service offerings [29]. As the introductory Amazon Web Services example illustrates, the analysis of big data is separated into different service modules (e.g., different analytic frameworks and databases). The different analytic frameworks (e.g., Amazon EMR, Amazon Elasticsearch Service) are combined with different databases (e.g., Amazon DynamoDB, Amazon RDS) to provide the service (i.e., big data analytics). As a consequence, Amazon Web Services achieve a high service flexibility through service modularity. Hence, we hypothesize:

H1: Service modularity is positively associated with service flexibility.

Literature has found a positive effect of service modularity on service flexibility both for services with a low [15] and high degree of digitalization [16], but neglects the

possible moderating role of service digitalization on the effect of service modularization. Conceptual literature on service digitalization keeps emphasizing that digital
services offer unpreceded possibilities of service flexibility [17] in comparison to
services provided by personnel. To address this equivocal relationship, we propose a
positive moderating effect of service digitalization on the relationship between service
modularization and service flexibility. Service digitalization reflects the degree to
which the service is provided by IT instead of personnel. Especially, services provided
by IT to a major part enable suppliers to rapidly recombine service modules without
sacrificing cost or quality [25]. The positive effect of service digitalization on the
relationship between service modularity and service flexibility (cf., H1) can, for
example and among others, be achieved by time- and location-independence of the
service provision, service scalability, and possibilities of automatic recombination of
the service modules. Scalability, which is inherent in digital services, facilitates the
provision of services for a growing number of customers without an increase in cost.
Hence, it increases the service flexibility for a given level of service modularization
[16]. Chan et al. [30] have shown that customer participation to foster service customization can create job stress when the services are provided by personnel because of
their loss of power and control, increased input uncertainties, and incompatible
demands and expectations. The same accounts for service modularity and leads to a
stronger increase in service flexibility of services provided mainly by IT in comparison
to services provided mainly by personnel. Hence, the effect of service modularity on
service flexibility is larger for services with a high degree of digitalization than for
services with a low degree of digitalization. This leads to our second hypothesis:

> H2: An increase in service digitalization positively moderates the effect of service
> modularity on service flexibility.

To succeed in the market, services have to generate value for the customers [5].
Service value reflects the superiority of a service in terms of its quality and benefits for
the customer [19]. As, especially in the B2B context, customer needs are heterogeneous
and change over time [20], service flexibility enables suppliers to offer superior services. This is achieved by the suppliers' increased responsiveness to misalignments
during service provisions and to new market opportunities [31]. Thus, we hypothesize:

> H3: Service flexibility is positively associated with service value.

4   Methodology

A survey among representatives of 147 IT suppliers was conducted in September and
October 2017. By addressing the companies' sales managers and consultants, we relied
on the key informant approach [32]. Sales manager and consultants, who are involved
in the marketing of the IT services and/or their provision, should be knowledgeable
about the characteristics of the offered services and business relationships as well as
about general company characteristics used as control variables.

4.1 Sample

At the beginning of the survey, the participants had to self-report their level of knowledge about the suppliers' marketing activities as well as offered services and only representatives with a sufficient level were included in the study. In cooperation with a market research firm, we collected responses from representatives of 147 suppliers of IT services. Out of these representatives, 51% were consultants, 27% sales managers, and 12% general managers. In average, the representatives were working in their respective position for 8 years. The main offerings of the companies were IT system integration services (40%), IT infrastructure services (26%), business process services (19%), and general consulting services (9%). That guarantees a variety of services regarding their degree of digitalization. The companies had at least 50 and in average 3,200 employees.

4.2 Construct Conceptualization

At the beginning of the survey, all participants were asked to choose a service that they had offered to a customer, which could have finally adopted the service or not, during the last six months and to describe the chosen service in detail. Afterwards they were instructed to answer all questions against the background of that service and customer relationship respectively.

For the measurement items, standard scale development procedures were applied, including the conduction of a comprehensive literature review. For all constructs except service digitalization existing measurement items were used which were modified or further developed when necessary to match the study's context (cf., Table 1 in the Appendix). As there is no established measurement for service digitalization in the literature, a new scale was developed for measuring that construct. Service digitalization, that is the degree of service provision which is done by IT, reflects a concrete service characteristic, which can be suitably measured with a single item [33, 34]. The measurement was inspired by literature [2, 3] and discussed as well as validated in multiple interviews with practitioners.

All items were pretested in interviews with twelve independent researchers and practitioners as well as by investigating the answers of the first 30 participants, which in combination ensured final clarity of the items' formulations.

5 Results

To analyze our data, we use the variance-based partial least squares (PLS) method. PLS is chosen as it is especially suited for exploratory research [35], which applies to our investigation of the effect of service digitalization on the relationship between service modularization and service flexibility.

5.1   Measurement Model Assessment

The assessment of the measurement model's psychometric properties includes testing of convergent validity, discriminant validity, and internal consistency reliability [36]. Additionally, we test for the common method bias [37].

To ensure convergent validity, the item loadings and the average variance extracted (AVE) are assessed (cf., Table 2 in the Appendix). In general, the outer loadings of the items on their respective construct should exceed 0.7 [35, 38] which is the case for all items. On the construct level, the AVE is considered to ensure convergent validity. The smallest AVE of the constructs is 0.580 (service value) and, hence, all AVE values exceed the recommended threshold of 0.5 [35].

Discriminant validity is given when the items' loadings are greater than their cross-loadings on other constructs [39], the Fornell-Larcker criterion is met [40], and the constructs' heterotrait-monotrait ratios (HTMT) do not exceed the given threshold of 0.90 [41]. The first criterion, investigating the item level, is established as all items load higher on their respective construct than on any other construct. On the construct level, the Fornell-Larcker criterion [40] and the HTMT are applied. As shown in Table 3 in the Appendix, the Fornell-Larcker criterion is fulfilled for all constructs. Additionally, all constructs' HTMT values (highest value of 0.405) meet the threshold of 0.85 [41]. In summary, discriminant validity can be assumed.

Internal consistency reliability comprises the assessment of Cronbach's α and composite reliability (CR). The values of Cronbach's α and CR should meet the lower threshold of 0.7 [39]. As this is the case for all constructs (cf., Table 2 in the Appendix), internal consistency reliability is also met.

Lastly, we include a marker variable in our survey to test for a potential common method bias [37]. The results of the correlation analysis do not indicate any significant correlations between the marker variable and the other variables and, hence, the test does not show any indication for the existence of the common method bias [42].

5.2   Structural Model Assessment

After ensuring the validity of the measurement model, first, the estimated structural model (cf., Fig. 1) is analyzed and, second, the hypothesized relationships are assessed. To address potential collinearity issues between the exogenous latent variables, the Variance Inflation Factors (VIF) are examined. All VIF values are well below the threshold of 5 [35, 43], with the highest VIF value in our data being 1.326. Hence, collinearity seems not to be an issue. To assess the significances of the path coefficients, a bootstrap procedure with 5,000 samples is performed. The impact of all control variables is not significant. In addition to testing the main model as described below, we perform supplemental analyses which consider a direct effect between service digitalization and service modularity as well as service value respectively. All results show to be robust towards these additional constraints.

The effect of service modularity on service flexibility equals 0.297 and is significant (p-value of 0.001). Hence, H1 is supported. Our RQ aims at investigating the significance of the moderating effect of service digitalization on the positive relationship between service modularity and service flexibility. Hence, by following the suggestions

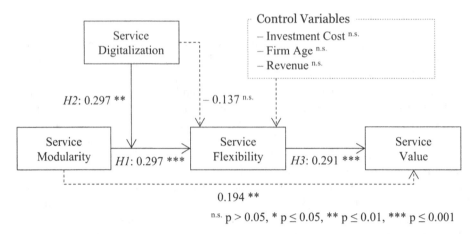

Fig. 1. Results of the estimated structural model.

of Henseler and Chin [44], the moderation (i.e., interaction term between service digitalization and service modularity) was modelled via the two-stage approach. To test the moderation (i.e., H2), we followed Hair et al. [36] and Sharma et al. [45]. The positive moderating effect of service digitalization is significant (p-value of 0.045). Following Henseler and Fassott [46], the f^2 value reflects the moderator's effect size. In our study, the f^2 effect size equals 0.090 which is considered as a large effect (threshold of 0.025 [36, 47]). The direct effect between service digitalization (i.e., the moderator and predictor variable) and service flexibility (i.e. the criterion variable) is not significant (p-value of 0.084). Hence, we can confirm a "pure moderator" [45], which fully supports the hypothesis (cf., H2) that service digitalization positively moderates the effect of service modularization on service flexibility. Furthermore, the effect of service flexibility on service value (i.e., H3) equals 0.291 and is significant (p-value of 0.001). Thus, our results strengthen the importance of service flexibility for service value and, in turn, the suppliers' market success.[1]

6   Discussion, Limitations, and Future Research

Based on an empirical model, we have shown that service digitalization positively moderates the positive relationship between service modularity and service flexibility and that service flexibility, again, is positively associated with service value. Hence, the results provide implications for research on service modularization and technology and innovation management research as well as managerial practice.

---

[1] Due to the significantly positive effect between service modularization and service value (cf., Fig. 1), we additionally can confirm a complementary mediation [48].

First, by considering service digitalization as a moderator for the effect of service modularization on service flexibility, we have extended the equivocal results of extant research stating that service modularity positively influences service flexibility both for services with a low degree of digitalization [15] and services with a high degree of digitalization [16]. Hence, in contrast to intuition, we showed that the advantages for service flexibility caused by digitalization outweigh the advantages for service flexibility caused by personnel. Additionally, by doing this, we have empirically underpinned the conceptual thoughts of Yoo [17] and Bharadwaj et al. [18] stressing that service digitalization offers unpreceded possibilities of service flexibility. Furthermore, we respond to Nambisan et al. [4], who call upon considering digitalization as an influencer of service modularity and not only as a mere context, and Iman [21], who recommends to investigate services in general instead of limiting the research to one degree of digitalization. Second, we respond to the call of Brax et al. [9], who ask whether service modularization influences the customers' service experience. By showing that service modularization positively influences service flexibility which, in turn, leads to an increase in service value, we investigate one aspect of service experience in detail. Third, our results yield important insights for practitioners offering services in the B2B context as well. By showing the importance of service modularity to achieve service flexibility, our results encourage suppliers to thoroughly check the modularity of their existing services as well as explicitly consider service modularity when creating new services. As our results have shown, this accounts especially for services with a high degree of digitalization. Hence, service modularity could be the key for suppliers to stay competitive in highly competitive B2B markets which are characterized by heterogeneous and rapidly changing customer needs [e.g., 12, 13, 20].

The findings of this research have to be interpreted in light of their limitations, which may provide an avenue for future research. Different effects are connected to service modularity (e.g., cost reduction, mitigation of risks). As we only focused on the effect of service modularity on service flexibility, future research should investigate the mechanisms of other outcomes of service modularity by also considering the possibly moderating influence of service digitalization. Due to our research design, service modularization only considers the service modules of one firm. Future research could broaden this view by considering service platforms which comprise modules of multiple firms. We assume that this might even further strengthens the moderating role of service digitalization.

In conclusion, we have offered new insights into the moderating influence of service digitalization on the relationship between service modularity and service flexibility.

Appendix

See Tables 1, 2, and 3.

Table 1. Measurement Items with respective loadings for the main constructs.

Service Modularity (SM)

Adapted from Vickery et al. [7]; reflective; 7-Point Likert scale with anchors 1 = "strongly disagree" and 7 = "strongly agree"

SM1	The service is composed of service modules (or self-contained processing units)	0.797
SM2	The service is broken down into service modules that can operate independently	0.788
SM3	Service modules can be added to or removed from the service without changing other service modules	0.882
SM4	The service is designed so that service modules can be added or removed without significant changes to other service modules	0.902
SM5	The service is designed to be rapidly disassembled and reconfigured	0.755

Service Flexibility (SF)

Adapted from Nelson et al. [14]; reflective; 7-Point Likert scale with anchors 1 = "strongly disagree" and 7 = "strongly agree"

SF1	The service can be adapted to meet a variety of needs	0.835
SF2	The service can flexibly adjust to new demands or conditions	0.905
SF3	The service is versatile in addressing needs as they arise	0.862

Service Value (SV)

Adapted from Stock and Zacharias [19]; reflective; 7-Point Likert scale with anchors 1 = "strongly disagree" and 7 = "strongly agree"

SV1	The service offers unique advantages to our customers	0.760
SV2	The service offers higher quality than services of our competitors	0.803
SV3	The service offers higher value than services of our competitors	0.766
SV4	The service solves the problems of our customers	0.724
SV5	The service delivers high benefits for our customers	0.753

Service Digitalization (SD)

Percentage scale between 0% and 100% with steps of 5%

SD1	Please rate the percentage of the service provision which is done by IT-systems	1.000

Investment Cost (IC)

Adapted from Benaroch et al. [49]; selection among predefined ranges in EUR with the boundaries 50,000, 0.1 million, 0.5 million, 1 million and 5 million as well as more than 5 million

IC1	What is the amount in EUR approved for this service?	1.000

Firm Age (FA)

Adapted from Demirkan et al. [50]

FA1	What is your companies age in years?	1.000

Revenue (R)

Selection among predefined ranges in EUR with the boundaries: 0.1 million, 1 million, 5 million, 10 million, 50 million, 100 million, 500 million, 1 billion and 1.5 billion as well as more than 1.5 billion

R1	What is your company's turnover in the past year?	1.000

Table 2. Indices for the assessment of internal consistency reliability.

Construct	Cr. α	CR	AVE
Service Modularity (SM)	0.886	0.915	0.684
Service Flexibility (SF)	0.835	0.901	0.753
Service Value (SV)	0.819	0.873	0.580
Service Digitalization (SD)	1.000	1.000	1.000
Investment Cost (IC)	1.000	1.000	1.000
Firm Age (FA)	1.000	1.000	1.000
Revenue (R)	1.000	1.000	1.000

Table 3. Correlations and Indices for the assessment of discriminant validity.

Construct	SM	SF	SV	SD	IC	FA	R
SM	0.827						
SF	0.252	0.868					
SV	0.267	0.340	0.761				
SD	0.244	−0.056	0.133	1.000			
IC	0.101	0.041	−0.012	0.077	1.000		
FA	−0.022	0.027	0.074	0.022	−0.028	1.000	
R	−0.057	−0.008	0.064	−0.011	0.463	0.067	1.000

References

1. Xue, L., Zhang, C., Ling, H., Zhao, X.: Risk mitigation in supply chain digitization: system modularity and information technology governance. J. Manag. Inf. Syst. **30**, 325–352 (2013)
2. Froehle, C.M., Roth, A.V.: New measurement scales for evaluating perceptions of the technology-mediated customer service experience. J. Oper. Manag. **22**, 1–21 (2004)
3. Böhmann, T., Junginger, M., Krcmar, H.: Modular service architectures: a concept and method for engineering it services. In: 36th Annual Hawaii International Conference on System Sciences, 10–19 (2003)
4. Nambisan, S., Lyytinen, K., Majchrzak, A., Song, M.: Digital innovation management: reinventing innovation management research in a digital world. MIS Q. **41**, 223–238 (2017)
5. Scherer, A., Wunderlich, N.V., von Wangenheim, F.: The value of self-service: long-term effects of technology-based self-service usage on customer retention. MIS Q. **39**, 177–200 (2015)
6. Baldwin, C.Y., Clark, K.B.: Managing in an age of modularity. Harvard Bus. Rev. **75**, 84–93 (1997)
7. Vickery, S.K., Koufteros, X., Dröge, C., Calantone, R.: Product modularity, process modularity, and new product introduction performance: does complexity matter? Prod. Oper. Manag. **25**, 751–770 (2016)
8. Voss, C.A., Hsuan, J.: Service architecture and modularity. Decis. Sci. **40**, 541–569 (2009)
9. Brax, S.A., Bask, A., Hsuan, J., Voss, C.: Service modularity and architecture – an overview and research agenda. Int. J. Oper. Prod. Manag. **37**, 686–702 (2017)

10. Dörbecker, R., Böhmann, T.: The concept and effects of service modularity – a literature review. In: 46th Annual Hawaii International Conference on System Sciences (HICSS-46), pp. 1357–1366 (2013)
11. Fixson, S.K.: Modularity and commonality research: past developments and future opportunities. Concurr. Eng. **15**, 85–111 (2007)
12. Rahikka, E., Ulkuniemi, P., Pekkarinen, S.: Developing the value perception of the business customer through service modularity. J. Bus. Ind. Mark. **26**, 357–367 (2011)
13. Martin, B., Stephan, K.: Providing a method for composing modular B2B services. J. Bus. Ind. Mark. **26**, 320–331 (2011)
14. Nelson, R.R., Todd, P.A., Wixom, B.H.: Antecedents of information and system quality: an empirical examination within the context of data warehousing. J. Manag. Inf. Syst. **21**, 199–235 (2005)
15. Moon, S.K., Shu, J., Simpson, T.W., Kumara, S.R.T.: A module-based service model for mass customization: service family design. IIE Trans. **43**, 153–163 (2010)
16. Lewis, M.O., Mathiassen, L., Rai, A.: Scalable growth in IT-enabled service provisioning: a sensemaking perspective. Eur. J. Inf. Syst. **20**, 285–302 (2011)
17. Yoo, Y.: The Tables have turned: how can the information systems field contribute to technology and innovation management research? J. Assoc. Inf. Syst. **14**, 227–236 (2013)
18. Bharadwaj, A., El Sawy, O.A., Pavlou, P.A., Venkatraman, N.: Digital business strategy: toward a next generation of insights. MIS Q. **37**, 471–482 (2013)
19. Stock, R.M., Zacharias, N.A.: Patterns and performance outcomes of innovation orientation. J. Acad. Mark. Sci. **39**, 870–888 (2011)
20. Chun-Hsien, L., Chu-Ching, W.: Formulating service business strategies with integrative services model from customer and provider perspectives. Eur. J. Mark. **44**, 1500–1527 (2010)
21. Iman, N.: Modularity matters: a critical review and synthesis of service modularity. Int. J. Qual. Serv. Sci. **8**, 38–52 (2016)
22. Tuunanen, T., Cassab, H.: Service process modularization: reuse versus variation in service extensions. J. Serv. Res. **14**, 340–354 (2011)
23. Sundbo, J.: The service economy: standardisation or customisation? Serv. Ind. J. **22**, 93–116 (2002)
24. Salvador, F.: Toward a product system modularity construct: literature review and reconceptualization. IEEE Trans. Eng. Manag. **54**, 219–240 (2007)
25. Yoo, Y., Henfridsson, O., Lyytinen, K.: Research commentary—the new organizing logic of digital innovation: an agenda for information systems research. Inf. Syst. Res. **21**, 724–735 (2010)
26. Tuunanen, T., Bask, A., Merisalo-Rantanen, H.: Typology for modular service design: review of literature. Int. J. Serv. Sci. Manag. Eng. Technol. **3**, 99–112 (2012)
27. Bask, A., Merisalo-Rantanen, H., Tuunanen, T.: Developing a modular service architecture for E-store supply chains: the small- and medium-sized enterprise perspective. Serv. Sci. **6**, 251–273 (2014)
28. Sambamurthy, V., Bharadwaj, A., Grover, V.: Shaping agility through digital options: reconceptualizing the role of information technology in contemporary firms. MIS Q. **27**, 237–263 (2003)
29. Henfridsson, O., Mathiassen, L., Svahn, F.: Managing technological change in the digital age: the role of architectural frames. J. Inf. Technol. **29**, 27–43 (2014)
30. Chan, K.W., Yim, C.K., Lam, S.S.K.: Is customer participation in value creation a double-edged sword? Evidence from professional financial services across cultures. J. Mark. **74**, 48–64 (2010)

31. Tiwana, A., Konsynski, B.: Complementarities between organizational IT architecture and governance structure. Inf. Syst. Res. **21**, 288–304 (2009)
32. Kumar, N., Stern, L.W., Anderson, J.C.: Conducting interorganizational research using key informants. Acad. Manag. J. **36**, 1633–1651 (1993)
33. Bergkvist, L.I., Rossiter, J.R.: The predictive validity of multiple-item versus single-item measures of the same constructs. J. Mark. Res. **44**, 175–184 (2007)
34. Rossiter, J.R.: The C-Oar-Se procedure for scale development in marketing. Int. J. Res. Mark. **19**, 305–335 (2002)
35. Hair, J.F., Ringle, C.M., Sarstedt, M.: Pls-Sem: indeed a silver bullet. J. Mark. Theory Pract. **19**, 139–152 (2011)
36. Hair, J.F., Hult, G.T.M., Ringle, C., Sarstedt, M.: A Primer on Partial Least Squares Structural Equation Modeling (Pls-Sem). Sage Publications, Los Angeles (2017)
37. Podsakoff, P.M., MacKenzie, S.B., Lee, J.-Y., Podsakoff, N.P.: Common method biases in behavioral research: a critical review of the literature and recommended remedies. J. Appl. Psychol. **88**, 879–903 (2003)
38. Bagozzi, R.P., Yi, Y., Phillips, L.W.: Assessing construct validity in organizational research. Adm. Sci. Q. **36**, 421–458 (1991)
39. Bagozzi, R.P., Yi, Y.: Specification, evaluation, and interpretation of structural equation models. J. Acad. Mark. Sci. **40**, 8–34 (2012)
40. Fornell, C., Larcker, D.F.: Evaluating structural equation models with unobservable variables and measurement error. J. Mark. Res. **18**, 39–50 (1981)
41. Henseler, J., Ringle, C.M., Sarstedt, M.: A new criterion for assessing discriminant validity in variance-based structural equation modeling. J. Acad. Mark. Sci. **43**, 115–135 (2015)
42. Lindell, M.K., Whitney, D.J.: Accounting for common method variance in cross-sectional research designs. J. Appl. Psychol. **86**, 114–121 (2001)
43. Bennett Thatcher, J., Perrewé, P.L.: An empirical examination of individual traits as antecedents to computer anxiety and computer self-efficacy. MIS Q. **26**, 381–396 (2002)
44. Henseler, J., Chin, W.W.: A comparison of approaches for the analysis of interaction effects between latent variables using partial least squares path modeling. Struct. Equ. Model.: Multidiscip. J. **17**, 82–109 (2010)
45. Sharma, S., Durand, R.M., Gur-Arie, O.: Identification and analysis of moderator variables. J. Mark. Res. **18**, 291–300 (1981)
46. Henseler, J., Fassott, G.: Testing moderating effects in PLS path models: an illustration of available procedures. In: Esposito, Vinzi V., Chin, W., Henseler, J., Wang, H. (eds.) Handbook of Partial Least Squares, pp. 713–735. Springer, Heidelberg (2010). https://doi.org/10.1007/978-3-540-32827-8_31
47. Kenny, D.A.: http://davidakenny.net/cm/moderation.htm. Accessed 02 Nov 2017
48. Zhao, X., Lynch, J.G., Chen, Q.: Reconsidering Baron and Kenny: myths and truths about mediation analysis. J. Consum. Res. **37**, 197–206 (2010)
49. Benaroch, M., Lichtenstein, Y., Robinson, K.: Real options in information technology risk management: an empirical validation of risk-option relationships. MIS Q. **30**, 827–864 (2006)
50. Demirkan, I., Deeds, D.L., Demirkan, S.: Exploring the role of network characteristics, knowledge quality, and inertia on the evolution of scientific networks. J. Manag. **39**, 1462–1489 (2013)

Modular Sales – Using Concepts of Modularity to Improve the Quotation Process for B2B Service Providers

Aleksander Lubarski[(✉)]

Research Group of Industrial Services, University of Bremen,
28359 Bremen, Germany
lubarski@uni-bremen.de

Abstract. The current trend of individualization forces industrial service providers to search for new ways of standardizing their internal processes without diminishing the flexibility of satisfying customer demands. This conflict of interests reaches its peak in the quotation process, where suppliers are spending a considerable amount of time and effort, often preparing their quotes from the scratch. The purpose of this paper is to apply the concept of service modularity to improve the efficiency of the quotation process on both operational and strategic levels. Using methods of qualitative research, I analyze the corresponding challenges of 19 German service providers and identify possible areas of improvement with the help of the service modularity. The result of the paper are 15 requirements for the appropriate IT-support to cover these identified challenges and enable the realization of the concept in practice. I hereby contribute to the ongoing discussion on service modularity by delivering empirical insights from the new area of the quotation process.

Keywords: Service modularity · Quotation process · B2B · IT requirements

1 Introduction

Over the past decades, the Business-to-Business (B2B) market and the supplier-customer interaction, in particular, have undergone many operational and strategic changes. The requirements of the industrial customers are becoming more and more individualized and complex, making it almost impossible to offer the same product or service twice [1]. Complex service projects involve close interaction with customers and supply-chain partners [2] and require a high degree of expertise often generated from tacit knowledge, assigned to certain employees [3]. Together with the overall globalization and increased information transparency, these changes force companies to optimize their internal processes via standardization in order to remain competitive. However, the standardization services is only possible as long as the flexibility to meet individual customer demands is not diminished [4].

This conflict of interest reaches its apogee in the quotation process that can last up to several months, involving different departments to determine technical and financial feasibility, often resulting in extremely long quotation documents [5]. While it is plausible for the customer to send his requests to numerous suppliers and thus apply

© Springer Nature Switzerland AG 2018
G. Satzger et al. (Eds.): IESS 2018, LNBIP 331, pp. 16–30, 2018.
https://doi.org/10.1007/978-3-030-00713-3_2

competitive tendering, the suppliers are often wasting considerable amount of their resources preparing the offer that they may not even get. The problem of such a "request-mentality" is intensified by the fact that quotation documents and price calculations are often created from the scratch due to seemingly unique customer requirements [6].

Motivated by the fact that the majority of B2B service providers are still relying solely on the experience of their senior sales managers with little to none supportive IT when preparing a quote [7], the purpose of this paper is to show how the concept of modularity can simplify the overall quotation process and prevent the loss of previous efforts put in the creation of the quotation documents. Here, the focus on "service modularity" as a more specific research theme of the general "modular system theory" is necessary, as it incorporates service characteristics such as immateriality, process nature, uno-actu principle and a high heterogeneity level especially in the B2B context along with the specificity of their sales process (e.g., service description or pricing of the composite service). While there exists sufficient amount of conceptual publications on service modularity, little is known on how it can improve company's efficiency in particular areas of operation such as the quotation process. Using methods of qualitative research, I identify the current operative and strategic challenges of 19 German industrial service providers and explore the potential of constructing quotation documents in a modular way. Based on these insights I derive 15 requirements for the appropriate IT-support acting as an enabler of the concept realization in practice. This paper answers existing research calls for new ways of standardization in the sales process via IT usage [7–10] and contributes to the ongoing discussion on service modularity by delivering empirical insights from the new area of the quotation process.

The remainder of this paper is structured as follows. After giving a quick overview of the theoretical background in Sect. 2, I propose a framework for the quotation process and its placement within a company's overall sales strategy and use it for the collection and structuring of the empirical data. Section 3 elaborates on the used methodology and interviewed companies. The results, consisting of the structure of the quotation document, identified strategic and operative challenges in the quotation process, as well as a derivation of 15 requirements for the appropriate IT-support are presented in Sect. 4. Finally, Sect. 5 concludes with a short summary of the finding, implications for the practitioners, research limitations, and future research pathways.

2 Theoretical Background

2.1 Service Modularity

The concept of modularity has been proposed as a possible solution to achieve standardization and simplification of services without diminishing firm's capability to meet individual customer demands [4]. In general, modularity can be seen as a principle of building a complex system from smaller parts (modules) that can be designed and improved independently, yet function together as a whole [11], as long as the interfaces are well-defined and a clear one-to-one matching of modules and functions exists [12].

For the past 50 years, modularity has influenced the design of products [13], organizational IT and processes [14], and recently – services. The rise of the service modularity as a distinct research direction started especially after Pekkarinen and Ulkuniemi [4], who regarded service modularity as a combination of modular processes, modular organization, and a certain level of customer integration. Industrial services in particular can be regarded as complex systems, which can be decomposed into individual process steps (modules) consisting of single activities (elements), although more levels of granularity are possible [15]. Unlike the physical interfaces within the product domain, service interfaces concentrate on the information flows, exclusion rules, documentation of the process steps, or assignment of the particular employee for the task completion [16]. A feasible combination of these pre-defined modules add up to a particular service variant, which is configured either by the sales employee based on the input of the customer, or directly by the customer, if he possesses sufficient technical knowhow.

An industrial example of a complex service from the wind energy sector is a maintenance plan for a power plant, which consists of different process steps such as video surveillance, oil exchange, parts replacement, and predictive maintenance. Naturally, all of these steps are independent from each other and can be added or configured using parameters such as power plant characteristics, speed of execution, or even qualifications of the service technicians. Such a modular service design and delivery offers an increased external service variety for the customer, while simultaneously decreasing internal process variety for the service provider [17]. Other claimed benefits of providing modular services include the reuse of components for future development [18], faster development process and decreased time-to-market [8], economies of scale and scope [19] as well as cost-efficiency in operations [20].

However, while concentrating on the efficient development and delivery of services, little attention has been awarded towards how a modular service portfolio can improve the sales function of the company. Exceptions include the study by Giannikis et al. [9], which pointed to the possible growth of operating profit by 30% in the field of legal services and called for more research on how to standardize the tender preparation process. Similarly, Schmidt [21] has published a survey among members of the Association of German Engineers (VDI) about quotation management in industrial companies. The results show that current quotation processes mainly arise from the practical experience of the upper management, but the trends concerning potential improvements move significantly towards standardization, modularization and the reduction of time between request and submission. In this regard, the desire for simplicity and systematic standards, as well as the need for controlling, is made clear.

2.2   The Notion of Quote Preparation

Currently, research activities on quote preparation and selling business services is rather scarce, even though practitioners have been calling for academic insights already for a while [21]. The fact that there exists no unified sales process amongst B2B providers due to the heterogeneity of their services makes it difficult to define where the creation of the quotation document actually begins and where it ends. Therefore, in order to collect empirical data and derive requirements for a supportive IT in a systematic way, I first propose a simplified framework of the quotation process and its placement within the overall strategic and operational sales activities of the service provider.

In their framework for service modularization, Lubarski and Pöppelbuß [22] categorize the long-term decisions regarding the service portfolio as a Build-Time phase, whereas everyday activities and the operative creation of quotation documents happen in the Run-Time phase. The decisions made during the *Build-Time* phase strongly depend on the underlying customer segmentation, which helps service providers to define their scope of activity and identify offerings for the customer [23]. Based on the customer segmentation and the understanding of customer's needs, providers are then able to define their service portfolio and develop new services [24]. Simultaneously, for cost efficiency reasons, the providers are forced to standardize their services, which is enabled through service modularity [19, 20]. However, the presented elements of the Build-Time phase should not be seen as a strict process, but rather exhibit a strong interdependence and can be performed in an arbitrary order.

The *Run-Time* phase, on the other hand, is a repetitive sequential process that is executed every time a quote has to be created and is based on the specifications from the Build-Time phase. For example, in their analysis of the procurement of logistic services, Söhnchen and Albers [25] derive a stepwise sales funnel for the industrial customer acquisition, where tender preparation covers the stages of "product presentation", followed by "design of the offer" and "handling objections". Similarly, the proposed model of Geiger and Krüger [26] begins with the rough clarification of customer requirements, over the detailed planning of the technical concept, overall resulting in the finalization of the calculations and submission of the tender. Finally, Rahikka et al. [23] identified three main challenges when developing value perception of the business customers – specification of services, estimation the co-creation efforts, and cost determination. The intersection of the presented models builds the so called Configure-Price-Quote (CPQ) process, which is a widely-used term in the B2B product industry [27].

However, I believe that these three steps alone cannot build the base for the company's overall ability to react to the customer's changing requirements, since there exists no reverse connection to the Build-Time phase. By ignoring former quote preparations, companies miss out the opportunities of reusing previous quotation documents both for operational processes (i.e. configuration, pricing and quotation), as well as continuous monitoring of strategic weak points (e.g. service redesign or innovation, identification of unprofitable services). Therefore, I add an additional step of Monitoring and Reporting, where the input from the quotation process (e.g. hit rate[1], the price offered for a certain service package) is gathered and analyzed ongoing, acting as an additional source of information for the future requests for quotations or the adjustment of the overall sales strategy. Figure 1 summarizes different perspectives of the sales process along with their interdependencies.

In order to avoid confusion and to emphasize the focus of this publication, I summarize the underlying concepts along with their interrelationships. First, service

---

[1] Hit rate is a metric traditionally associated with sales, which represents a relative success of a sales employee or department. In the context of this publication, I define hit rate as a proportion of successful quotes or tenders to the number of the overall attempts.

modularity is an overarching concept that envisages the decomposition of the provider's service portfolio into functional independent units (modules). Second, a modular quotation process is the operationalization of the concept from the sales perspective as it builds on the existing modular portfolio and allows customer-specific (internal or external) configuration of the desired solution. Finally, this configured service variant is communicated to the customer in a modular quotation document. In this publication I assume that provider's service portfolio has been (at least to some extent) already modularized (more information on the steps of the typical modularization process can be found in [22]) and focus on the requirements for the appropriate IT-support that would enable the creation of a modular quotation document.

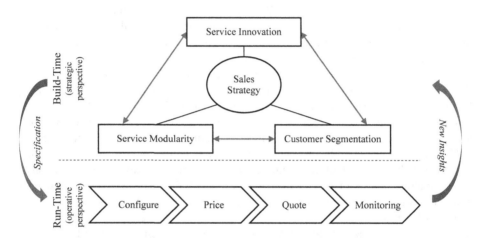

Fig. 1. Placement of the quotation process within the overall sales activities

3 Empirical Data

To understand the current challenges in the quotation process amongst B2B service providers and gather requirements for potential IT support, I employed methods of qualitative research [28]. Qualitative research is especially suitable when the research area is still emerging and not controllable by the investigators, which is the case both for the quotation process and service modularity. I applied semi-structured expert interviews as the approach for data collection. The questionnaires were divided into thematic sections according to the framework from Fig. 1. The interviews were conducted in July to October 2017 and lasted 45 min on average. All interview partners requested to remain anonymous. The analysis of the data was supported by using professional software for qualitative data analysis (MAXQDA 12), which eased the process of arranging, discussing and synthesizing the input into greater units of analysis. When choosing the interview partners, I searched for B2B companies who face challenges of addressing highly individual customer demands while simultaneously trying to cope with variant management by introducing standardization. The empirical data came from 19 B2B companies, which can be clustered into three distinctive groups.

Product Manufacturers (N = 7). Even though service modularity is primarily of interest for service providers, I also included product manufacturers to the sample, who, in addition to their individualized products (e.g., vacuum components, heating systems, warehousing equipment), offer supplementary services like leasing, maintenance or logistic solutions. The majority of this group are well-established big companies with over 1000 employees (three of the interviewed companies have over 10000), offering up to 10^{30} product variations and dealing with up to 5000 quotation requests per year. All of the companies from the sample were already using a certain level of IT support for quote preparation, but have stated to have difficulties with assigning services to their individualized products and drawing conclusions from their historic quotation documents (e.g. strategic price setting based on the hit rate).

Contract Service Providers (N = 6). The companies in this group are B2B service providers offering long-term projects ranging from financial or consulting services up to software development and contract logistics. These providers do not receive as many requests for quotation but put much time both in selection of the customers and the creation of the quotation documents. Unlike the first category, where product parts and their use are well-defined, a common step for this group is a feasibility check regarding the required resources (e.g., man-hour) and the estimation of how much of these resources are needed. In most cases, such an estimation is impossible without the expertise of the senior management unless it is well-documented within the company.

Product-Enabled Service Providers (N = 6). The third group can be seen as a combination of the first two as it consists of companies offering product-enabled services such as marketing campaigns, container rental, and transportation. Unlike Group 1, here the value for the customer is service-dominant with the products being solely a medium for transmission of the value [24]. The difference to the second group is a relatively high number of quotation requests to deal with as well as customers' demands for transparency, forcing them to both accelerate their sales process while still giving detailed information about price calculation and delivery times. However, as of now most of the interviewed providers still rely strongly on the expertise of their top management, thus making sales employee very dependent on the non-documented internal knowledge.

I deliberately picked a diversified sample, in order to derive generic requirements for the IT support of a modular quotation process that applicable to a wide range of industrial service providers, irrespective of their individual characteristics. In addition to the expert interviews, the companies also provided 18 quotation documents on which I was able to perform document analysis that involved skimming (superficial examination), reading (thorough examination) and interpretation [29]. By combining data from interviews with the documentary evidence, I hereby minimize bias and establish credibility.

4 Results

In the following, I will present the research results in three steps. First, I analyze the structure of the provided quotation documents and try to find common logic and constituents. Using the input from the interviews and the quotation documents, I identify current challenges in the quotation process amongst industrial service provi-ders. Finally, based on these results I propose a set of requirements for an appropriate IT support, covering both the operative and strategic perspectives of the sales process.

4.1   The Structure of the Quotation Document

Despite the differences in service offerings of the interviewed companies, it was still possible to identify overlaps and derive a typical structure of the quotation document (Fig. 2), which can also be regarded as a composed system consisting of modular substitutable elements [4]. However, due to a high individuality of the customer requests in the B2B sector, I believe that not all modules are fully exchangeable. Therefore, following the ideas of different interface types for manufacturing firms [30], I distinguish between *standardized, adaptable,* or *custom-made.* The first type (Fig. 2, blocks without outlines) describes elements that already come constant and complete and can be reused unlimitedly until they are changed on the strategic level (cf. Fig. 1). In contrast, adaptable elements (Fig. 2, dashed outlines) need adjustments or man-agerial approval before they can be reused in the new quotation document. Finally, when the difference to the former quotation documents is too vast to apply previous elements, the module has to be custom-made (Fig. 2, solid outlines).

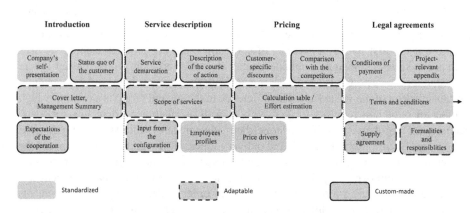

Fig. 2.   The modular structure of the quotation document

Based on the provided data, a typical quotation document consists of four inter-connected thematic blocks, which are created anew for each customer. Based on the provided quotation documents, modules in the middle lane can be seen as a lowest common denominator, while the remaining ones are optional. The block *Introduction* begins with the adaptable management summary and standardized company's self-presentation. In addition, using the input from the customer, the status quo before service provision is analyzed and expectations towards the upcoming cooperation are documented serving as a basis for the subsequent CPQ-process.

The second block *Service description* concentrates on the actual sales object. Here the scope of the service and its demarcation are adaptable, since they are (at least to some extent) assembled from a set of pre-defined service sub-components and, if needed, are then adjusted or changed based on the customer's requirements from the previous block. In some cases, the company also provides profiles of the involved employees (typically standardized, although project-specific adaptation is also possible) and a general description of the course of action, which has to be developed together with the customer.

Based on the service configuration, the anticipated *Price* is described in the next section, although the level of detail and transparency strongly depend on company's sales strategy and the level of uncertainty of the sales object. The price itself can be presented either in form of a well-structured and binding calculation table (e.g., maintenance costs of a certain machine for Group 1), or a rough effort estimation (e.g., software development for Group 2) that can be updated during the actual service provision. Depending on the complexity and uniqueness of the offered service, the price cannot always be based on pre-defined elements or historic quotation documents, but requires managerial approval, making it adaptable. The underlying price drivers (e.g., workforce, material costs) along with customer-specific discounts (based on the customer segmentation, cf. Fig. 1) are usually standardized, whereas comparison with the competitors (if desired) has to be updated each time a quote is created.

Finally, the document ends with *Legal agreements* and other project-related formalities. This block covers not only the legal aspect and penalties of not fulfilling parts of the contract but also other terms of execution like acceptable travel and catering expenses, delivery times, payment types, etc. While some of the modules are similar for all quotation documents (conditions of payment, terms of delivery), others are adapted based on the customer requirements (supply agreement, formalities, and responsibilities) or are created from scratch (project-relevant appendix)

4.2   Identified Challenges

The analyzed structure of a typical quotation document can now be combined with the responses of the expert interviews to identify current challenges in the quotation process. Following the logic of the sales framework (cf. Fig. 1), these challenges can be assigned either to the strategic or to the operational perspective.

Table 1. Interview quotes (selection)

ID	Industry	Quote
#1	Finance	New colleagues sometimes complain about not really knowing what we can deliver and thus have difficulties explaining it to the customers
#2	Software engineering	Sometimes the customer describes requirements that make no technical sense at all or are simply not feasible. This shows that the customer does not really know where the problem is, only that there is one
#3	Logistics	The estimation is further complicated by the fact that offer documents are not stored in any central storage location that could serve as a basis for research for inexperienced sales employees

(continued)

Table 1. (*continued*)

ID	Industry	Quote
#4	Consulting	From a sales perspective it is quite critical if the customer sends his requirements and no one knows becomes responsible for it. There is often a long search process combined with a message "ping-pong game"
#5	Consulting	You can also make the mistake by assuming upfront that you will get the deal and block the necessary resources. However, if in the end the quote is declined – the company loses money
#6	Automotive	You go through the individual service modules of an offer. They are put together like a jigsaw puzzle, which, if lucky, finally adds up to a price
#7	Software engineering	Every time you are making an offer – this requires time and effort. This involves, let us say, 10 employees for three weeks. And this adds up to 10.000€, 20.000€, or even 30.000€
#8	Manufacturing	We have a new tool that promises high efficiency. However, while the quotes are indeed created faster, but their quality is not checked at any point. Doing something suboptimal faster is not really our goal

Strategic Perspective. The first challenge lies in a proper *Structuring of the service portfolio and its external representation* (Table 1, #1). This appears to be problematic especially for companies from the Group 2, who describe their services only in general terms, leaving their customers unaware of the scope of their expertise. This results in increased marketing and pre-sales expenses involving on-site visits and workshops to present own service portfolio and start a dialogue with the customer. Closely interrelated is the next challenge of *Understanding the customer needs*, which can be seen as an extension of the customer segmentation (Table 1, #2). While companies from Group 1 minimize the level of interaction with the customer in the pre-sales stage (partly due to the high number of requests), Group 2 and 3 point out that early customer involvement is inevitable both for analyzing the status quo and for joint solution development. However, this procedure can take a long time until an customer involvement is inevitable it, while the success is not guaranteed and the service provider is not being paid during this time.

Another strategic challenge arises from overall difficulties in *Knowledge documentation and sharing* within the company (Table 1, #3). Although some templates are already used during service definition and price calculation by some of the companies of Group 1 and 3 – the quote preparation is still mainly based on the employee's experience, which exists solely in their heads. As a result, the quotation documents are not constructed uniformly as the content is individually formulated each time, with little to none predefined sections for recurring services. This complicates reuse of historic documents and leads to redundant and timewasting *Unstructured communication* between the departments as well as with the customer (Table 1, #4). Moreover, some of the interviewed companies addressed the problem of knowledge sharing especially in the context of new employees, who find it very difficult to acquire knowledge about service composition without a well-structured and constantly maintained reference database.

Operational Perspective. One of the main identified operational challenges that influences the whole CPQ process is the *Allocation of experts and resources* when staffing the team for an incoming opportunity (Table 1, # 5). Providers from Group 2 and 3 mention that an overview of employees' profiles (cf. Fig. 2) can even act as an important criterion for selecting a provider. Hence, this allocation has a direct influence on the feasibility and profitability of the quotation request, which brings us to the next challenge of *Pricing and effort estimation* (Table 1, #6). Due to individual nature of service requests, prices have to be calculated for each customer individually, although there exist certain price drivers. The introduction of appropriate price lists and standards together with the consideration of the previous calculation could accelerate and simplify this process.

Another critical challenge that was mentioned by the majority of the interviewed companies were the *Delays and time-consuming procedures* (Table 1, #7). On the one hand, these delays result from the pricing negotiations with the customers due to very different perceptions of risks and benefits. On the other hand, the quotation process is decelerated due to additional approval processes and communication within the company, since no central decision-support system is available. Finally, the ability to constantly *Identify weak points* and draw insights from previous quotation documents is considered another long-term success factor for service providers dealing with a high number of quotation requests (Group 1 and 3) since the efficiency in the quotation preparation alone does not necessarily mean the efficiency in winning the tender. By looking at the hit rate and other controlling characteristics, providers can draw conclusions both on the operative (e.g., price adjustments) and on the strategic levels (e.g., identifying potential candidates for removing).

4.3 Approaches to Overcome the Challenges

Based on existing literature, I now propose how the identified challenges can be solved by three structural changes – modular service offerings, customer integration, and modularity in organizations [4], thus overall simplifying and accelerating the quotation process. The first challenge of making services visible to the customer can be solved by a modular service offering [4]. It has been shown that modular service offerings can help to better specify the services, which leads to a facilitation of managing complex services and a higher understanding on the customer side. Moreover, modularity enables the partly replication of service offerings thus avoiding repetitive effort estimations or pricing processes [9]. The accurate definition and the common composition of service module bundles also make prices transparent and more traceable.

However, although such a modular portfolio helps to present service offerings in a compact and comprehensible way, this alone would not suffice to fully understand the customer needs. Therefore, companies should also integrate their customers into the quotation process, thus enabling value co-creation [23] and minimize subsequent clarifications and negations. Only if it is transparent which efforts arise for the customer during service provision, the promised service can be performed satisfactorily and a price satisfying both sides can be determined.

The mentioned difficulties in gathering expert knowledge and the assignment of the responsibilities during quotation process could be overcome with the help of a modular organization [4]. On the one hand, such an unbounded organization would make tacit knowledge explicit and understandable for colleagues [31]. On the other hand, it would provide standard channels for the communication (including communication with the customer) and bring the right people together to share and combine knowledge [3]. In this way, teams can be built and reorganized easier and faster in response to new business opportunities or customer-driven customization requests.

Lastly, the ability to learn from previous quotation documents is indirectly supported by a modular service portfolio (e.g., it is possible to analyze the popularity of a certain service module or a price of a service package) and a close customer interaction (e.g., detailed feedback loops, especially after a rejection of the quotation). An overview of the proposed approaches together with the mapped challenges can be found in Table 2.

Table 2. Overcoming challenges by modular approaches

	Identified challenges	Approaches to overcome challenges		
		Modular service offerings	Modularity in organizations	Customer integration
Strategic	Portfolio structure and representation	x		
	Understanding the customer needs	(x)		x
	Difficulties in knowledge sharing		x	
	Unstructured communication		x	(x)
Operative	Allocation of experts and resources		x	
	Complex pricing and effort estimation	x		
	Delays and time-consuming procedures	x	x	
	Identification of the weak points	(x)		(x)

4.4 Requirements for the Appropriate IT-Support

With the growing number of service variations and quotation requests each year (as reported in the interviews), the realization of the upper mentioned approaches becomes possible only with the use of an appropriate IT support. Therefore, in addition to the identified challenges, the interviewees were also explicitly asked about what to consider when introducing such solution. Since a real implementation strongly depends on an individual use case and company's sales strategy, the goal of this paper was not to develop a concrete action plan but to start a discussion of the most critical aspects. For overview purposes, the derived 15 requirements were separated into Functional, Non-Functional, and Domain Requirements, which is a common categorization within

Requirements Engineering literature [32]. *Functional requirements* focus on what functionality the system should provide to its users and how it should behave in particular situations [33]. In addition to the classic steps of the CPQ process (configuration of the service modules, pricing of the final service and generating a quotation document), the interviewed companies stated the access to employee profiles to be an important feature for project management. Similarly, to allow direct interaction and value co-creation between the business units as well as with the customer, structured communication channels, and project-dependent right management should be ensured.

Non-functional requirements define system properties and constraints and affect overall architecture rather than individual components [34]. First, all the steps of the quotation process should be integrated and executed in one place. Second, a central point of data accessible to authorized users is needed to simplify knowledge sharing and shorten approval processes during the creation of quotation documents. This can work only if the data is consistent and constantly maintained and the system is accepted by its users. Finally, since the overall motivation of the use of the IT support and service modularity is performance improvement and foreseeable cost-efficiency, the advantages of the standardized quotation process (in particular shorter completion time and overall higher success rate) should exceed the initial investment in the restructuring of the service portfolio and software implementation.

Lastly, the *Domain Requirements* are imposed by the operating environment [32]. Especially in the context of B2B markets, the system should have an interface with the customer both during quotation preparation and service provision, although this integration should not be solely IT-based in order to maintain a necessary level of personal touch and customization ability. Furthermore, the software should also have a reporting unit connected to the central point of data, in order to make reuse of previous quotation documents and learning loops possible. Finally, such a software should not act as a stand-alone solution, but being able to integrate in existing enterprise systems to ensure data consistency. Each of the presented requirements is aimed to realize the proposed approaches (i.e. modular service offerings, modular organizations and customer integration) as well as ensure overall performance of the software (Fig. 3).

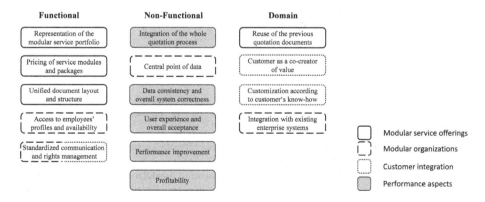

Fig. 3. Requirements for the IT-support of the modular quotation process

5 Conclusion

With the current trend of individualization and quickly changing market environments, companies are searching for new ways to go that extra mile for their customers, while still standardizing their internal processes for cost-efficiency reasons. This conflict of interest reaches its apogee in the quotation process, which requires a considerable amount of time and effort, while often still resulting in the rejection of the offer. With this background, the concept of service modularity can be seen as a possible solution to achieve a sound balance between customer-driven individualization and company-driven standardization ambitions. In order to derive a solution empirically, I applied methods of qualitative research and presented the results in three consecutive steps. First, I analyzed the structure of the quotation documents and discussed different types of modules within this document as well as their potential for exchangeability. Second, current strategic and operative challenges were mapped against three different architectural approaches – modular service offerings, modularity in organizations and customer integration. Finally, these results served as a basis for the derivation of overall 15 requirements for the appropriate IT-support to realize the full potential of service modularity in the context of the quotation process. Thereby I contributed to the ongoing discussion on service modularity by delivering empirical insights from the new area of quotation process.

However, the results of this paper do not come without limitations. Due to a relatively small sample of 19 companies, the results of the paper cannot provide universal validity, thus calling for more quantitative research in this direction. Furthermore, the identified structure of quotation documents and the derived requirements for the IT-support of service modularity are meant to start a discussion of the most critical aspects, rather than claim completeness and universality. In this regard, more case studies are needed that would concentrate on the specificity of the operating environment and provided services, thus resulting in additional or more detailed requirements. Finally, the transition to a modular quotation process and the implementation of a prototype would also deliver valuable insights and reveal practical difficulties, which can best be seen when conducting a longitudinal action research. In conclusion, it should be said that the topic of modularized quotation process is still in its infancy and yet to attract more attention of researchers from both marketing and the IS.

References

1. Lindemann, U., Reichwald, R., Zäh, M.F. (eds.): Individualisierte Produkte: Komplexität beherrschen, in Entwicklung und Produktion. Springer, Berlin (2006). https://doi.org/10.1007/3-540-34274-5
2. Andersson, D., Norrman, A.: Procurement of logistics services—a minutes work or a multi-year project? Eur. J. Purch. Supply Manag. **8**, 3–14 (2002)
3. Nätti, S., Ulkuniemi, P., Pekkarinen, S.: Implementing modularization in professional services—the influence of varied knowledge environments. Knowl. Process Manag. **24**, 125–138 (2017)

4. Pekkarinen, S., Ulkuniemi, P.: Modularity in developing business services by platform approach. Int. J. Logist. Manag. **19**, 84–103 (2008)
5. Agndal, H., Axelsson, B., Lindberg, N., Nordin, F.: Trends in service sourcing practices. J. Bus. Mark. Manag. **1**, 187–208 (2007)
6. Lubarski, A., Pöppelbuß, J.: Vertrieb industrienaher Dienstleistungen - Erkenntnisse aus der Windenergie- und Logistikbranche. Staats- und Universitätsbibliothek Bremen, Bremen (2017)
7. Kindström, D., Kowalkowski, C., Alejandro, T.B.: Adding services to product-based portfolios: an exploration of the implications for the sales function. J. Serv. Manag. **26**, 372–393 (2015)
8. Böttcher, M., Klingner, S.: Providing a method for composing modular B2B services. J. Bus. Ind. Mark. **26**, 320–331 (2011)
9. Giannikis, M., Mee, D., Doran, D., Papadopulos, T.: The design and delivery of modular professional services: implications for operations strategy. In: Proceedings of the 6th International Seminar on Service Modularity, Helsinki, Finland (2015)
10. Lindberg, N., Nordin, F.: From products to services and back again: towards a new service procurement logic. Ind. Mark. Manag. **37**, 292–300 (2008)
11. Baldwin, C.Y., Clark, K.B.: Design Rules: The Power of Modularity. MIT Press, Cambridge (2000)
12. Arnheiter, E.D., Harren, H.: A typology to unleash the potential of modularity. J. Manuf. Technol. Manag. **16**, 699–711 (2005)
13. Sanchez, R., Mahoney, J.T.: Modularity, flexibility, and knowledge management in product and organization design. Strateg. Manag. J. **17**, 63–76 (1996)
14. Malone, T.W., Crowston, K., Herman, G.A. (eds.): Organizing Business Knowledge: The MIT Process Handbook. MIT Press, Cambridge (2003)
15. Lin, Y., Pekkarinen, S.: QFD-based modular logistics service design. J. Bus. Ind. Mark. **26**, 344–356 (2011)
16. de Blok, C., Meijboom, B., Luijkx, K., Schols, J.: Interfaces in service modularity: a typology developed in modular health care provision. J. Oper. Manag. **32**, 175–189 (2014)
17. Yang, L., Shan, M.: Process analysis of service modularization based on cluster arithmetic, April 2009
18. Carlborg, P., Kindström, D.: Service process modularization and modular strategies. J. Bus. Ind. Mark. **29**, 313–323 (2014)
19. Tuunanen, T., Bask, A., Merisalo-Rantanen, H.: Typology for modular service design: review of literature. Int. J. Serv. Sci. Manag. Eng. Technol. **3**, 99–112 (2012)
20. Bask, A., Lipponen, M., Rajahonka, M., Tinnilä, M.: The concept of modularity: diffusion from manufacturing to service production. J. Manuf. Technol. Manag. **21**, 355–375 (2010)
21. Schmidt, D.H.: Studie zum Angebotsmanagement, p. 34
22. Lubarski, A., Pöppelbuß, J.: Methods for service modularization - a systematization framework. In: Proceedings of the Pacific Asia Conference on Information Systems, Chiayi, Taiwan (2016)
23. Rahikka, E., Ulkuniemi, P., Pekkarinen, S.: Developing the value perception of the business customer through service modularity. J. Bus. Ind. Mark. **26**, 357–367 (2011)
24. Lusch, R.F., Nambisan, S.: Service innovation: a service-dominant logic perspective. MIS Q. **39**, 155–176 (2015)
25. Söhnchen, F., Albers, S.: Pipeline management for the acquisition of industrial projects. Ind. Mark. Manag. **39**, 1356–1364 (2010)
26. Geiger, I., Krüger, S.: Anfragenbewertung und Angebotserstellung. In: Kleinaltenkamp, M., Plinke, W., Geiger, I. (eds.) Auftrags-und Projektmanagement, pp. 59–89. Springer, Wiesbaden (2013). https://doi.org/10.1007/978-3-658-01352-3_2

27. Gartner Inc.: Configure, Price, Quote (CPQ) Application Suites (2017). https://www.gartner.com/it-glossary/configure-price-quote-cpq-application-suites
28. Yin, R.K.: Case Study Research: Design and Methods. SAGE Publications, Thousand Oaks (2013)
29. Bowen, G.A.: Document analysis as a qualitative research method. Qual. Res. J. **9**, 27–40 (2009)
30. Ulrich, K.: The role of product architecture in the manufacturing firm. Res. Policy **24**, 419–440 (1995)
31. Langlois, R.N.: Modularity in technology and organization. J. Econ. Behav. Organ. **49**, 19–37 (2002)
32. Sommerville, I., Sawyer, P.: Requirements Engineering: A Good Practice Guide. Wiley, Hoboken (1997)
33. Kotonya, G., Sommerville, I.: Requirements Engineering: Processes and Techniques. Wiley, Hoboken (1998)
34. Glinz, M.: On non-functional requirements. In: 15th IEEE International Requirements Engineering Conference, RE 2007, pp. 21–26. IEEE (2007)

Omni-Channel Service Architectures in a Technology-Based Business Network: An Empirical Insight

João Reis[1(⊠)], Marlene Amorim[2], and Nuno Melão[3]

[1] Department of Military Science and CINAMIL/CISD, Military Academy,
Lisbon, Portugal
joao.reis@academiamilitar.pt
[2] Department of Economics, Management and Industrial Engineering
and Tourism, and GOVCOPP, Aveiro University, Aveiro, Portugal
mamorim@ua.pt
[3] Department of Management and CI&DETS,
School of Technology and Management of Viseu, Polytechnic Institute of Viseu,
Viseu, Portugal
nmelao@estgv.ipv.pt

Abstract. This article investigates the existing omni-channel service architectures in the front-office of technology-based business networks. It discusses the implications from the existing alignment between the network-preferred channel with other channels and clients. The methodological approach is qualitative, exploratory in nature, and employs case study research in a large private retail bank in Portugal. It includes multiple sources of data collection for corroboration purposes, including semi-structured interviews, direct observation and institutional documents. Although we have identified four types of omni-channel architectures in a business network context, the case analysis revealed that only two of them meet all the requirements, namely: the mixed services and pure virtual services. For academics this is the first attempt to discuss a growing topic in the operations management literature. Thus, this study may also help practitioners to understand the challenges they may have to deal with an omni-channel strategy in a business network context.

Keywords: Technology-based business networks · Empirical research
Service operations · Service architecture · Case study · Omni-channel services
Front-office

1 Introduction

Services have become an integral part of modern society [1–3], in a continuously evolving context stimulated by diverse factors such as new technologies and online channels. The integration of online service delivery channels, employing self-service technologies and interfaces (e.g. self-service checkout systems) are enabling firms to change and optimize the design of service encounters in order to meet customer requirements and convenience at an unprecedented pace [4, 5]. This happens along

© Springer Nature Switzerland AG 2018
G. Satzger et al. (Eds.): IESS 2018, LNBIP 331, pp. 31–44, 2018.
https://doi.org/10.1007/978-3-030-00713-3_3

with an unprecedented level of customer connection and empowerment [6], that is enabling customers to exhibit preferences towards the existing channel options when conducting processes for purchasing goods and services [7–9]. In this context of technically equipped, empowered and knowledgeable customers' organizations are compelled to continuously adapt their service approach, and evolving from a multi- to omni-channel strategies [10], with profound implications for the management of service systems and operations. The adoption and operation of new channels is a challenging task that requires for huge efforts of channel coordination and integrating with the other existing resources from the firm [11]. One viable option, that organizations can consider when they have different priorities or limited capacity to quickly add new channels and technologies to their portfolio, is to look for synergies with other providers with already established specific capabilities or resources that offer opportunities to customers and organizations [12]. Recent work on competitiveness has also emphasized the importance of business networking, the evidence illustrating that those firms which do not co-operate reduce their ability to enter into exchange relationships [13] and lose the capacity to share markets [14]. The underlying argument here is that we are witnessing the evolution of some companies that previously operated solely in a business-to-consumer (B2C) context and are now moving to a technology-based business network, with other providers, a move that allows them to be more competitive and flexible for providing quick and agile answers to customers' evolving demand for rich and technology infused service encounters. However, it is timely to ask about how familiarized are the companies that are joining these business networks with the front-office service architectures that result from the combination of channels and providers, to address customer interactions? Are companies prepared for the challenge of working in an omni-channel network environment, involving the alignment of the service experience along multiple providers? This is of particular concern since firms can only be properly managed if practitioners understand the omni-channel service architectures and how they function on a business network context. The next sections aim to provide some answers to these questions.

2 Literature Review

Service business aiming for an effective adoption of multiple delivery channels, need to invest in the deployment of adequate operational capabilities. This process involves different stages of maturity in what concerns the level of interconnection and process integration for the different channels [15]. Many companies expand their business from one single channel to a service configuration employing multiple channels [10, 16]. Several concepts have been advanced to label these different channel strategies, (multi–, cross–, omni–channels), and whereas they are often used indistinctively in the academic literature, researchers continuously try to determine boundaries to avoid overlaps. For this reason, it is important to put forward some defining elements to frame each concept, and support our study. Channels are often defined as a customer contact point through which a firm interacts with their customers [17, p. 96]. Sousa and Voss [18, p. 357] distinguish virtual from physical channels: where "virtual channels consist of a means of communication using advanced telecommunications, information, and multimedia technologies (e.g. self-service checkout)" and "physical channels consist of a means of communications with customer employing a physical ("bricks-and-mortar") infrastructure". From a service process standpoint, channels can be addressed as the sum of routes or paths for customer-provider interaction employed by a company to deliver its products, services, or exchange information with recipients [19]. As customers are increasingly offered different channel alternatives and modes of communication with service providers [20], multi-channel strategies raise major technological and organizational concerns [21]. Jeanpert and Paché [21] stress that, in the literature dedicated to this theme, the emphasis is being placed on a global and combined management of all channels offered to consumers in terms of their coordination or even their integration. Sousa and Voss [18] define multi-channel service as a service involving components (physical and/or virtual) that are delivered through two or more channels and it may comprise a combination of virtual and physical services. A virtual service is defined as the "pure information component of a customer's service experience provided in an automated fashion (without human intervention) through a given virtual channel" [18, p. 357], whereas a physical service is defined as the "portion of a customer's service experience provided in a non-automated fashion, requiring some degree of human intervention, either through a virtual or a physical channel" [18, p. 358]. Cross-channel service emerges as a set of integrated activities that involve the use of a widespread set of channels to offer accessible services or products in-store and/or on the Internet, whereby the customer can trigger partial channel interaction and/or the retailer/service provider controls partial channel integration [21, 22]. In a cross-channel context, the complementarity and compatible nature of channels is a significant consideration for managers [23] as this is considered a crucial characteristic of this strategy [24]. Lastly, omni-channel services have been defined as a seamless and integrated shopping experience across all channels that blurs the distinction between physical and online stores, and culminates in an integrated brand experience [25, 26]. Although researchers have witnessed a continued evolution of channel strategies, there is already academic evidence of organizational synergies that are beyond the omni-channel concept, as they bring together a mix of channels and providers that need to be articulated in a seamless interaction with the customers [12]. These synergies may be interpreted as technology-based business networks, which are formalized as business-specific linkages negotiated between individual organizations [27] providing services delivered through a set of advanced technologies [28]. Each participating partner is mutually dependent upon resources controlled by the other, so that certain goals only become attained when their divided resources are combined [29]. Empirically, Reis *et al.* [30] found that these technology-based business networks (Tb2N) involve two or more companies in partnership, combine more than one channel and more than one service, which the customer perceives as an integrated network of brand experiences or multi-brand experiences (Fig. 1).

Technology-based business networks are introducing new dynamics; this strategy can be considered the next evolutionary step of omni-channel services, supporting the argument that this channel strategy arrived to stay.

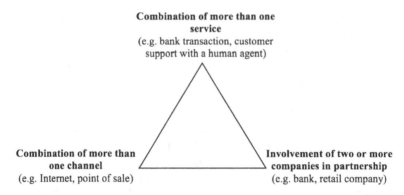

Fig. 1. Integrated network of brand experiences [30]

3 Methodology

This article employs a qualitative case study approach as the omni-channel service phenomenon should be studied in its natural setting; also because the case study methodology lends itself to early, exploratory investigations, where the variables are still unknown and the phenomenon not at all understood [31–33]. This methodology consisted on the analysis of multiple sources of data collection, including 7 semi-structured interviews, direct observations and analysis of institutional documents from a large private retail bank in Portugal. This study was conducted in Portugal because the banking industry offers a rich setting of increasing employment of omni-channel services. Over the past three decades' banks have been pioneers in adopting new information communication technologies [34] and in adopting new technology-enabled services [35], with the goal of improving customer relationships by empowering customers [36]. The prime source of data in this case study are semi-structured interviews, which is perhaps the most common type of interview used in qualitative social research [37]. Convenient and snowball sampling were used to select interviewees. The researchers made use of their personal network of contact inside the bank to identify the respondents who were in a best position to provide replies to the interview protocol. Subsequently, they were asked to nominate other employees, from different functional areas and different levels of responsibility, at the bank's physical branch. This process continued until theoretical saturation was achieved [38]. Direct observation as a source of evidence, contributes to the development of robust case studies, since it is an appropriate way to measure reality and generate truth about the world [39]. This technique allowed for the collection of aspects of everyday activities that may go unreported by participants, and gave the researchers direct experience of the phenomena being studied, while providing the opportunity to see and listen what was happening in the social setting as opposed to the focus on solely narrative descriptions of participants [40, 41]. For a reliable and accurate observation, field notes were taken [42]. These field notes served to document real life phenomenon events, serendipitous moments [43] and informal conversations with the interviewees [44]. The institutional documents are generally produced by organizations for communications or record-keeping purposes [41] and are a source of exceptional data collection because they were accessible and record the organizations' day-to-day activities. Official documents included organizational newsletters [42], internal records and reports available from the official bank website. The data was analyzed according to the technique of content analysis [45]. The textual data was categorized into codes or categories to identify consistent patterns and relationships between variables to reduce data and making sense of them [39]. NVivo 11 software was used to implement the data analysis procedure described, thus contributing to the robustness of the chain of evidence [46]. The reliability and validity of the case research data was enhanced by a well-designed research protocol [47], it was improved by using the multiple sources of evidence and by double-checking the transcripts and interview analyzes by participants.

4 Findings

This section provides an empirical summary that includes real-life statements, collected from the bank employees, direct observations and documental analysis.

4.1 Case Study General Overview

SIBS Group has been providing payment services worldwide over the last three decades. In Portugal, it performs a central role as a technology operator in the payments sector. In particular, with respect to the banking services, it manages the ATM network and the latest MB WAY brand [48]. The MB WAY has a vast number of adherents, including banks and retail stores, that have merged into a business network. With this solution it is possible to purchase services employing mobile payments (m-payments) in retail stores or online. SIBS, as a well established technology-based business firm, is managing the network, recruiting other companies and linking these companies by using a network-preferred channel (MB WAY). In this research we conducted a case study within a bank that joined the network and uses the network's preferred channel to do business. Our case is a well established private retail bank operating in Portugal, it uses a vast array of its own physical and virtual channels that are available to customers for service provision purposes. The bank mainly interacts with their customers using direct channels: (1) *bank mail*; (2) *bank website*, which includes two communication icons (click to call/chat) – *click to call*, a virtual icon that allows customers to receive contacts from the bank, and the *click to chat*, a virtual icon that allows customers to interact with the bank using a chat box; (3) *call center*; (4) *brick-and-mortar*; (5) *social networks*. As the bank has already joined the business network, this poses an immense challenge to operations management, since the bank will have to interact with the channels of other companies. Although there are no systematic metrics that allows us to accurately measure the degree of omni-channel intensity of an organization, we observed that the bank services have omni-channel characteristics. For instance, the bank employees have presented a mortgage loan as an omni-channel service, showing that the service purchase can go through all of the bank's direct channels, from the initial consultation to the provision of the loan. The purpose of this case study was to analyze if the omni-channel concept can be extended beyond the portfolio of channels owned and managed directly by an individual organization, i.e. integrate an entire business network, and the mix of proprietary channels and those of partner companies and providers, as the case of MB WAY. With this assumption, we are going to investigate the existing omni-channel service architectures in the technology-based business network (Tb2N).

4.2 Omni-Channel Architectures in a Technology-Based Business Network

The MB WAY concept, developed by the SIBS group, works as the network-preferred channel. This solution allows the bank addressed in the case study to connect with several retail companies that have joined the network. The network-preferred channel is a solution for mobile payments that enables immediate transfers and payments for purchase in several channels via mobile device and can thus combine an act of physical purchase with a virtual payment. The continuous evolution of wireless technologies, in combination with the widespread use of mobile devices, has paved the way for fast evolution of mobile commerce settings [49]. Several employees reported that there is an involvement of mixed services in a technology-based business network. From the interview data, it was possible to ascertain that mixed service encounters occur when two different and heterogeneous companies are involved, encompassing, simultaneously, a physical and a virtual purchase with a virtual payment to deliver a service to a customer. The direct observations corroborated the aforementioned arguments, with the witnessing of settings where a customer may choose a retail store from the business network to purchase a physical service, the purchase can be paid by e.g. mobile devices (m-payment), which connects the bank with the retail store. These mobile devices have advantages known as queue avoidance, immediacy, ease of use and low cost [49]. This experience can be considered as an omni-channel mixed service architecture in a technology-based business network; it comprehends a virtual payment to acquire a physical purchase. The Fig. 2 illustrates this real-life situation.

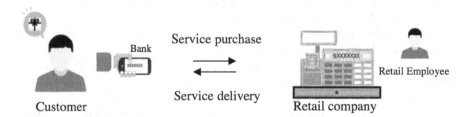

Fig. 2. Omni-channel mixed service architecture in a Tb2N context

Recent studies are in line with this concept in the sense that the Tb2N presuppose a trilogy crossing heterogeneous companies-channels-services; Reis *et al.* [30] argue that in a Tb2N context there is the involvement of two or more companies in a partnership (e.g. bank, retail store), the combination of more than one channel (e.g. smartphone, point of sale) and more than one service (e.g. bank transaction, customer support –

human agent), which is perceived by the customer as an integrated network of brand experiences or multi-brand experience. For those reasons, this omni-channel service architecture is a full Tb2N experience. Notwithstanding, the interview data also distinguished the abovementioned concept from other traditional mixed services. Traditional mixed services are offered when a single provider employs more than one channel, to offer different services, i.e. not involving more than one organization. The bank employees stated that, for instance, many of those customers that opt to open an account online, first search for additional information through the call center, and then upload all the necessary documents to open the account through the virtual channel, using their mobile device (e.g. table, smartphone). To finish the account opening process, they must perform a monetary transfer e.g. through the network-preferred channel, to officially start using the new account and complete the process. This situation involves the combination of more than one service (account opening and bank transfer), more than one channel (virtual channel and physical channel), but is missing the involvement of another company in the process, which is a pillar of the multi-brand experience triangle [30]. For that reason, we consider this as an incomplete architecture with regard to the Tb2N, although we recognize that traditional omni-channel architectures, working within organizations, may eventually be generalized in the future to the entire network.

Data analysis also reported the existence of pure virtual services in a technology-based business network. From the interview data it was possible to determine that pure virtual services comprise two different companies (e.g. bank and SIBS), it also comprehends different channels (e.g. mobile devices and ATM), and services (e.g. withdraw and balance inquiry). Official documents corroborated this information, and mention that customers may establish electronic bridges from their bank and automatic teller machines that are managed by SIBS. This connection is performed by mobile device, using the network-preferred channel (e.g. MB WAY), in order to withdraw money without any human intervention. We consider this experience as omni-channel pure virtual service architecture in a technology-based business network; it comprehends a virtual purchase to collect a physical service delivery. The Fig. 3 illustrates this real-life situation.

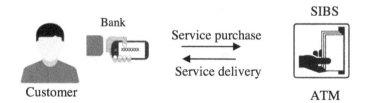

Fig. 3. Omni-channel pure virtual service architecture in a Tb2N context

Traditionally, these services do not involve two companies, e.g. transactions using debit/credit cards, which reinforces the idea that Tb2N is different because it includes more than one company in the process. In the Fig. 3 it is involved the user's bank (e.g. an adherent bank) and the company that manages the ATM network (i.e. SIBS). What underlies is that customers are not just debit/credit card holders, they make part of the process as self-service buyers and the purchase is encompassed by virtual services, and is virtually involving two companies. As self-service buyers, these customers also have other features that further distinguish this service from the traditional one. Customers are increasingly involved in the service, being able to interact with other customers, e.g. by sending a code, so that a member of the MB WAY can withdraw money from an ATM in real-time and geographically elsewhere.

Other example can be offered concerning the virtual pure services when customers goes to a retail store and pay their purchases at the self-checkout (employing what is known as self-service technologies), scanning their acquisitions and making the payment without any or low human intervention. In this example, we may include the use of a self-checkout, owned by the retail store (e.g. hypermarket), and the MB WAY application, that represents the customer's bank (network adherent bank). By using these self-service technologies, customers perform the service, or part of the service, traditionally performed by the service provider [50]. Recent research is in line with this concept: as self-service checkout systems intend to improve the efficiency of checkout operations and minimize the negative effects of traditional checkout service (e.g. minimize waiting experiences) [5], as this promote the expansion of virtual services through a business network. For these reasons, this omni-channel service architecture is a full Tb2N experience, which unlike the previous examples, fulfills the full requirements of the multi-brand experience triangle.

Lastly, the front-line employees also reported that there can be also an involvement of pure physical services in a technology-based business network. From the interview data we learned that the pure physical is a traditional service that comprises only one company (e.g. bank), it includes more than one channel (e.g. call center, bank employee) and services (monetary transfers, account opening). The bank employees mentioned a situation similar to one previously identified, where customers that open an account, first they search additional information through the call center (cf. Fig. 3, A), just then they start the process at the branch (cf. Fig. 3, B). Similarly, to the traditional mixed services, we consider this experience as omni-channel pure physical service architecture. The Fig. 4 illustrates this real-life situation.

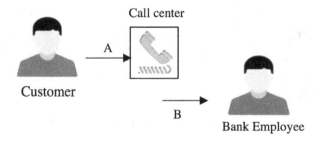

Fig. 4. Omni-channel pure physical service architecture

Extant theory is in line with the aforementioned concept of physical service [vide 18]. Human intervention can take place in the front office, the back office, or both [18], in this case we just considered the front-office. For the above reasons, we consider this experience as an incomplete architecture with regard to the Tb2N, since it does not fulfill the requirements of the multi-brand experience triangle. Table 1 summarizes this section.

Table 1. Omni-channel service architectures in a Tb2N

Omni-channel service architecture		Triangle of Tb2N multi-brand experience (full experience)			
		Combination of more than one service	Combination of more than one channel	Involvement of more than one company in partnership	Is this architecture a full Tb2N experience?
Mixed service encounter	Tb2N	X	X	X	Yes
	Traditional	X	X		No
Pure virtual service		X	X	X	Yes
Pure physical service		X	X		No

The existing business ecosystem (Table 1) is as complex as unique, due to its own specific relationships and technological features; therefore, the results should not be generalized to other contexts, although there are resemblances with similar networks (e.g. Apple Pay). We do acknowledge the traditional and physical banking services may include more than one company in partnership, which seems to contradict our results. This remark depends on the adjustment of the scope for a specific service, any of them can have a complete/incomplete architecture. To avoid misconceptions, we did map all the existent banking services and channels, thus we did not find any signs of business-to-business (B2B) relations that could call into question the incompleteness of both architectures (i.e. traditional mixed service encounter and pure physical service). The bank under study actually does establish B2B relations with other firms (i.e. by outsourcing services), but none of those firms operates inside the ecosystem. We do not exclude the possibility of one of those firms to join the ecosystem, or the bank to establish new and direct partnerships within the network. If all the modalities discovered in the case would satisfy the multi-brand experience triangle, our arguments would be strengthened and of greater value to achieve effective and adaptable omni-channel service architectures.

4.3 Bringing Operations Management to a Successful Channel Integration

A few years ago we observed a move from the adoption of multi- towards omni-channel strategies [10]. Launching an omni-channel strategy raises the challenge of managing and integrating the operations between several business partners. Previously, during the transition from brick-and-mortar to bricks-and-clicks, we noticed that channels were developed and managed separately by companies with limited integration [10, 51]. The multi-channel shopping revolutionized retail operations. For traditional retailers the growing importance of online sales meant creation of new business models, which required a solid understanding of the operational processes [52]. Companies had to redesign their processes within their premises. But, with the need to integrate operations within business networks, there is a need to redesign processes among several companies and, therefore, retailers' success often depended on the alignment between operations and the brand image [53]. The expanding range of online offerings and mobile devices are significantly changing firms' structures; for many firms, this implies completely new operational processes [52]. Thus, the implementation of an omni-channel strategy requires different companies to stop attempting to improve their own processes independently, in order to achieve a global benefit [54]. A parallelism can be made when retailers recognized the difficulty inherent in "going it alone" in the transition to an online environment [55]. One way of overcoming these difficulties was through joining an established partnership and integrates different interests among the business networks. The use of a common language among several companies has been fundamental for understanding each other. Nowadays, the challenge of technological integration and the consequent implementation of an omni-channel strategy pose challenges as well. For instance, Hansen and Sia [56] described how Hummel, a European sports fashion company, overcame the challenges and successfully transitioned toward omni-channel retailing. These authors noted that companies must focus on changes in technology infrastructures and organizational practices to successfully transform towards an omni-channel strategy [24]. Hansen and Sia [56] identified four lessons: (1) *embrace your channel partners in the omni-channel strategy*, which requires a continuous clarification of its omni-channel focus, specially in a business network; (2) *Recognize that a successful omni-channel strategy requires deep change*, it is not just about adding channels, it requires integration; (3) *leverage the strategic role of chief digital officer*, to establish a common mind-set across the organization; (4) *evolve the role of chief information officer (CIO) in enabling an omni-channel strategy*, to continue to be relevant in a digital transformation and in updating technical competencies in managing more front-end, customer facing IT systems. As depicted in Table 1, not all architectures are integrated as a full Tb2N experience. In other words, there are traditional architectures that use the network-preferred channel, but are not yet operating all their services within the network. Although the evolution from multi- to an omni-channel strategy within a company is far from being straightforward, and it might require a transition, inherent to the complexity that a move from a strategy to another implies [26]; we believe the same process takes place within the business network. Through direct observation we verified that not all the network organizations have the same omni-channel maturity and are not integrated into the business network to the same extent. The move to a Tb2N involves new challenges as it requires a transition based on operations management to allow these firms to adapt their processes and channels, in order to be able to collaborate in a heterogeneous network of firms [30]. Data analysis has shown that not all companies are fully prepared for the challenges of working in an omni-channel network environment, since they are not aware of the service architectures potential enabled by the entire technology-based business network. One possible solution is to invest in operations management, in a way that would be possible to integrate all the channels of a certain organization that join the network. The bank's employees were peremptory, companies should not manage the omni-channel services alone, this concept must be extended to the entire network, which means that the omni-channel strategy should go beyond individual organizations and should integrate the entire business network. Only in this way managers can move to the next level and integrate all the architectures of the omni-channel service as a full Tb2N experience, which is currently incomplete (cf. Table 1).

5 Concluding Remarks

The case study analysis revealed two types of omni-channel services architectures in the front-office of technology-based business networks, namely: mixed services and pure virtual services. Although we found four types of omni-channel architectures not all of them fulfilled the requirements of the multi-brand experience triangle and for that reason they were excluded. It is likely that this study will help practitioners to understand the challenges they will have to overcome within a Tb2N. However, it is possible that this article does not provide a complete classification of the omni-channel service architectures, since the ones that have been presented are limited to one case, and the theme is taking the first steps. Despite the exploratory nature of the article, we intend to fill a gap in the literature, for which this study may be a relevant contribution. Due to confidentiality reasons we do not provide any information about key informants and the respective organization, as the researcher is responsible not only for maintaining the confidentiality of all information but also for information that might affect the privacy of the research participants [57]. A number of avenues for future research arise from our study: first, empirical work is needed to enrich the Tb2N, as it may be interesting to conduct a study within the Tb2N that would focus not only on collecting data from one company but from the entire network; second, there is a need of measurement instruments to determine the maturity level for omni-channel banking services, so as to measure the degree of omni-channel intensity of those organizations.

Acknowledgment. This work is partially financed by funds from CINAMIL/CISD research and development center. Furthermore, we would like to thank the continuous support of the Polytechnic Institute of Viseu and CI&DETS.

References

1. Cook, D., Goh, C., Chung, C.: Service typologies: a state of the art survey. Prod. Oper. Manag. **8**(3), 318–338 (1999)
2. Vargo, S., Maglio, P., Akaka, M.: On value and value co-creation: a service systems and service logic perspective. Eur. Manag. J. **26**(3), 145–152 (2008)
3. Lusch, R., Nambisan, S.: Service innovation: a service-dominant logic perspective. MIS Q. **39**(1), 155–175 (2015)
4. Meuter, M., Ostrom, A., Roundtree, R., Bitner, M.: Self-service technologies: understanding customer satisfaction with technology-based service encounter. J. Mark. **64**(3), 50–64 (2000)
5. Morimura, F., Nishioka, K.: Waiting in exit-stage operations: expectation for self-checkout systems and overall satisfaction. J. Mark. Channels **23**(4), 241–254 (2016)

6. Chou, S., Shen, G., Chiu, H., Chou, Y.: Multichannel service providers strategy: understanding customers' switching and free-riding behavior. J. Bus. Res. **69**(6), 2226–2232 (2016)
7. Chiou, J., Wu, L., Chou, S.: You do the service but they take the order. J. Bus. Res. **65**(7), 883–889 (2012)
8. Chiu, H., Hsieh, Y., Roan, J., Tseng, K., Hsieh, J.: The challenge for multichannel services: cross-channel free-riding behaviour. Electron. Commer. Res. Appl. **10**(2), 268–277 (2011)
9. Verhoef, P., Neslin, S., Vroomen, B.: Multichannel customer management: understanding the research-shopper phenomenon. Int. J. Res. Mark. **24**(2), 129–148 (2007)
10. Verhoef, P., Kannan, P., Inman, J.: From multi-channel retailing to omni-channel retailing: introduction to the special issue on multi-channel retailing. J. Retail. **91**(2), 174–181 (2015)
11. Sousa, R., Amorim, M., Pinto, G., Magalhães, A.: Multi-channel deployment: a methodology for the design of multi-channel service processes. Prod. Plan. Control **27**(4), 312–327 (2016)
12. Reis, J., Amorim, M., Melão, N.: New ways to deal with omni-channel services: opening the door to synergies, or problems in the horizon? In: Za, S., Drăgoicea, M., Cavallari, M. (eds.) IESS 2017. LNBIP, vol. 279, pp. 51–63. Springer, Cham (2017). https://doi.org/10.1007/978-3-319-56925-3_5
13. Pittaway, L., Robertson, M., Munir, K., Denyer, D., Neely, A.: Networking and innovation: a systematic review of the evidence. Int. J. Manag. Rev. **5/6**(3/4), 137–168 (2004)
14. Holm, D., Eriksson, K., Johanson, J.: Creating value through mutual commitment to business network relationships. Strateg. Manag. J. **20**, 467–486 (1999)
15. Hübner, A., Wollenburg, J., Holzapfel, A.: Retail logistics in the transition from multi-channel to omni-channel. Int. J. Phys. Distrib. Logist. Manag. **46**(6/7), 562–583 (2016)
16. Brynjolfsson, E., Hu, Y., Rahman, M.: Competing in the age of omnichannel retailing. MIT Sloan Manag. Rev. **54**(4), 23–29 (2013)
17. Neslin, S., et al.: Challenges and opportunities in multichannel customer management. J. Serv. Res. **9**(2), 95–112 (2006)
18. Sousa, R., Voss, C.: Service quality in multichannel services employing virtual channels. J. Serv. Res. **8**(4), 356–371 (2006)
19. Mehta, R., Dubinsky, A., Anderson, R.: Marketing channel management and the sales manager. Ind. Mark. Manag. **31**(5), 429–439 (2002)
20. Kotarba, M.: New factors inducing changes in the retail banking customer relationship management (CRM) and their exploration by the fintech industry. Found. Manag. **8**(1), 69–78 (2016)
21. Jeanpert, S., Paché, G.: Successful multi-channel strategy: mixing marketing and logistical issues. J. Bus. Strateg. **37**(2), 12–19 (2016)
22. Beck, N., Rygl, D.: Categorization of multiple channel retailing in multi-, cross-, and omni-channel retailing for retailers and retailing. J. Retail. Consum. Serv. **27**, 170–178 (2015)
23. De Faultrier, B., Boulay, J., Feenstra, F., Muzellec, L.: Defining a retailer's channel strategy applied to young consumers. Int. J. Retail Distrib. Manag. **42**(11/12), 953–973 (2014)
24. Mirsch, T., Lehrer, C., Jung, R.: Channel integration towards omnichannel management: a literature review. In: Proceeding of the 20th Pacific Asia Conference on Information Systems (PACIS) (2016)
25. Aradhana, G.: Technological profile of retailers in India. Indian J. Sci. Technol. **9**(15), 1–16 (2016)
26. Picot-Coupey, K., Huré, E., Piveteau, L.: Channel design to enrich customers' shopping experiences. synchronizing clicks with bricks in an omni-channel perspective – the direct optic case. Int. J. Retail Distrib. Manag. **44**(3), 336–368 (2016)

27. Jackson, K., Matsumoto, S.: Business networks in Japan: the impact of exposure to overseas markets. In: Business Networks in East Asia Capitalisms: Enduring Trends, Emerging Patterns, p. 143 (2016)
28. Palo, T., Tahtinen, J.: A network perspective on business models for emerging technology-based services. J. Bus. Ind. Mark. **26**(5), 377–388 (2011)
29. Willms, W., Flirber, U., Hardt, U., Jung, H.: Konzept fur eine regionale Infrastrukturpolitik im Raum der gemeinsamen Landesplanung Bremen/Niedersachsen. Band II: Wissenschaft und Forschung. In: Gemeinsame Landesplanung Bremen Niedersachsen Schriftenreihe, pp. 5–94. Hannover, Bremen (1994)
30. Reis, J., Amorim, M., Melão, N.: Omni-channel services operations: building technology-based business networks. Service operations and logistics, and informatics (SOLI). In: IEEE International Conference, Bari, Italy, pp. 96–101 (2017)
31. Meredith, J.: Building operations management theory through case and field research. J. Oper. Manag. **16**, 441–454 (1998)
32. Benbasat, I., Goldstein, D., Mead, M.: The case research strategy in studies of information systems. MIS Q. **11**, 369–386 (1987)
33. Voss, C., Tsikriktsis, N., Frohlich, M.: Case research in operations management. Int. J. Oper. Prod. Manag. **22**(2), 15–219 (2002)
34. Cortiñas, L., Chocarro, R., Villanueva, M.: Understanding multi-channel banking customers. J. Bus. Res. **63**(11), 1215–1221 (2010)
35. Sousa, R., Amorim, M.: A framework for the design of multichannel services. Project for the Foundation for Science and Technology, under grant number PTDC/GES/68139/2006 (2009)
36. Elgahwash, F., Freeman, M.: Self-service technology banking preferences: comparing libyans' behaviour in developing and developed countries. In: Banking Finance, and Accounting: Concepts, Methodologies, Tools, and Applications, pp. 714–727. IGI Global (2015)
37. Dawson, C.: Practical Research Methods: A User-friendly Guide to Mastering Research Techniques and Projects. How to books Ltd., Oxford (2002)
38. Saunders, M., Townsend, K.: Reporting and justifying the number interview participants in organizational and workplace research. J. Manag. **27**(4), 836–852 (2016)
39. Given, L.: The Sage Encyclopedia of Qualitative Research Methods. Sage Publications, Thousand Oaks (2008)
40. Patton, M.: Qualitative Research and Evaluation. Sage, Thousand Oaks (2002)
41. Mills, A., Durepos, G., Wiebe, E.: Encyclopedia of Case Study Research, vol. 1. Sage, Thousand Oaks (2010). Index, L.-Z.
42. Berg, B., Lune, H., Lune, H.: Qualitative Research Methods for the Social Sciences. Pearson, Boston (2004)
43. Fisher, C.: Researching and Writing a Dissertation: A Guidebook for Business Students. Pearson Education, London (2007)
44. Hancock, D., Algozzine, B.: Doing Case Study Research: A Practical Guide for Beginning Researchers. Teachers College Press, New York (2015)
45. Marvasti, A.: Qualitative Research in Sociology. Sage, Thousand Oaks (2003)
46. Bazeley, P., Jackson, K.: Qualitative Data Analysis with NVivo. Sage Publications, London (2013)
47. Yin, R.: Case Study Research: Design and Methods. Sage Publications, Thousand Oaks (2003)
48. SIBS International Worldwide Payment Solutions. https://www.sibs-international.com

49. Kousaridas, A., Parissis, G., Apostolopoulos, T.: An open financial services architecture based on the use of intelligent mobile devices. Electron. Commer. Res. Appl. **7**(2), 232–246 (2008)
50. Demoulin, N., Djelassi, S.: An integrated model of self-service technology (SST) usage in a retail context. Int. J. Retail Distrib. Manag. **44**(5), 540–559 (2016)
51. Verhoef, P.: Multi-channel customer management strategy. In: Handbook of Marketing Strategy, pp. 130–150 (2012)
52. Hübner, A., Holzapfel, A., Kuhn, H.: Operations management in multi-channel retailing: an exploratory study. Oper. Manag. Res. **8**(3–4), 84–100 (2015)
53. Brun, A., Castelli, C.: Supply chain strategy in the fashion industry: developing a portfolio model depending on product, retail channel and brand. Int. J. Prod. Econ. **116**(2), 169–181 (2008)
54. Bagchi, P., Skjoett-Larsen, T.: Supply chain integration: a european survey. Int. J. Logist. Manag. **16**(2), 275–294 (2005)
55. Kennedy, A., Coughlan, J.: Online shopping portals: an option for traditional retailers? Int. J. Retail Distrib. Manag. **34**(7), 516–528 (2006)
56. Hansen, R., Sia, S.: Hummel's digital transformation toward omnichannel retailing: key lessons learned. MIS Q. Execut. **14**(2), 51–66 (2015)
57. Marczyk, G., DeMatteo, D., Festinger, D.: Essentials of Research Design and Methodology. Wiley, Hoboken (2005)

An Approach for Customer-Centered Smart Service Innovation Based on Customer Data Management

Katharina Blöcher[✉] and Rainer Alt

Information Systems Institute, University of Leipzig, Leipzig, Germany
{katharina.bloecher,rainer.alt}@uni-leipzig.de

Abstract. The purpose of this paper is the development of a customer-centered understanding of smart services by utilizing customer-dominant logic as a theoretical view. It extends current perspectives on smart services and proposes a definition highlighting the relevance of customer centricity and customer data management for service engineering. For this purpose, smart service dimensions will be deduced from current literature. These dimensions represent the foundation for a procedure model that will be developed and tested within a research initiative focusing on the restaurant industry. This procedure model is intended to support the discovery of new ideas based on data as well as the management of data requirements to successfully design smart services. The restaurant sector serves in this context as an innovative application field for smart service prototyping. The paper represents current research in progress, outlines the objective of the research initiative and demonstrates first empirical research results.

Keywords: Smart service engineering · Customer-dominant logic
Personalization · Customer data management · Procedure model

1 Introduction

Driven by digitalization, whole economies are transforming into digital service industries. In the time of big data, the Internet of Things and servitization, there is an unprecedented potential for service innovation based on emerging technologies and new volumes of data [1, 2]. Rising amounts of sensor data, social data and more, new levels of information exchange and intelligent algorithms create the foundation for new services that offer exciting value propositions to customers [2, 3].

Smart services have emerged and spread into a variety of different industries, e.g. smart home, smart mobility or smart tourism [4–6]. Characterized by the interconnectedness of their ecosystem, they make use of available data and offer close customer interaction and intelligent, context-based solutions [5, 7]. In this way, they follow increasing customer expectations, in terms of convenience, participation or transparency in our information-driven society [8].

After the financial, automotive and electricity sector, the food industry has started its digital transformation [9]. As one important cornerstone of the branch, the ongoing digital revolution of the restaurant sector shows very clearly how digitalization is

© Springer Nature Switzerland AG 2018
G. Satzger et al. (Eds.): IESS 2018, LNBIP 331, pp. 45–58, 2018.
https://doi.org/10.1007/978-3-030-00713-3_4

changing traditional service industries, e.g. [10]. Service processes, such as reservations, food delivery, or payments are increasingly supported by IT solutions [11]. Social Media, chatbots, and voice technology transform communication processes and the Internet of Things has brought innovative sensor-based digital contact points, such as interactive tables, bars or digital signages [12]. This does more than creating new experiences of interaction and co-creation with customers [13]. Data tracks produced at every digital contact point also offer new opportunities for intelligent data-based service innovation, such as personal menu recommendations [3, 14]. Even if this represents an exciting playground for innovation, it seems reasonable to make use of systematic approaches to start the journey of tapping into this potential due to massive volumes of data, an increasing amount of diverse service providers and rising complexity, e.g. [15].

The research area that focuses on the structured development of services is the field of service engineering. Being part of the service science discipline [16], it is an interdisciplinary approach that aims to systematically design service value propositions using frameworks, methods and tools [17, 18]. With increasing demands for personalized offers in the food industry [19, 20] just as in other sectors, customer centricity and personalized customer solutions have gradually gained in importance in service engineering [21, 22]. Services need to be tailored more strongly to the individual, which means that detailed customer knowledge has become more valuable than never before [23]. For this reason, personal customer data has been highlighted recently as a key concept in service innovation [15, 23]. With the aid of personal customer data, a broader customer understanding can be developed [3, 15] and innovative value propositions, higher-quality services and deeper customer relationships are created based on personalization [5, 23]. However, managing personal customer data is also associated with several challenges – especially in a world of growing amounts of (big) data. Data needs to be aggregated and processed from different contact points and disparate external and internal systems – often even in real-time to generate value to customers [6, 24, 25]. New data formats need to be handled [23] and privacy and data ownership issues ensured [15, 25]. With recent changes, such as the new European GDPR regulation [26], these issues become even more crucial.

As a result, this work points out that the management of customer data for smart service innovation requires a specific focus. It is postulated that efficient innovation processes to rapidly design novel prototypes based on customer data require new approaches due to their specific requirements. For this reason, this research aims to further develop the perspective on smart service innovation to a strong focus on personalization and personal customer data management.

2 Research Objective and Paper Contribution

As mentioned above, recent digital movements result in an increasing significance of systematic service engineering [27, 28]. However, even if more and more researchers emphasize the importance of personal customer data [23, 29], there is currently less research, e.g. [28, 29], that focuses specifically on the role of data as a resource for smart service innovation, while simultaneously addressing relevant data management aspects associated with, for example, rising volumes of data or increasing interdependencies of service partners.

As a consequence, this research initiative has the objective to develop a new procedure model with a focus on customer data management in order to evolve the research field of smart service engineering. For this purpose, the following research questions have been formulated:

RQ1: How can smart service engineering benefit from a focus on customer data management?

RQ2: Which requirements need to be considered in smart service engineering that are related to personal customer data?

This procedure model is intended to support the discovery of innovative ideas based on customer data as well as the management of data requirements by offering appropriate guidelines and methods [30]. The restaurant industry serves in this context as innovative application field.

Following a design science approach, the research will be divided into three different phases taking advantage of various methods, such as prototyping, interviews, workshops and systematic literature reviews [31, 32]. In the foundation phase, the application field will be analyzed with the aim to understand current challenges of the transforming restaurant industry. Furthermore, a systematic literature review evaluates existing models and assembles relevant criteria that are related to customer data management. Based on these results, a procedure model will be developed.

The second research phase aims to apply the model in a practical environment. For this purpose, a research project has been set-up in cooperation with TNC Group in Leipzig and a consortium of stakeholders, such as the food and beverage industry (e.g. Coca Cola, Nestlé), tool providers (e.g. OpenTable, Vectron, etc.) and leading restaurant owners, that constitute a good representation of the German restaurant landscape. Workshops and interviews are being organized around this consortium in order to test relevant concepts and methods and to generate learnings for model refinement. A third evaluation phase simultaneously assesses the use of models and methods to further improve the artefacts by means of surveys and interviews.

This paper provides theoretical groundwork for the development of the procedure model and can therefore be located in the primary research phase. It will present a customer-centered smart service definition as a base for systematic smart service engineering [30]. Such a definition of the essence of smart services is currently missing due to little theoretical knowledge [15, 33, 34] and no common understanding of smart services [33, 35].

By outlining relevant research contributions in (1) service engineering and (2) smart service innovation, smart service characteristics will be linked to a theoretical perspective focusing on customer centricity and customer knowledge: the customer-dominant logic that is grounded in the research tradition of the Nordic school of service management [36–38].

Using this customer-centered approach, the role of customer data management will be emphasized. Simultaneously, a broader view on smart services will be developed [39, 40] in order to guide smart service innovation, for example in the restaurant sector, in wider directions to create more encompassing customer solutions. Finally, this paper summarizes the first empirical results utilizing the definition for first idea management and requirements analysis and rounds off with an outlook to further research activities.

3 Theoretical Foundation

3.1 Service System Engineering

Service engineering (SE) approaches are based on the conviction that systematic service design needs to consider the specific characteristics of services [41] that differentiate them from products [30]. Furthermore, service-dominant logic (SDL) has significantly influenced the SE discipline. SDL specifically highlights the customer's role as well as the concept of value-in-use utilizing available resources for value creation (e.g. labor, knowledge). In addition, it describes services as a mutual process of co-creation and collaboration between service provider and customers [42, 43].

Services are increasingly created in service value networks with multiple agents that cooperate and exchange resources [44, 45]. This view on service systems offers new possibilities for innovation based on resource sharing, but it also signifies that the network of multiple agents has to be taken into account in service engineering [46].

Due to digitalization, ICT plays an enormous role in service systems and changes the way interaction and information exchange take place [1, 47]. There are, however, still major differences in the degree of IT usage [48]. On the one hand, interconnected objects, new forms of product-service systems or cyber-physical systems [3, 49] have emerged, triggering automatic service processes and work without any human interaction, e.g. in smart factories. On the other hand, there are industries, such as the restaurant sector, where services are strongly focused on personal interaction. Even if these human-centered service systems are transforming and a rising share of "eservices" are supported by ICT [50], they still require special attention on personal interaction [51, 52].

With the widespread presence of eservices, digital contact points and the resulting rise of information exchange, information-intense services or knowledge-based services offer new potential for service innovation, e.g. econsulting, elearning [2]. In this context, successful SE needs to consider the management of data sources, data collection, analysis of data or the information delivery, etc. [29]. Additionally, if information is allocated across different service systems, e.g. in diverse IT systems, SE needs to take care of the systematic management of data within the whole network [15, 29]. Here, multi-sided service platforms have gained in importance in order to simplify centralization and access to data [6, 15, 23].

To conclude, services have specific characteristics that need to be considered. However, the understanding of services has recently changed dramatically due to the significance of ICT in service provision. ICT and new volumes of data do not only transform service industries, such as the restaurant sector, they also offer novel opportunities for new service innovation [5].

3.2   Smart Service Innovation

Smart service innovation is on the agenda of a lot of different research streams, e.g. business literature [53], service management & marketing [4], IS literature [5] as well as in practice [7]. It is understood as a transformation of existing service systems, a recombination of existing service elements [28] or even the establishment of new service systems [54].

Research in this field often takes place in specific industries, e.g. smart home [55], smart mobility and smart cities [56] as well as smart tourism and hospitality [33]. Even if there is no mutually defined understanding of smart services [33, 35], there are particular characteristics that differentiate them from previous forms of service provision. Key characteristics are (1) the use of data (e.g. web logs, sensor networks, social networks, etc.) [3, 7], (2) the ability to infer intelligence and actions based on algorithms and machine learning [33, 35], and finally (3) the interconnectivity to objects and products as well as the synchronization of different technologies [55]. As a consequence, they can be described as extremely information-intensive or knowledge-based services [15, 29]. Smart services can significantly differ in their visibility and the associated level of customer participation [57]. Smart interactive services, for example, are characterized by a high level of interaction. In contrast, self-regulating services [35], such as smart home applications for temperature regulation, do not require any active customer participation. Arising research themes are, for example, user acceptancy or further success factors for smart service usage, e.g. embeddedness in daily routines, visibility, user engagement, etc. [4, 55, 56]. Additionally, the field of product-service systems [58] and cyber-physical systems [3, 49] as a combination of product-service bundles has received a lot of attention recently.

Furthermore, first important frameworks have been formulated. Nadj et al. [35], for example, have categorized specific activities (sensing, decision making and learning) and abilities of smart services (e.g. autonomy, proactivity, etc.). Klein [34] has examined barriers of implementation and has formulated specific design guidelines for smart service development. Kleinschmid et al. [52] have developed design principles to align service system engineering with business modelling in human-centered service systems. The specific role of personal data for smart service innovation has been highlighted by certain researchers [15, 29]. Increasing volumes of data offer opportunities for a new customer understanding [3, 23], but need to be managed systematically across service systems as mentioned above. The ability to do so also depends on the appropriate infrastructure and organizational capabilities, e.g. [6, 29]. User context or user profiles need to be built for personalization [22] and data privacy must be guaranteed [3, 15].

4   A Customer-Centered Smart Service Definition

As mentioned, smart service research is a relatively young research field with little theoretical foundation. Even if the existing body of research describes particular characteristics of smart services [35, 54], research is often marked by the dependency of domains, use cases or specific research issues, e.g. user acceptancy [34, 55]. As a consequence, there is a lack of comprehensive understanding of the essence of smart services [33, 35]. Even if a strong emphasis on customer orientation and data is embedded in the understanding of smart services, so far barely any research could be identified that focuses specifically on the significance of personal customer data and analyses associated requirements for smart service design.

Nonetheless, the development of a procedure model for smart service engineering requires a definition that formulates and conceptualizes relevant service dimensions and focuses on customer data management. For this purpose, this paper aims to develop a theory-grounded definition that uses customer-dominant logic (CDL) [36, 37] as a theoretical lense.

This work postulates that this perspective is not only suitable for emphasizing strong customer orientation in smart service design [39]. It also provides new perspectives for innovative customer solutions based on successful customer data management, thus responding to rising challenges in terms of customer empowerment and competition [38]. Smart service characteristics are conceptualized in the definition below and distinct dimensions are worked out with the aim to guide the analysis of requirements and the engineering process of smart services. In the following part, the fundamental principles of CDL will be described. Service literature has been analyzed in order to synthesize perspectives for the definition of smart service dimensions.

Customer-dominant logic (CDL) represents a managerial perspective on service management that originated from the business and marketing field [36, 37]. In contrast to service-dominant logic, CDL does not focus on the single service situation and interaction process between service provider and customer. Instead, it positions the individual customer and the understanding of their daily life and objectives in the center of activities.

Personalized: Heinonen et al. [36] address the necessity of considering the customer's specific individual situation, their objectives, daily routines and experiences. Their activities are based on their own individual logic driven by their aspirations, tasks, and previous experiences [38]. Heinonen and Strandvic insist that value formation is also dynamic and varies depending on the individual customer. Consequently, services need to adapt to the customer's individual situation.

Smart service literature also states: smart services should assist customers in their daily tasks by understanding the customer's individual context [5, 7].

Applying this to a smart service definition, smart services need to be personalized and tailored to the individual customer/customer's logic and provide in this way a personal value to them, e.g. finding the right restaurant for the right situation and place.

Data-Based: CDL states that the consideration of the customer's individual context requires deep knowledge for building up understanding of activities, practices and experiences. Understanding the customer's logic should be the center of the service offering [37]. By taking advantage of growing information sources, e.g. sensors, social media, etc., technology and data can be utilized to create such a deep understanding [22, 23]. Generating this knowledge within the customer's service system [36] is, however, related to the analysis of relevant contact points, e.g. inside and outside the restaurant, describing customer activities and processes as well as the collection, management and aggregation of valuable customer data. This also depends on possibilities of smoothly identifying the customer across different contact points (e.g. via NFC-technology or face recognition).

Intelligent: Comprehensive knowledge then allows "to participate in and support customer's processes" with the aid of individual service offerings [36].

Analytics can, thereby, support businesses to build up this knowledge about the customers' reasoning and logic. According to Heinonen and Strandvik [37], an important assumption is that the customer's logic can be identified and classified into "a usable number of groups". Analytics and algorithms can discover these patterns in customers' activities, to group logics together and to build up valuable knowledge about the customer.

According to current literature, smart services tailor assistance by making use of analytics and logics on how to best assist customers with relevant activities (e.g. recommendations, decision-supporting, etc.), e.g. [35].

Consequently, it can be stated that smart services are intelligent services that take advantage of analytics and algorithms with the aim to build up deep knowledge about the customer's logic.

Solution Bundle: According to CDL, value creation never takes place in a vacuum of single service situations. Customers need to handle different problems in their daily activities so that services should support the orchestration of these activities. Heinonen et al. [36] postulate that customers create their value independently in different service situations and orchestrate them to reach their objectives. Services should facilitate this orchestration. This indicates that activities should not be perceived separately. For a successful restaurant experience, for example, guests need to reserve or look for a table, they must choose a meal, order, pay, etc. These activities should be interpreted as a network that needs to be supported by solution bundles instead of separate services [59, 60]. This understanding can also be found in smart service contributions, defining smart services as solution bundles combining products and services in encompassing product-service systems [5].

Consequently, smart services need to bundle different service components with the aim to simplify the orchestration of customer activities and to offer valuable solution bundles to customers [7].

Interconnected: Paying attention to daily routines and activities of the customer also means that services need to be integrated into the customer's life. Heinonen and Standvic [37, 38] postulate that (service) providers need to be "present" and involved in customers' lives and service offerings should be embedded smoothly in customers' activities in order to create value.

New cyber-physical applications (screens, tables, smart devices, etc.) go beyond digital services, such as smartphone or desktop apps, and allow for a much deeper positioning of services in daily routines. Consequently, smart services are able to support daily activities at their best when technologies are deeply embedded in customers' lives [4, 55]. This also requires convenient forms of information provision in the context of the service provision [29]. Furthermore, based on cloud computing, virtualization and synchronization of different technologies [55] smart services need to be device-independent so that the services can be used on every suitable device [55].

Control Features: CDL further develops the role of the customer and emphasizes that customers are not only co-producers like in SDL [42]. Instead, they are the initiators that utilize and control service processes for pursuing their objectives [36]. This signifies that they are also fully in control of their resources. Transferring this critical point to a customer-centered definition, this demonstrates that smart services should do more than ensuring visibility and interactivity [4]. Appropriate instruments are also necessary to enable customers to initiate and control services as well as the active exchange of resources, including personal data. Smart services must simultaneously allow customers to transparently look at their data and to actively control the provision or the "not-provision" of their information. This is also in line with available research [4] as well as privacy regulations (EU GDPR).

The following Table 1 displays relevant service dimensions related to the main statements of the customer-dominant logic (CDL) perspective.

Table 1. Overview: smart service dimensions.

CDL [36–38]	Smart service dimensions
Personal value in customer lives.	**Dimension I: Personalized** Smart services are customer-centered services offering personalized value by assisting customers in their activities to reach their individual objectives.
Deep knowledge at the center of service offerings.	**Dimension II: Data-based** Using available information sources in the customer's ecosystem, smart services are based on customer data by integrating & connecting data to a single customer with the objective to reach an holistic view of them.
Customer logic can be classified into groups.	**Dimension III: Intelligent** Smart services make use of analytics to build up knowledge on how to best understand the customers' logic and to assist them with relevant activities (e.g. recommendations, decision-supporting, etc.).
Services must facilitate the orchestration of activities.	**Dimension IV: Solution Bundle** As digital assistants, they offer solution bundles that facilitate the orchestration of services for customers' daily life tasks.
Embeddedness in daily routines.	**Dimension V: Interconnected** With the aid of technologies (applications, cloud computing, virtualization, etc.), smart services are interconnected with relevant objects to integrate service provision as deeply as possible in daily life activities. Simultaneously, they can be used across contact points or devices so that customers can make use of the service at every relevant device or contact point.
As initiators, customers control service processes.	**Dimension VI: Control Features** Smart services are interactive services enabling customers to control service processes as well as relevant resources, especially personal data, at any time.

5 Empirical Results

In empirical research, the smart service definition has already been applied to offer guidance for the work on smart service innovation. A first workshop served as starting point to build up practical knowledge of the application field, to understand current stakeholders, technical systems and current collaboration structures as well as to discuss current challenges and opportunities for the restaurant industry. Furthermore, it has been utilized to analyze necessary requirements for smart service engineering from a practical point of view.

For this purpose, design thinking methods, such as personas and customer journey mapping were chosen. Design thinking approaches with a strong focus on user centricity, customer integration and multidisciplinary collaboration have recently become highly relevant in SE [61] driven by the growing emphasis on customer orientation and service systems. Consequently, these methods have been highly suitable for the customer-centered development of smart service ideas. In preparation for this workshop, a systematic review of current market studies (restaurant/food industry) was conducted. Relevant consumer studies were utilized (e.g. [19, 20]) in order to describe six personas that summarize actual trends in customer needs and behaviour, e.g. the need for information transparency, customization and convenience during the restaurant experience [19]. Furthermore, journey mapping was used to display a typical visit at a restaurant with typical tasks and activities, such as table reservation, waiting for food, etc. In this way, specific customer needs were deduced that served as starting point for idea generation.

The workshop resulted in four smart service ideas, e.g. a health and nutrition platform for customers that demonstrate the opportunities of smart services in the restaurant industry. By developing these ideas, characteristics of smart services and related requirements were derived from the discussion how to bundle different services and to manage personal customer data for personalized offerings. The results are displayed in Table 2. Requirements related to customer data management are, for example, organizational issues, e.g. the development of new business models, the clarification of data ownership, or process issues, such as channel management as well as the set-up of interorganizational platform architectures to ensure information exchange. Even if these aspects only represent the beginning of a systematic requirements analysis, similar points could be already identified within current literature, e.g. [6, 15, 25].

Table 2. Workshop results (Nestlé Competence Center, Frankfurt am Main, September 2017).

Dimensions	Workshop Results Structured by Smart Service Dimensions
Dimension I: Personalized	**Personalized Value Propositions** • Personalized service in restaurants • High-quality consulting in health care and nutritional matters • Exclusive invitations for restaurants and events and entertainment services • Shopping functionalities and home delivery for favorite products (e.g. wine) • Ordering processes within specific activities, such as gaming or driving • Personalized offerings and gratification processes **Discussed Requirements** • Mutual business model for all service providers within the service system • Ownership of infrastructure • Novel business potentials through cooperation and data sharing • Payment options on customer side vs. service provider side
Dimension II: Data-based	**Customer Data** • Contact information, wine preferences, car brand, allergies, hobbies, profession, average meal size, last orders, collected points for gratification system, health information, favorite restaurants, Youtube stars/influencer **Discussed Requirements** • Necessity of contact information & permission management • Necessary reason for data provision on customer side • Open APIs and interconnectedness of available systems (e.g. reservation systems, POS systems, etc.) • Identification method: One central log-in on customer-side, other identification methods, such as face recognition • Central management of customer data on service provider side • Central management of all data sources for data integration
Dimension III: Intelligent	**Analytics** • Recommendation engine: content or entertainment services, restaurants, menus, etc. • Location-based services
Dimension IV: Solution Bundle	**Bundling Features** • Platform as possibility to bundle different services across various industries **Discussed Requirements** • Open access & cooperation models
Dimension V: Interconnected	**Possible Contact Points** • Alexa, Siri, smartphone apps, tables, instore signages, sms, car, call-center assistant, waiter, chatbot, augmented reality, digital ear-prompter, Youtube, smart TV, wearables **Discussed Requirements** • Independency of devices or contact points; virtualization • Form of information provision (voice-assistant vs. sms vs. online games)
Dimension VI: Control Features	**Possible Control Features** • One central portal for self-management of personal data • Search engine to receive relevant recommendations based on profile information and search queries • Configurators, filter options, etc.

6 Future Research

As mentioned above, the customer-centered definition of smart services serves as starting point for further research. This paper displays current research in progress and intends to expand research knowledge on smart service design. Defined dimensions will be used for the development of a procedure model that guides the engineering process by focusing on personal customer data management. A preliminary exemplified model is displayed in Fig. 1. The model contains more than the relevant steps along the engineering process, e.g. idea management. It also forms a guide to new potentials and summarizes relevant requirements by focusing on customer data management. The dimensions of the smart services definition represent relevant guiding perspectives for service innovation, e.g. regarding the innovation potentials that arise from information control/provision features, or data requirements related to the bundling of services. In this way, the model aims to guide the engineering process by identifying new potentials for service innovation (RQ1) as well as managing relevant issues (RQ2) around customer data.

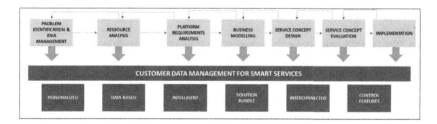

Fig. 1. Reduced draft of a procedure model for smart service engineering [28, 29].

Based on a comprehensive literature review, further requirements for smart service engineering will be assembled. Existing procedure models will be compared and a detailed criteria catalogue will be developed. For this purpose, insights from different research fields such as CRM, information management, network theory and service system literature will be utilized. The objective is to structure particular aspects and to develop guiding principles as part of the procedure model that will be continuously discussed and supplemented with the aid of further empirical and theoretical work.

In future workshops, further methods will be tested that highlight the potential of customer data for service innovation and guide in this way the innovation process.

References

1. Barrett, M., Davidson, E., Prabhu, J., Vargo, S.L.: Service innovation in the digital age: key contributions and future directions. MIS Q. **39**(1), 135–154 (2015)
2. Lusch, R.F., Nambisan, S.: Service innovation: a service-dominant logic perspective. MIS Q. **39**(1), 155–175 (2015)

3. Demirkan, H., Bess, C., Spohrer, J., Rayes, A., Allen, D., Moghaddam, Y.: Innovations with smart service systems: analytics, big data, cognitive assistance, and the internet of everything. Commun. Assoc. Inf. Syst. **37**(35), 734–752 (2015)
4. Wünderlich, N., et al.: Futurizing smart service: implications for service researchers and managers. J. Serv. Mark. **29**(6/7), 442–447 (2015)
5. Thomas, O., Nüttgens, M., Fellmann, M.: Smart Service Engineering: Konzepte und Anwendungsszenarien für die digitale Transformation. Springer, Heidelberg (2016). https://doi.org/10.1007/978-3-658-16262-7
6. Buhalis, D., Leung, R.: Smart hospitality - interconnectivity and interoperability towards an ecosystem. Int. J. Hosp. Manag. **71**, 41–50 (2018)
7. Kagermann, H.: Smart Service Welt - Umsetzungsempfehlungen für das Zukunftsprojekt Internetbasierte Dienste für die Wirtschaft. Acatech Deutsche Akademie der Technikwissenschaften, München (2015)
8. Euromonitor: The Impact of Millennials' Consumer Behaviour on Global Markets. Euromonitor International, London (2015)
9. Horizont. http://www.horizont.net/marketing/nachrichten/Nestl-Studie-Wie-die-digitale-Transformation-die-Lebensmittelbranche-veraendert-138258. Accessed 14 Jan 2016
10. New York Times. https://www.nytimes.com/2017/08/25/dining/restaurant-software-analytics-data-mining.html. Accessed 25 Aug 2017
11. Ruiz-Molina, M., Gil-Saurab, I., Berenguer-Contr, G.: Information and communication technology as a differentiation tool in restaurants. J. Foodserv. Bus. Res. **17**(5), 410–428 (2014)
12. Coca Cola. http://www.coca-colacompany.com/stories/coca-cola-bringing-google-powered-digital-signage-system-to-reta. Accessed 12 Apr 2017
13. Neuhofer, B., Buhalis, D., Ladkin, A.: High tech for high touch experiences: a case study from the hospitality industry. In: Cantoni, L., Xiang, Z. (eds.) Information and Communication Technologies in Tourism 2013, pp. 290–301. Springer, Heidelberg (2013). https://doi.org/10.1007/978-3-642-36309-2_25
14. IBM Chef Watson. https://www.ibmchefwatson.com/community. Accessed 09 Mar 2018
15. Peters, C., et al.: Emerging digital frontiers for service innovation. Commun. Assoc. Inf. Syst. **1**(39), 136–149 (2016)
16. Chesbrough, H., Spohrer, J.: A research manifesto for services science. Commun. ACM **49**(7), 35–40 (2006)
17. Bullinger, H.-J., Scheer, A.-W. (eds.): Service Engineering. Springer, Berlin (2006). https://doi.org/10.1007/978-3-662-09871-4
18. Maglio, P.P., Spohrer, J.: Fundamentals of service science. J. Acad. Mark. Sci. **36**(1), 18–20 (2008)
19. Nestlé: Nestlé Future Study. How Germany will be& eat in 2030. Nestlé, Frankfurt (2016)
20. Deloitte: The Restaurant of the Future. Creating the Next-generation Customer Experience. Deloitte, New York (2016)
21. Leimeister, J., Österle, H.: Individualization & Consumerization. In: Proceedings der 11. Internationalen Tagung Wirtschaftsinformatik (WI2013), Leipzig, vol. 1, pp. 7–8 (2013)
22. Becker, M., Klingner, S.: Konzepte zur kundenspezifischen Anpassung von Dienstleistungen. In: Thomas, O., Nüttgens, M., Fellmann, M. (eds.) Smart Service Engineering, pp. 2–28. Springer, Wiesbaden (2017). https://doi.org/10.1007/978-3-658-16262-7_1
23. Ostrom, A.L., Parasuraman, A., Bowen, D.E., Patricio, L., Voss, C.: Service research priorities in a rapidly changing context. J. Serv. Res. **18**(2), 127–159 (2015)
24. Payne, A., Frow, P.: A strategic framework for customer relationship management. J. Mark. **69**(4), 167–176 (2005)

25. Kunz, W., et al.: Customer engagement in a big data world. J. Serv. Mark. **31**(2), 161–171 (2017)
26. European GDPR. https://www.eugdpr.org/. Accessed 09 Mar 2018
27. Fähnrich, K.P., Opitz, M.: Service Engineering—Entwicklungspfad und Bild einer jungen Disziplin. In: Bullinger, H.J., Scheer, A.W. (eds.) Service Engineering. Springer, Heidelberg (2003). https://doi.org/10.1007/978-3-662-09871-4_4
28. Beverungen, D., Lüttenberg, H., Wolf, V.: Recombinant service system engineering. In: Leimeister, J.M., Brenner, W. (eds.) Proceedings der 13. Internationalen Tagung Wirtschaftsinformatik (WI 2017), St. Gallen, pp. 136–150 (2017)
29. Lim, C., Kim, K.H., Kim, M.J., Heo, J.Y., Kim, K.J., Maglio, P.P.: From data to value: a nine-factor framework for data-based value creation in information-intensive services. Int. J. Inf. Manage. **39**, 121–135 (2018)
30. Bullinger, H.-J., Schreiner, P.: Service Engineering: Ein Rahmenkonzept für die systematische Entwicklung von Dienstleistungen. In: Bullinger, H.-J., Scheer, A.-W. (eds.) Service Engineering, pp. 53–84. Springer, Berlin (2006). https://doi.org/10.1007/3-540-29473-2_3
31. Hevner, A.R.: A three cycle view of design science research. Scand. J. Inf. Syst. **19**(2), 87–92 (2007)
32. Peffers, K., Tuunanen, T., Rothenberger, M.A., Chatterjee, S.: A design science research methodology for information systems research. J. Manag. Inf. Syst. **24**(3), 45–77 (2007)
33. Gretzel, U., Sigala, M., Xiang, Z., Koo, C.: Smart tourism: foundations and developments. Electron. Mark. **25**(3), 179–188 (2015)
34. Klein, M.: Design rules for smart services. Doctoral dissertation, University of St. Gallen (2017)
35. Nadj, M., Haueßler, F., Wenzel, S., Maedche, A.: The smart mobile application framework (SMAF) – Exploratory evaluation in the smart city context. In: Nissen, V., Stelzer, D. Straßburger S., Fischer D. (eds.) Multikonferenz Wirtschaftsinformatik (MKWI), pp. 1367–1378. Universitätsverlag Ilmenau (2016)
36. Heinonen, K., Strandvik, T., Mickelsson, K., Edvardsson, B., Sundström, E., Andersson, P.: A customer dominant logic of service. J. Serv. Manag. **21**(4), 531–548 (2010)
37. Heinonen, K., Strandvik, T.: Customer-dominant logic: foundations and implications. J. Serv. Mark. **29**(6/7), 472–484 (2015)
38. Heinonen, K., Strandvik, T.: Reflections on customers' primary role in markets. Eur. Manag. J. **36**(1), 1–11 (2017)
39. Alt, R.: Electronic markets on customer-orientation. Electron. Mark. **26**(3), 195–198 (2016)
40. Baron, S., Russell-Bennett, R.: Editorial: the changing nature of data. J. Serv. Mark. **30**(7), 673–675 (2016)
41. Parasuraman, A., Zeithaml, V.A., Berry, L.L.: A conceptual model of service quality and its implications for future research. J. Mark. **49**, 41–50 (1985)
42. Vargo, S.L., Lusch, R.F.: Evolving to a new dominant logic for marketing. J. Mark. **68**(1), 1–17 (2004)
43. Vargo, S.L., Lusch, R.F.: Why service? J. Acad. Mark. Sci. **36**, 25–38 (2008)
44. Spohrer, J., Maglio, P.P., Bailey, J., Gruhl, D.: Steps toward a science of service systems. Computer **40**(1), 71–77 (2007)
45. Vargo, S.L., Lusch, R.F.: Institutions and axioms: an extension and update of service-dominant logic. J. Acad. Mark. Sci. **44**(1), 1–19 (2016)
46. Maglio, P.P., Vargo, S.L., Caswell, N., Spohrer, J.: The service system is the basic abstraction of service science. IseB **7**, 395–406 (2009)
47. Breidbach, C.F., Maglio, P.P.: A service science perspective on the role of ICT in service innovation. ECIS Research-in-Progress Papers (2015)

48. Fitzsimmons, J.A., Fitzsimmons, M.J. (eds.): New Service Development, pp. 1–32. SAGE Publications, Thousand Oaks (2000)
49. Mikusz, M.: Towards a conceptual framework for cyber-physical systems from the service-dominant logic perspective. In: Twenty-First Americas Conference on Information Systems, pp. 1–13. Puerto Rico (2015)
50. Leimeister, J.M.: Dienstleistungsengineering und- management. Springer-Verlag, Heidelberg (2012). https://doi.org/10.1007/978-3-642-27983-6
51. Maglio, P.P., Kwan, S.K., Spohrer, J.: Commentary - toward a research agenda for human-centered service system innovation. Serv. Sci. **7**(1), 1–10 (2015)
52. Kleinschmidt, S., Burkhard, B., Hess, M., Peters, C., Leimeister, J.M.: Towards design principles for aligning human-centered service systems and corresponding business models. In: International Conference on Information Systems (ICIS), Dublin (2016)
53. Allmendinger, G., Lombreglia, R.: Four strategies for the age of smart services. Harv. Bus. Rev. **83**(10), 131–134 (2005)
54. Maglio, P.P., Lim, C.: Innovation and big data in smart service systems. J. Innov. Manag. **4**(1), 11–21 (2016)
55. Woodall, T., Rosborough, J., Harvey, J.: Proposal, project, practice, pause: developing a framework for evaluating smart domestic product engagement. AMS Rev. **8**, 1–17 (2017)
56. Höjer, M., Wangel, J.: Smart sustainable cities: definition and challenges. In: Hilty, L.M., Aebischer, B. (eds.) ICT Innovations for Sustainability. AISC, vol. 310, pp. 333–349. Springer, Cham (2015). https://doi.org/10.1007/978-3-319-09228-7_20
57. Wünderlich, N.V., von Wangenheim, F., Bitner, M.J.: High tech and high touch: a framework for understanding user attitudes and behaviors related to smart interactive services. J. Serv. Res. **16**(1), 3–20 (2013)
58. Becker, J., Beverungen, D.F., Knackstedt, R.: The challenge of conceptual modeling for product–service systems: status-quo and perspectives for reference models and modeling languages. IseB **8**(1), 33–66 (2010)
59. Winter, A., et al.: Manifest Kundeninduzierte Orchestrierung komplexer Dienstleistungen. Informationspektrum **35**(6), 399–408 (2012)
60. Sachse, S., Alt, R.: Kundenzentrierte Komposition komplexer Dienstleistungen - Eine empirische Untersuchung der Vorteile kundenzentrierter Servicekomposition. In: Nissen, V., Stelzer, D., Straßburger, S., Fischer, D. (eds.) Multi-Konferenz Wirtschaftsinformatik (MKWI), pp. 1379–1390. Technische Universität Ilmenau, Ilmenau (2016)
61. Schallmo, D.R.A.: Vorgehensmodell des design thinking. In: Design Thinking erfolgreich anwenden, pp. 41–60. Springer, Wiesbaden (2017). https://doi.org/10.1007/978-3-658-12523-3_4

Closing the Gap Between Research and Practice – A Study on the Usage of Service Engineering Development Methods in German Enterprises

Simon Hagen$^{(\boxtimes)}$, Sven Jannaber, and Oliver Thomas

Chair of Information Management and Information Systems,
University of Osnabrueck, Katharinenstr. 3, 49074 Osnabrueck, Germany
{simon.hagen, sven.jannaber,
oliver.thomas}@uni-osnabrueck.de

Abstract. Service Engineering (SE) evolved in the mid-1990s and has become a popular and interdisciplinary field of research in service science since then. However, the diffusion of SE research results into practice is still rare. This is especially crucial, since structured SE methodologies are required to support businesses with the ongoing digitalization of their services. To help closing the gap between research and practice, we conducted 13 semi-structured interviews with experts from eight enterprises in Lower Saxony, Germany, that are involved in (technical) services. The results reveal several requirements and barriers, which hinder companies from implementing and using structured SE methodologies. The findings can be used to help researchers developing industry-friendly approaches and practitioners to set up their enterprises for future-oriented (smart) service engineering.

Keywords: Service engineering · Practice · Science · Empirical analysis
SME

1 Introduction

The opportunities and the importance to offer services as a company to stay competitive grows [1], which is particularly true for small and medium sized enterprises (SME) [2]. This trend, often referred to as servitisation or service innovation, can be capitalized on by using Service Engineering (SE) methodologies, which gained increased attention in this regard since associated methods strive for structured approaches to support companies transition towards digitalized services [3, 4]. Primarily producing companies have shown an increasing interest in servitisation due to changing business models and profit margins and, therefore, have evolved a demand for SE methodologies over the recent years [5].

To support enterprises in implementing and applying systematic SE, an increasing amount of procedure models has been developed and published [5, 6]. However, although many models are extensively discussed in relevant literature, only few SE procedure models are actually implemented and used in practice [7]. Not only does this

© Springer Nature Switzerland AG 2018
G. Satzger et al. (Eds.): IESS 2018, LNBIP 331, pp. 59–71, 2018.
https://doi.org/10.1007/978-3-030-00713-3_5

cast a critical glance at the practical applicability and usefulness of these models, but also bears significant disadvantages for enterprises, since the usage of standardized structures and processes facilitates an increased growth in productivity and quality [8]. In previous studies, the usage of SE methods in practice has already been investigated: Meiren showed that 70% of their interviewed companies have no or little formalized service development procedures [9]. The project "KoProServ" (run-time 2009 to 2012) conducted a similar analysis [7] and found that there is a huge demand in managing service offerings and IT tools in companies. Similarly, Spath et al. stated that "one of the main gaps in the PSS [(Product-Service System)] design and SE is the absence of industrial tested methodologies [...]" [10]. Both PSS and SE methods aim at the same goal, developing integrated solution offerings containing products and services [11].

In opposite to service development, many producing companies already use structured and digitalized methods to develop e.g. their product models by means of CAD applications [12]. Therefore, knowledge about the benefits of using structured methods and procedures in general can be assumed. Since the previous studies in the field of SE are no longer current and partially do not address SME, we determine the following three research questions (RQ) for this contribution:

1. What is the state-of-the-art with respect to the usage and acceptance of SE procedure models in practice?
2. Which barriers exist that hinder the adoption of SE procedure models?
3. Which requirements do SME have regarding SE procedure models?

To address these research questions, 13 interviews with experts from eight different companies have been conducted with primary focus on SME. Using the interview results, we gained insights into the gap between scientific efforts in terms of SE on the one hand, and the requirements of practitioners working in practice on the other hand. The look into both perspectives enables us to propose solutions to scientists and practitioners to close the gap, which may facilitate companies to adapt SE methods that focus on their specific needs and allow them to increase their ability for a successful servitisation and therefore a more sustainable business model.

With respect to the research questions, the paper is structured as follows. First, a brief introduction to SE (Sect. 2) and a detailed explanation of the research method used is given (Sect. 3). The results of the interviews are presented in accordance to the research questions stated above in Sects. 4.1 to 4.3. In addition, Sect. 4.3 proposes guidelines, which support future research in developing new SE procedure models. Finally, the results are discussed and reviewed in Sect. 5.

2 Service Engineering

Since the 1960s the impact of services on economies can be found in scientific literature [13] and are firstly examined in depth in the "new service development" phase in the 1970s and 1980s [14] which resulted in the emergence of service engineering in the 1990s [15]. Since then the interest in this topic decreased, which changed with the arise of concepts such as internet of things (IoT) or smart products as well as their academic pendant "(smart) service systems" [16, 17] and ultimately led to the "second wave" of SE [10].

SE aims at a "systematic development and design of services using suitable models, methods and tools as well as the management of service development processes" [18]. The tools and methods usually compile from other, engineering related research fields like product, process or software engineering or operations research [19]. However, todays markets do not focus on products or services solely, but strive for the integration and thus the integrated development, which is also referred to as service systems or product-service systems (PSS) engineering [20]. Therefore, SE increasingly seeks for integrating (or integration into) development approaches from other disciplines to develop the service aspect in these systems with respect and interfaces to the other components (e.g. product or IT). By doing so, companies can achieve a reduction of complexity, reusable artefacts and usage of efficiency potentials [19] which enables them to provide more efficient and higher qualitative offers to their customer by means of service, in which producing companies as well as pure service providers are interested in [21].

To support companies in adapting SE, structured methods to guide the designers through the process have been proposed in relevant research [22]. Such SE methods assist enterprises in executing the development process in a structured manner and reduce the complexity of the project by ex-ante defining the different phases [23]. Well-known examples of SE methods are for example published by Pezzotta et al. [24] or Bullinger and Schreiner [25].

3 Research Method

We conducted semi-structured expert interviews in accordance to [26, 27] to obtain qualitative information from practitioners regarding the usage of SE. Semi-structured interviews have been chosen as a methodology to address the stated research questions, since they promise good practical data. As a preparatory step, an interview guideline has been prepared to ensure quality, integrity and topic-related information [28]. In addition, we evaluated the first iteration of the guideline within a pre-test with other researchers to prevent ambiguity and complexity of used phrases and terms and to gain first insights about parameters such as time spent and complexity of terms. The interview guideline is divided into three sections: (1) the interview starts with a short introductory part including information about the interview and general questions regarding practical background of the interviewee. Beside the interesting insides about the interviewee, it introduces them with basic questions to the situation and reduces barriers in terms of the relationship between the interviewer and the interviewee. Subsequently, the main part (2) focuses on the usage of SE procedure models, as well as advantages and drawbacks, involved people and requirements for an application of such models in practice. Closing with (3) final comments and personal opinions of the interviewees, the interviews took an average of 30–40 min. The questions and the structure respect the guidelines from Porst [29], for example the introductive questions mentioned above and the avoidance of hypothetical or disjunct questions.

The interviewees are primarily employees of SME and work in the middle and higher management. However, employees from a larger enterprise is included as well to allow a comparison of the results. In total 13 experts from eight different companies were interviewed. Employees from the same company are from different departments to obtain a richer understanding of the whole company. To get a variety of insights, companies have been chosen that are spread across different industries, from machine construction through mobility to pure (IT) service provider. A list of all companies and their interviewed employees including additional information is given in Table 1.

To evaluate the transcribed interviews, we follow the quantitative content analysis approach from Mayring [30], which is a three-step-process that aims for reduction of the results to a set of relevant statements. First, the statements are paraphrased and, in a second step, generalized towards a generic abstraction level. Following this, Mayring proposes a reduction of the generalization to the most important messages in two additional steps, which has been applied with respect to RQ2 and RQ3. In terms of RQ1, the interviews are assessed regarding statements that concern the usage of SE. All steps are performed by two researchers independently and merged afterwards to ensure a common understanding. The results of our analysis are presented and explained in the following.

Table 1. Interviewees (NOE: Number of employees)

#	Company industry	Revenue	NOE	Interview experts & position
1	Agricultural engineering	4 Bill. €	11.500	1. Company & product development
2	HVAC industry	68 Mill. €	720	2. Manager smart service 3. Manager sales 4. Team leader marketing
3	Electrical engineering	23 Mill. €	160	5. Division manager 6. CEO 7. Technical manager
4	Agricultural engineering	45 Mill. €	125	8. Technical manager
5	IT & software provider	4 Mill. €	60	9. Manager marketing
6	Mobility	121 T. €	8	10. Authorized representative 11. Product development
7	Energy	–	3	12. CEO
8	Consulting	–	1	13. Manager/consultant

4   Results

4.1   Usage and Acceptance of Systematic SE Development Methods (RQ1)

The interviews revealed five categories of SE method usages that are plotted in Fig. 1, describing the state-of-the-art for the adoption and acceptance of systematic SE methods in practice. The results show that currently no experts use or are aware of SE procedure models ($a = 0\%$). Only one respondent stated that their company applies a defined procedure, which is, however, adopted from other fields such as engineering and thus did not primary originate in the field of SE (b = 7%). The majority of experts (c = 77%, 10 respondents) indicated that they do not use a method to develop their service offering at the moment, but are aware of the opportunities and advantages a systematic procedure may yield for their company. The remaining interviewees are split equally over the remaining categories (d, e = 8%, one respondent each) and therefore either do not see advantages of using a systematic method for SE or were not able to contribute to this topic.

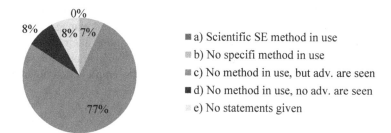

Fig. 1. Proportional distribution of usages of SE methods in practice

Resulting from the insights given, RQ2 and 3 are set to shed additional light into reasons that lead to SE adoption respectively rejection. Especially the results of category (c) have appeared very promising in this regard, since 77% of interviewed experts are aware of potential benefits that SE methods may come along with, but multiple reasons seem to have prevented an implementation yet. Henceforth, Sect. 4.2 deals with issues that are currently considered as a hindrance for SE method adoption. Subsequently, Sect. 4.3 reveals requirements that a SE method needs to incorporate in order to be practically applicable.

4.2 Barriers for Companies to Adopt and Use SE Procedure Models (RQ2)

Figure 1 highlights shortcomings within the adoption of SE procedure models in practice. Hence, RQ2 specifically addresses barriers and obstacles expressed explicitly by the interviewees. An overview is given in Fig. 2, which is explained below.

Reviewing the results, particularly one problem stands out, which is the *information gap* (45%) with respect to SE and corresponding models. Almost half of the experts state that they are not aware of the existence of SE methods or rather which of them are applicable for their specific industry and how the enterprise should use and benefit from them (e.g. "I haven't thought about whether a method exists. I don't really think so now.").

Further stated barriers (with 27% each) are product focus, missing perceived value and customer invocation. *Product focus* refers to the current focus on physical products, rather than a combination of product and services, in most of the enterprises interviewed within this paper. Therefore, many methods that originate in related fields such as engineering or product development are applied, without the use of additional service-related methods. In one case, related methodologies have been modified in order to support the development of the service as well. (c.f. category (b) in RQ1). Naturally, these results are influenced by the industries interviewed, which are in this case German SMEs that are primarily manufacturing-intensive firms. However, this also indicates the need for integrated product/service systems (c.f. Sect. 2). Another barrier is a *missing perceived value* for using systematic methods to develop new services. Many interviewees stated that, now, experienced domain experts are involved in developing services in their field. Hence, the perceived added value of adhering to a structured procedure for service development is seen as rather low, since the domain experts are already knowledgeable in SE. However, interviewed managers state that in case these experts may leave the company in the near future, their knowledge will mostly be lost, which can lead to a situation where new methods for subsequent employees are needed (e.g. "It will be useful, when the employer turnover will be higher").

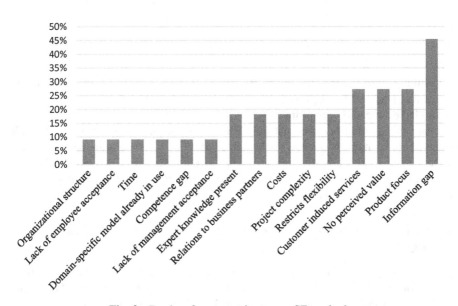

Fig. 2. Barriers for companies to use SE methods

Customer invocation refers to the situation that many services are often *invoked by the customer* himself. Therefore, such services are treated at the moment of occurrence and thus are not planned and developed in comparison to regular service offerings. This might also relate to the industries examined, since in manufacturing companies service offerings usually are synonymous to maintenance and repair.

The last two aspects are related to statements concerning *expert knowledge present*. The employees seemingly have enough knowledge and thus most companies do not see the need to offer a more structured service, so that customer requests are executed "spontaneous" ("I can't imagine an example where we have to document what we have to do"). *Limited flexibility* is another barrier that has been mentioned throughout the interviews. If the service offering is planned and structured too much in advance, companies express fear to lose a certain degree of flexibility and adaptability in terms of service features and its execution. In addition, this is related to *project complexity*, which is, according to the interviews, another barrier for SE adoption. For smaller projects or service offerings, there is no need for a systematic design, since the costs exceed the benefits ("[…] depends on how big the project is"). The *costs* themselves have also been referred to as barrier and have often been associated with the expenses needed to establish the necessary structure, e.g. *organizational structures* ("[…] sometimes a matter of budget"). The last major point that has been stated more than once during the interview is the *relation to the business partners* of the interviewed enterprises. The explanations thereby were based in the service structure of the companies, which is partly outsourced to firms who act as resale and maintenance provider. To prevent cannibalizing their own partners and due to the not required own structures, no systematic methods for the service development were used.

Furthermore, few barriers were stated which affect the *time* needed (related to the cost-barrier), the *competencies necessary are not established* yet and that the *management does not seem the necessity* for using structured procedures.

4.3 Guidelines to Develop Practical Applicable SE Procedure Models (RQ3)

Addressing RQ3, the conducted interviews revealed 30 requirements that were mentioned by the questioned experts with respect to a practical applicability of procedure models specifically tailored towards the development of service within their company. In addition, three requirements have been deducted by the authors from assessing the stated barriers of adopting SE procedure models. Henceforth, in total 33 identified requirements are summarized in Table 2. Within this table, the identified requirements have been differentiated regarding their source ("I" = Interviews; "B" = deducted from barriers, see Sect. 4.2) and are clustered and consolidated (column "C") according to their general point of concern into ten requirement clusters, described below.

Cluster A summarizes requirements that concern embedding structured SE into an organization. Here, interviewees stated the necessity of a SE procedure model to be adaptable to the organizational structure ("We need […] a constant flow of information between product development department and sales department"). Exemplarily, many businesses have not implemented a separate corporate entity specifically for SE purposes. Therefore, SE is interwoven into different departments, such as product development and sales. Similarly, the SE procedure model should be able to be applied across different functional domains, without being limited to the service department only.

Within *Cluster B*, the interviews have revealed a number of requirements that address the wish to consider various forms of information in service development ("We have a treasure of experience in our customer relations department, especially regarding services"). Thus, this cluster encompasses for example non-process related information such as roles and responsibilities, but also the integration of information flows and IS interfaces.

Table 2. Requirements for SE procedure models (C = Clusters)

#	C		SE procedure model reqs.	#	C		SE procedure model requirements
1	A	I	SE method should fit the organizational structure	17	F	I	Integrate cost view
2		I	Applicability in different functional domains	18		I	Integrate market analysis
3	B	I	Add non-process relevant information	19		I	Integrate project management
4		I	Integrate corporate knowledge base	20		I	Integrate resource planning
5		I	Integrate information flows	21		I	Plan service scope and vision
6			Integrate IS interfaces	22	G	I	Define USPs
7	C	I	Customizable	23		I	Ensure service quality
8		I	Consist of distinct phases	24	H	I	Broad domain respectively company acceptance
9		I	Ensure validation	25		I	Follow compliance rules
10		I	Provide different entry points	26		I	Standard conformity
11		I	Structured procedure model	27	I	I	Handbook/guidelines/documentation
12		I	Incorporate continuous feedback loop	28		B	Promotion/marketing/best practice
13		I	SE procedure model is of high pragmatic quality	29		B	Guidelines for consistent SE
14		B	SE procedure model is of low complexity	30	J	I	Customer integration
15	D	I	Service development considers technical SotA	31		I	Employee integration
16	E	I	Incorporate/include service object	32		I	Maintain flexibility of employees
				33		I	Partner integration

Cluster C primarily focuses on the SE procedure model itself. In the interviews it has been noted that a structured, iterative cycle is preferred as an SE model by practitioners ("We work with defined milestones for each project, with a separate idea phase, a concept phase [...]"). The cycle should consist of distinct phases and needs to be customizable towards specific organizational needs. Furthermore, the SE procedure model needs to incorporate a validation phase to ensure service quality. In general, the model should be of low complexity while providing a high degree of pragmatic quality to ensure comprehensiveness. *Cluster D* contains a single requirement that reflects the expressed need to ensure that the SE procedure model considers the current technological state-of-the-art ("We need to ensure quality and general applicability of used technology before entering the market"). Hence, services that result from applying the model for service development are built with technological advances in mind, such as improved sensor technology or mobile devices.

Similarly, *Cluster E* represents a single requirement that puts focus on the service object on which the service activity is performed. During the interviews, statements have been made that specifically addressed the need to center a SE approach to the service object offered by enterprises, as well as the desire to integrate service object-specific procedure models into SE ("Services are rather a co-product resulting from our product development cycle"). This has become predominantly apparent in the manufacturing industry, where the development of products such as machinery and corresponding services, e.g. maintenance, is closely tied and often follows specific procedure models originating in the engineering domain.

Cluster F primarily comprises requirements that concern the integration of preparatory planning phases ("[SE] is a matter of budget"). In the interviews, practitioners have expressed the necessity for a sophisticated market analysis and resource planning, which need to precede the service development. Additionally, an SE procedure model needs to incorporate aspects of project management as well as maintain a holistic cost view across each phase to allow for target costing a service offering.

Cluster G summarizes requirements that concern the service itself as a result of SE procedures. Here, practitioners express the demand to clearly determine a service's USP within its development cycle ("We need to know that the [service's] USPs are"). Similarly, measures should be established within a SE method that ensure a certain degree of quality of the result.

In *Cluster H*, requirements primarily concern rules and regulations that influence service development coming from both external and internal authorities. This includes compliance rules on the one hand ("We need to consider NDAs [...]"), but also standards and norms to be followed within service development and service delivery.

Cluster I addresses requirements for a successful implementation of the SE procedure model. During the interviews, practitioners highlighted the importance of a holistic and comprehensive documentation and guidelines on how to apply the SE procedure model within their enterprise ("We need to ensure a consistent procedure and consistent quality"). Furthermore, best practices need to be given in order to facilitate the implementation process.

Lastly, *Cluster J* subsumes different stakeholder perspectives that need to be integrated into the service development procedure. Specific emphasis is put on customer, employee and partner integration, which includes e.g. customer touch points and entry points for partner participation, but also the consideration of specific stakeholder demands during both service development and service execution ("I would strongly recommend integrating service technicians in service documentation and development").

Based on the stated requirements and the derived clusters, we propose ten guidelines below that can be used to steer SE development towards a more practical applicability. Herby, each cluster is associated to a specific guideline:

(A) SE procedure models need to be able to adapt to the specific organizational structure of an enterprise
(B) SE procedure models should be designed to incorporate various forms of information flow, as well as information objects processes related interfaces to external IS

(C) SE procedure models should be designed as a structured, iterative cycle that consists of distinct phases, which are customizable towards the enterprise's needs

(D) SE procedure models need to consider the current state-of-the-art in terms of technological advances

(E) SE procedure models need to put primary focus on the service object on which the service is performed

(F) SE procedure models should integrate a systematic planning and strategy phase within their cycle that precedes service development

(G) SE procedure models need to explicitly determine a service's USP and ultimately ensure service quality

(H) SE procedure models must incorporate external and internal rules as well as guidelines across different domains

(I) SE procedure models need to be comprehensively documented and provide guidelines of how to successful implement the model

(J) SE procedure models must integrate each service stakeholder into the service development process and reflect their specific necessities

Ultimately, the proposed guidelines may serve as a blueprint and evaluation schema for developing SE procedure models that incorporate the need of SE domain experts and thus facilitate the usage of structured service development approaches in practice.

5 Discussion, Conclusion and Outlook

The results of our analysis reveal that only few experts do not know about SE methods or acknowledge the benefits of using specific methods, but have not been able to implement it within their firm yet. In terms of the advantages of SE for businesses and their business models, this result has to be valued positive. Beneficial for both research and practice, one of the most dominant barrier stated, the information gap with respect to SE methodologies, can be tackled with additional information provision and training only. Additionally, interviews also reveal a growing trend of enterprises further investing in SE, in particular structured procedure models for service development. Still, obstacles have to be overcome:

From a practical point of view, regular education and trainings, especially in the field of SE and the development of service offerings, need to be implemented in order to build a sustainable knowledge base about advantages, benefits and best-practices in SE. In addition, management executives need to propagate these benefits of SE methodologies, for example in terms of new and innovative business model respectively value generation within their enterprise. Top level management is advised to strengthen the integration of external partners into their business model or to incorporate external businesses as well as to invest in SE by engaging employees, increase transparency, establish designated service departments and hire employees specifically educated in SE. The overall objective needs to be the integration of the service perspective into the product sales and its development.

From a research perspective, additional focus is needed on the interconnection of product and service development, since the results demonstrate that both perspectives are still mostly perceived as distinct discipline. Additionally, SE research needs to be aware of a more differentiated view on various stakeholder groups, which should be covered within an appropriate method. Furthermore, existing project management methods should build the base of the procedure model to improve familiarity and reduce complexity. The awareness of further service details, which are important in practice for service development, such as interfaces to internal or external IS, links or specific phases, and the provision of a detailed documentation and guidelines, are equally important success factors mentioned by SE experts. Lastly, one should elaborate on strengthening transparency on the corporate benefits obtained by SE methods.

However, limitations of this study need to be considered: First, the findings are based on 13 interviews within German SMEs. Further research needs to verify the results within a larger scope of experts. Hence, expanding the interview towards a larger set of interviewees coming from a background that is more diverse seems to be a fruitful direction for subsequent work. Furthermore, comparing the results with data from additional sources like literature would strengthen the study, e.g. [11]. Third, primary focus of this contribution has been put on SE procedure models. Therefore, other forms of support for service development, for example tool support, have not been covered in the conducted interviews. Nevertheless, the obtained results yield a valuable insight into the practical application of SE methodologies within German enterprises. No major contradictions occurred during the interviews, the knowledge and quality of experts have been perceived as high. In addition, the results of the different research questions confirm each other, which supports the validity of the findings. Therefore, this contribution can confirm the findings of previous research (see Sect. 1), that offering services is often not linked to the main business model (usually a physical product) of a firm. In addition, our findings extend previous studies by barriers and requirements specifically for SMEs.

Therefore, the need for further research in SE, especially for practitioners, is given. Based on the findings and to assist the development the next steps in the research are to evaluate, how the proposed guidelines (Sect. 4.3) can be implement and to adapt existing SE procedure models. In addition, they need to be evaluate with practitioners to be able to develop e.g. a platform, which can be used by companies to develop their service offering according to a specific procedure model.

Acknowledgements. This research was conducted in the scope of the research project "SmartHybrid – Service Engineering" (ID: 6-85003236), which is partly funded by the European Regional Development Fund (ERDF) and the State of Lower Saxony (Investitions- und Förderbank Niedersachsen – NBank). We like to thank them for their support.

References

1. Patrício, L., Gustafsson, A., Fisk, R.: Upframing service design and innovation for research impact. J. Serv. Res. **21**, 3–16 (2018)
2. Leyh, C., Bley, K., Schäffer, T.: Digitization of German enterprises in the production sector – do they know how "digitized" they are? In: AMCIS (2016)
3. Haller, S.: Dienstleistungsmanagement. Gabler Verlag, Wiesbaden (2012)

4. Tan, A.R.R., Matzen, D., McAloone, T.C.C., Evans, S.: Strategies for designing and developing services for manufacturing firms. CIRP J. Manuf. Sci. Technol. **3**, 90–97 (2010)
5. Pezzotta, G., Cavalieri, S., Romero, D.: Engineering value co-creation in PSS: processes, methods, and tools. In: Rozens, S., Cohen, Y. (eds.) Handbook of Research on Strategic Alliances and Value Co-creation in the Service Industry, pp. 22–36. IGI Global, Hilliard (2017)
6. Schneider, K., Daun, C., Behrens, H., Wagner, D.: Vorgehensmodelle und Standards zur systematischen Entwicklung von Dienstleistungen. Service Engineering Entwicklung und Gestaltung Innovativer Dienstleistungen, pp. 113–138 (2006)
7. Klingner, S., Meiren, T., Becker, M.: Produktivitätsorientiertes Service Engineering. LIV, Leipzig (2013)
8. Hahn, A., Geuter, J.J.J.: Towards an executable sociotechnical model for product development and engineering systems. In: ECMS 2014, pp. 47–53. Cambridge (2014)
9. Meiren, T.: Service Engineering im Trend. Stuttgart (2006)
10. Spath, D., Meiren, T., Lamberth, S.: Trends und Perspektiven des Service Engineering. ZWF Zeitschrift für wirtschaftlichen Fabrikbetr **108**, 193–194 (2013)
11. Hagen, S., Jannaber, S., Thomas, O.: Towards practical applicability of service engineering: a literature review as starting point for SE method design. In: 9th International Workshop on Enterprise Modeling and Information Systems Architecture, pp. 90–94 (2018)
12. Gembarski, P.C., Lachmayer, R.: Mass customization und product-service-systems: Vergleich der Unternehmenstypen und der Entwicklungsumgebungen. In: Thomas, O., Nüttgens, M., Fellmann, M. (eds.) Smart Service Engineering, pp. 214–232. Springer, Wiesbaden (2017). https://doi.org/10.1007/978-3-658-16262-7_10
13. Cowell, D.W.: The Marketing of Services. Heinemann, Portsmouth (1984)
14. Bullinger, H.J., Fähnrich, K.P., Meiren, T.: Service engineering - methodical development of new service products. Int. J. Prod. Econ. **85**, 275–287 (2003)
15. Cavalieri, S., Pezzotta, G.: Product-service systems engineering: state of the art and research challenges. Comput. Ind. **63**, 278–288 (2012)
16. Alter, S.: Metamodel for service design and service innovation: integrating service activities, service systems, and value constellations. In: ICIS 2011 Proceedings, Shanghai (2011)
17. Beverungen, D., Müller, O., Matzner, M., Mendling, J., vom Brocke, J.: Conceptualizing smart service systems. In: Electronic Markets (2017)
18. Pezzotta, G., Cavalieri, S., Gaiardelli, P.: Product-service engineering: state of the art and future directions. IFAC Proc. **42**, 1346–1351 (2009)
19. Burr, W.: Markt- und Unternehmensstrukturen bei technischen Dienstleistungen. Springer Fachmedien Wiesbaden, Wiesbaden (2014). https://doi.org/10.1007/978-3-658-02286-0
20. Acatech: Smart Service Welt - Umsetzungsempfehlung für das Zukunftsprojekt Internet-basierte Dienste für die Wirtschaft, Berlin (2015)
21. Bullinger, H.-J., Scheer, A.-W.: Service Engineering: Entwicklung und Gestaltung innovativer Dienstleistungen. Springer, Berlin (2006). https://doi.org/10.1007/3-540-29473-2
22. Fähnrich, K.-P., Opitz, M.: Service engineering—Entwicklungspfad und Bild einer jungen Disziplin. In: Service Engineering, pp. 83–115 (2003)
23. Stickel, E., Groffmann, H.-D., Rau, K.-H.: Gabler Wirtschaftsinformatik Lexikon. Gabler Verlag, Wiesbaden (1997)
24. Pezzotta, G., Pinto, R., Pirola, F., Ouertani, M.Z.: Balancing product-service provider's performance and customer's value: the SErvice Engineering Methodology (SEEM). Procedia CIRP **16**, 50–55 (2014)

25. Bullinger, H.-J., Schreiner, P.: Service Engineering: Ein Rahmenkonzept für die systematische Entwicklung von Dienstleistungen. In: Service Engineering Entwicklung und Gestaltung Innov. Dienstleistungen, pp. 53–84 (2006)
26. Bogner, A., Littig, B., Menz, W.: Interviews mit Experten: Eine praxisorientierte Einführung. Springer Fachmedien Wiesbaden, Wiesbaden (2014). https://doi.org/10.1007/978-3-531-19416-5
27. Myers, M.D., Newman, M.: The qualitative interview in IS research: examining the craft. Inf. Organ. **17**, 2–26 (2007)
28. Meuser, M., Nagel, U.: Das Experteninterview—konzeptionelle Grundlagen und methodische Anlage. In: Pickel, S., Pickel, G., Lauth, H.-J., Jahn, D. (eds.) Methoden der vergleichenden Politik- und Sozialwissenschaft, pp. 465–479. VS, Wiesbaden (2009)
29. Porst, R.: Frageformulierung. In: Baur, N., Blasius, J. (eds.) Handbuch Methoden der empirischen Sozialforschung, pp. 687–699. Springer Fachmedien, Wiesbaden (2014). https://doi.org/10.1007/978-3-531-18939-0_50
30. Mayring, P.: Qualitative Inhaltsanalyse. In: Handbuch Qualitative Forschung in der Psychologie, pp. 601–613. VS Verlag für Sozialwissenschaften, Wiesbaden (2010)

Customers Input via Social Media for New Service Development

Intekhab (Ian) Alam[(✉)]

School of Business, State University of New York (SUNY),
Geneseo, NY 14454, USA
alam@geneseo.edu

Abstract. The objective of this research is to investigate the use of social media for customer interaction for developing new services. To collect data we adopted a multi-phase research approach used in several recent ethnographic studies. The findings suggest that a firm must use social media to interact with the customers while developing new services. Other traditional methods of customer interaction also play a key role in developing a successful new service. The article has implications for the financial service firms interested in marketing new services in the United States and other developed countries.

Keywords: Customer input · New service development · Social media

1 Introduction

New Service Development (NSD) is the primary driver of growth in a firm and therefore, it is an important area of research [8, 55]. Despite the emergent robustness of the NSD literature, scholars have argued that more needs to be done because the extant literature has failed to offer consistent solutions to many problems faced by the service managers [8, 54]. One such complex problem is that new service failure rate continues to be very high [61]. The literature on new service and product success and failure has invariably suggested that capturing voice of customers during the NSD process is a key to success [e.g., 42]. This suggestion for capturing the voice of customers has led many firms to increasingly interact with the customers at various stages of new product and NSD process [18, 25].

However, the literature of customer interaction has focused mainly on the tangible product domain as shown in a recent study by Bosch-Sijtsema and Bosch [9]. To fill this gap, scholars have called for more research on customer interaction that apply specifically to the services area [17, 46] The somewhat growing literature base of NSD and customer interaction points to two key challenges that are involved in customer interaction. One such challenge is that the NSD literature has taken a traditional approach in which it suggest that a firm must interact with their customers via interviews, focus group, group discussion and so on [3]. However, as Lemon and Verhoef [38] have argued, the customers interact with a firm via multiple channels including the emerging touch point of social media. Indeed, over the last decade the digital media have revolutionized the overall marketing practice by offering new ways to reach out to

© Springer Nature Switzerland AG 2018
G. Satzger et al. (Eds.): IESS 2018, LNBIP 331, pp. 72–87, 2018.
https://doi.org/10.1007/978-3-030-00713-3_6

customers [37]. As a result, during this time frame a number of articles have been published that relate to the use of social media platforms for various marketing activities [27, 39, 58, 60, 62]. More specifically, the literature related to the use of social media for engaging customers is rather robust [e.g. 32, 36]. Acknowledging the growing importance of social media, Alam [1] proposed a NSD model that contains the use of social media for customer input. Yet, how to engage customers via social media for obtaining their input for NSD is largely unknown.

Against the above backdrop, we investigate the main research question: How to use social media to obtain customer input for NSD? By answering this question we propose a process of obtaining customer input via social media that can help service firms address various tactical issues when developing new services. To achieve this goal a case study of NSD was conducted in a financial service firm in the USA. The case study investigates the development of new services with inputs from company's customers. This research is set in the financial services industry in which we investigate the business-to-business (B2B) services. We selected financial services industry because of the increasing focus on the innovation related activities in that industry resulting from several structural and technological changes [22, 63, 64]. In addition, we were motivated to select B2B services because of the growing importance of the B2B service industries [26], the increasing use of social media by many B2B firms [52] and the dearth of social media research in the B2B context [52]. The rest of the article is organized as follows. First, we discuss the literature related to NSD and customer interaction. Next we also review the literature related to social media as customer touch points. We then outline the research methodology. In the subsequent section we discuss findings and theoretical and managerial implications. The articles concludes with a discussion of research limitations and future research agenda.

2   Literature Review

The extant literature has paid increasing attention to the NSD research that can broadly be categorized into three research themes. The first research theme has focused on the process of NSD or service innovation. In line with [8, 17], for this research we use the term service innovation and NSD interchangeably. The second research theme has focused on exploring why new services fail and the role of customer interaction during NSD in avoiding new service failure [16, 47]. More recently, a third research theme has focused its attention on the emerging role of social media in product and service innovation. We review these three research themes next.

2.1   New Service Development Process

Innovation literature is strongly biased towards tangible products because new product development has been studied for several decades resulting in a large body of knowledge about new product development process [e.g., 10, 14, 30, 31]. Taking the lead from this rich literature base, service researchers have proposed several NSD models over a period of last 30 years starting from Barras [6] and Bowers [11, 12] to Alam [1]. Notable among them is the NSD model proposed by Scheuing and Johnson

[56], in which they suggested a structured 15 step model for developing new services. After over a decade, Alam and Perry [4] proposed a model containing 10 development stages including strategic planning, idea generation, idea screening, business analysis, formation of cross-functional team, product design and process/system design, personnel training, product testing and pilot run, test marketing, and commercialization and added that customers' input into the new services should be obtained throughout the development process. Later, emphasizing the importance of involvement of customers, frontline employees and other stakeholders, Kindstrom and Kowalkowski [35] offered a four stage NSD model including, market sensing, development, sales and delivery. Similarly, Song et al. [59] also proposed a four stage model containing the stages of business and market opportunity analysis, service design, service testing and service launch. More recently, Alam [1] challenged the applicability of the above models in the current era of technological advancement and social media, and proposed a much simplified and improved model containing only four phases of NSD: initiation phase, comprehension phase, corroboration phase and execution phase. He claims that these four phases are the better representation of NSD process because they take into account the influence of social media and digital technologies. Consequently, in this research, we use this model to analyze the customer interaction process. Since, customer interaction during NSD is an important construct and the focal concept of this study, we review the literature of customer interaction in NSD next.

2.2   Customer Interaction in NSD

Both scholars and practitioners agree that customer interaction during product/service development is a source of competitive advantage [18, 21, 40]. Several empirical studies have investigated the benefits of customer interaction in both new product and service development and reported that customer input might lead to high-quality innovations [3, 24]. As Cui and Wu [20] have argued there are three forms of customer involvement: customer involvement as an information source, customer involvement as co-developers and customer involvement as innovators. We focus on first two types as these are directly related to the main domain of this study. As buyers of current and future new services, the customers may contribute to all the stages of NSD, from idea generation to commercialization. For instance, Hoyer et al. [34] studied customer interaction in new service development and reported that customer interaction during the early stages of a development process significantly influenced the performance of new services products. Therefore, they argued for intense interaction between customers and product/service developers during the innovation process. Likewise in the case of NSD, Alam [3] suggested that the intensity of service producer-customer interactions during the idea generation stage should be higher than all other stages. He also noted that customer interaction results in important benefits such as reduced cycle time, superior services and customer education. Extending his research on customer interaction, Alam [2] further reports that customer interaction during the fuzzy front end, i.e. the first three stages of idea generation, idea screening, and concept development are more important than other later stages. However, Chang and Taylor [18] report that many firms do not interact with their customers on a regular basis and thereby miss a significant opportunity of developing successful new products and services. They assert the need for more research on customer interaction.

2.3   Social Media for Customer Interaction

For this research we define social media as "any web based interaction platform that can be deployed to generate content (such as online postings on a firm's various social media platforms including Facebook, microsites etc.) and develop networks (such as online communities)". The literature dealing with social media and their roles in various business issues is robust [37]. To make a coherent contribution to our understanding of the social media for NSD, we must first define our focal concept, which is the use of social media for customer interaction for NSD. Therefore we restrict our review as it relates only to customer interaction and new product and service idea generations using social media particularly a "microsite". A microsite is website which is distinct and separate from a firm's main website and which is created for a short time only for a specific purpose such as a promotion or an event.

As customers are increasingly relying on social media to gather information about products and services, many firms are also using social media tools to gain customer knowledge and behavior [43]. A recent IBM study suggests that getting closer to the customers is a top priority of the CEOs and therefore many firms are building social media platforms to interact with their customers [5]. Scultz and Peltier [57] have argued that marketers need to move beyond the short term goal of promotion to a more long term goal of using social media for regular customer engagement and interaction.

As reported previously, idea generation from customers using the traditional modes of interaction has been the subject of intense scrutiny both in new product [e.g., 25, 34] and new service development literature [e.g., 3, 40]. The literature has also started to emphasize the importance of social media in customer interaction by suggesting the fact that it has provided a new mean of researching the market and customers. For example, social media facilitate active customer involvement that results in co-creation of products and services [45, 65]. In addition, social media can be an inexpensive interactive marketing tool and a means for getting customer information rather quickly [49]. As a result, for the last few years the use of social media for generating new product and service ideas has been gaining momentum in many firms [39]. For example, Starbucks, Dell and Walmart have used microsites such as Mystarbuck-sidea.com, Ideastorm.com and getontheshelf.com respectively to generate new product ideas from their customers [7, 19]. Similarly, several open online platforms such as innocentive.com and ideascale.com develop a community of customers who are willing to offer new product ideas via social media [39]. More recently, based on their study of the online brand communities Hajli et al. [29] have argued that social media platform of online communities provide an important platform for customer interaction and co-creation of products and services. Yet, despite the increasing attention on the use of various social media, many firms are unsure about how to convert data collected via social media posts into actionable business strategies [28, 58]. In conclusion, social media can provide a platform for both types of customer interaction and involvement: customer involvement as an information source, customer involvement as co-developers [44]. Yet, we are not aware of any research that directly investigates the process of customer interaction for NSD using social media.

3   Research Methodology

To discover how a firm interacts with its customers for NSD, we adopted a single case setting that would allow an in depth understanding of the pattern of customer interaction. For this purpose, we intended to identify a case firm engaged in NSD in the B2B financial services industry. After negotiating with several firms, we gained access to one of the leading US based firms Ameri Corp Inc. (a pseudonym) that employs 400 people and markets mostly the B2B services. Our research design adopted a multiphase research approach used in several recent ethnographic studies [e.g., 13, 50]. The data were collected over a period of 9 months in 2017 by using several qualitative data collection techniques including in-depth interviews [41, 51], ethnographic observation [53] and netnography [50]. Over the nine month period the author along with several company managers ethnographically observed and analysed NSD activities. We also followed an abductive research method of moving back and forth between literature and the data that emerged during our research [23, 26]. In the first phase of the research, we interviewed the managers of the service firms. In the second phase, we observed customers' interaction via social media postings. In the third and final phase, we interviewed the business customers who were involved in the NSD process. We coded and analysed the data by adopting the theoretical sampling method recommended by Bryman and Bell [15]. We were motivated to develop this research design because of a recent call for more research on NSD using qualitative research approaches, including in-depth interviews, action research, ethnographic research and participant observations [8].

In phase one of the research we interviewed the mangers of Ameri Corp. We relied on a theoretical sampling procedure to identify respondents across various functions, provided they were involved in NSD activities and had significant knowledge about the NSD initiatives in the firm. Initially we selected 10 managers but found that only 6 managers were directly involved in NSD and customer interaction strategies. The main objective of the in-depth interviews with the managers was to investigate how the service managers had obtained the customer input for their NSD projects in the past and how they plan to proceed with the similar projects in the future. Using an interview protocol we posed several questions to probe the overall NSD process used by the company and the managers' overall perception about customer interaction methods employed by them. In phase two, we contacted 70 customers and requested their participation in the research project. After prequalification, we selected only 59 customers who had the richest information to offer; yet only 38 customers agreed to participate in the research. Following Homburg et al.'s [33] approach, in selecting the customer respondents, we attempted to maximize diversity among the participants so that we could discover a variety of new service concepts and ideas. These customers then joined the research team at various phases of the NSD to provide input. The details of both manager and customer respondents are summarized in Tables 1 and 2 respectively.

Table 1. Detail of the manager respondents

First name	Position in the Firm	Experience (no. of years)	Experience in NSD (no. of years)	No. of projects completed
David	Marketing Manager	16	06	21
Isaac	Marketing Supervisor	13	09	11
Peter	District Sales Manager	09	03	07
Sarah	Vice-President Sales	07	02	05
Marilyn	New Product Manager	11	06	07
Madison	Customers Service Manager	12	04	02

Table 2. Detail of the customer respondents

Type of customer firms	Experience no. of years in their firms	No. of years. dealing with Ameri Corp.	Number of respondents
Handicrafts exporter	7	5	2
Insurance firm	8	5	3
Automobile dealer	8	4	6
Leather goods manufacturer	5	3	6
Construction company	12	8	3
Travel agent	2	2	2
Software firm	4	1	4
Small retailer	5	3	3
Manufacturer of paper products	9	6	2
Leather tannery	4	2	4
Tire manufacturer	4	2	1
Exporter of textiles	10	4	2
Total	Avg. 6.5 years	Avg. 3.75 years	38

During this second phase of the research we also started a microsite www. Ameriideas.com (now closed) for collecting new service ideas from the customers. We invited all the 38 customers to log in to the site and provide input on various issues and factors related to the NSD. The main objective of this microsite was to facilitate frank discussion about customer needs and obtain solutions for their needs. The microsite had all social media tools needed to encourage interactions among customers including, discussion forums, online chats, product and service reviews, product and service ratings, likes and dislikes, blogging features, video postings etc. We adopted several methods from the literature to facilitate customer interaction over the social media and instructed the customers to freely post their ideas and offer critiques of the ideas suggested by other users [e.g., 13, 29, 39]. The customers could also offer 1–5 star rating of the ideas suggested by other users. Consistent with the previous studies customers were free to submit as many or as few ides the desired and there was no time restriction on their postings. Customers were also instructed to post the details of their unmet needs and overall dissatisfaction with the existing services being offered in the market. Any user that shares the same concern can simply use the "Like" feature similar to the Facebook to support the concerns raised by their peers. Any user could ask any questions and other users could freely answered the questions. No incentives, either monetary or non-monetary were offered and customers agreed to participate voluntarily. Based on the content analysis of the microsite social media posts, we recorded a total of 336 posts and 219 "Likes" during the nine months period. We selected the individual discussion and post as the unit of analysis. To discover emerging themes in the data we conducted a thorough text analysis of these posts using the NVivo 9.0 software. To code the data we used the triangulation method suggested by Bryman and Bell [15] in which three different research assistants coded the data separately and compared their coding with each other. After the comparison and successive readings we modified the interpretations and codes. During the coding process we looked for distinct data patterns and common elements and differences.

In phase three, we interviewed all the 38 customers who had posted their responses on the microsite. The objective for the customer interviews was to obtain corroboratory evidence. All the interviews were tape-recorded and details notes were taken. We started each interview with a grand tour question asking informants to describe their overall experience [41]. Next we structured the interviews around the following questions: (1) what input did you provide to this firm for their efforts in developing new services? (2) how would you describe your experience with this project both in online and offline settings? and (3) given this opportunity again in the future what is your suggestions in regard to customer interaction? To avoid active listening we carefully phrased the questions [41]. Each interviews lasted 60–90 min and were tape recorded and subsequently transcribed resulting in 207 pages of data. During the interviews we took extensive notes. We also consulted several company documents and archival records for data triangulation purpose. These customer interviews served as an effective means of data triangulations and improving the credibility of findings. We analysed the data line by line using NVivo 9.0 software and looked for broader data themes. Using the method of abduction, we compared and contrasted the data themes with the literature related to NSD, customer interaction and social media marketing discussed above.

4   Findings and Discussions

The interviews of the managers revealed three emergent themes: (a) process of customer interaction using the traditional research methods (b) various stages of customer interaction and, (c) the future of customer interaction and how the social media can be employed for the purpose of obtaining customer input. During the managers' interviews the respondents went to great lengths to convey the importance of customer input for NSD particularly in the financial services industry. One manager described his feeling this way: *Financial service market is complex and the customer needs are evolving everyday due to regulatory and other environmental changes. I can't imagine developing new services without customer inputs. Although, customer interaction and input is a new strategy, it has become a must in our organization.*

After the completion of manager interviews, the research team that included the author and six managers approached the customers and asked for their input and suggestions at various stages of the NSD. During the nine months period we conducted focus group interviews and in-depth interviews of all the 38 customers and sought their input for the new services. In addition, the firm organized several innovation retreats, informal mixers in which several customers actively participated and provided input into various aspects of the new services. Some of the active customers also worked as part of the NSD team. In regard to the effectiveness of each mode of interaction the managers reported that in-depth interviews provided very rich and usable information because they could elicit rather detailed responses from the customers related to the new service ideas. One manager explained the benefits of interviews this way: "*Any type of interaction is good and effective, yet I believe the very nature of the interviews provides more opportunities to inquire, assess and probe the customer inputs. It facilitates frank discussion about customer needs and requirements. Customers open up more in one on one meetings*". However the customer respondents provided a different perspective on the modes of interaction. They suggested that the working as the NSD team worked very well for them because it was very enriching experience for them and that they provided input that are workable and usable, as illustrated by a customer respondent: "*As part of the team, I suggested many ideas that some liked and a few disliked, yet we could still work on these ideas after going back and forth on the merit of each ideas. The team dynamics and discussion format provided a platform where we all could analyse and synthesize the new service ideas and the final outcome of the new services*".

The theme generated from interviews of both service managers and customers supported the notion that social media particularly a micro site can produce very important insights into the new service ideas. A manager respondent illustrated his preference for social media this way: "*We have been observing some of our customers for several years now. They are avid social media users. They blog, tweet and post on social media regularly. It dawned upon us why not use social media for interaction and your (the researcher) involvement further boosted our confidence in the process.*" The customer interviews further substantiated the managers' claims that many customers prefer a more modern approach of interaction particularly via social media because it provides them more flexibility. This comment of a customer clearly outlines the customers' preference: *In the past the company will call us and ask our availability to talk or they call me and say can we meet. I suggested to them that why don't your start a website. Give me a link and I will post my responses to your questions whenever I am free. It was easy. I forwarded the link to many of my acquaintances who were also the customers of this company.* In summary, the need for social media was driven by the customers who demanded the use of digital media. Thus, we discuss the data collected via social media next.

4.1 Social Media Interactions

While conducting customer interviews, we simultaneously started the microsite for the interaction purpose. During the research period, all the 38 selected customer actively participated in the research by posting their responses. Some of the more active customers even contributed to the blog feature on the site itself in which they wrote detailed analysis of their needs. Discussion about customer need related topics were the most prevalent area of discussion comprising of 80% of the postings. A majority of customer also provided detailed critiques of the suggested ideas that helped us screen the new service ideas. However, some of the postings about new service ideas were rather vague and not specific enough. Yet, a few customers posted a very detailed and specific analysis of their needs and how the company can best solve their needs by a developing a new services. We probed this contradictory finding while interviewing the customers. One customer's remark explains the situation this was: *There is a lot I can add to the discussion about the needs. However, I am not a professional. I have been the customer of the company for the last 4 years and have some ideas, yet can't tell you how they should solve my problem. I can provide insight; they have to turn that insight into a product.* In an interview with another participant 2 weeks later the respondent stated: *"I had mentioned to one of the managers of the company about an idea for a new pension product for my employees. I had provided all the details and how the pension will work. I am pretty savvy with the financial businesses, yet my business is not financial services. I am an exporter of leather garments. Yet, I am sure I and a lot of others will definitely adopt this service if offered"*.

Table 3. Customer input from the social media at various stages of NSD

NSD Stages [1]	Input from social media postings	No. of posts (comments)	No. of likes
Initiation stage	Suggest desired features, benefits and preference for a new service	41	56
	Identify customers' problems not solved by the existing services	37	32
	Provide a new service wish list and key attributes	39	31
	Offer critiques of the existing services available in the market	31	12
Comprehension stage	Rate the liking, preference and purchase intents of all the new service ideas suggested in the previous NSD stage	27	9

(*continued*)

Table 3. (*continued*)

NSD Stages [1]	Input from social media postings	No. of posts (comments)	No. of likes
Corroboration stage	Suggest improvements by identifying fail points in service delivery	21	21
	Compare their wish list with the proposed blue prints of the service	23	12
Execution stage	Comment on various aspects of the marketing plan for the new services	24	18
	Comment on their satisfaction with marketing mixes including pricing and distribution system	27	7
	Suggest desired improvements to the final new services	37	7
	Promote the new service ideas to other potential users within their respective networks	29	14
Total		336	219

The above findings suggest that all customers are not equal. Each bring different sets of skills and resources. It is the responsibility of the service managers to use their input appropriately and involve the customers based on their skills and talents. Also a considerable amount of discussion occurred about the overall marketability of the new service ideas in which customers commented about overall service delivery process, pricing and performance issues. One post clearly describes how the customers took some of the issues rather seriously. *"It is a price sensitive market. There are many banks who want my business. I would say in this market one should look extremely carefully at the fees and charges that are levied on some of the financial products your company wants ti impose'*. This one post attracted the single largest "Likes" from other participants. Further analysis of data resulted in the identification of actual input that the customers can provide at various stages of the NSD. All input/posts are summarized in Table 3. The data summary in Table 3 also points to the fact that the first and the last phases of NSD recorded the largest number of posts. Although, social media interactions provided rich information about the new services, the managers reported using several traditional modes of interaction as well including interviews, innovation retreats, customers inducted into the NSD teams and informal customer-managers mixer as mentioned in previous research [3]. This comment of a manager clearly illustrates the benefits of the combined use of social media and traditional methods of interaction: *"We had used several methods of interactions in the past. We bring our key customers to our head office regularly and conduct mini conferences, innovation seminars and idea clinics and even organize picnics and barbeques. These events have been extremely productive. The social media site that we added for this purpose provided us an additional method for obtaining customer input"*. The customer input data obtained from other modes of interaction is summarized in Table 4.

Table 4. Summary of customer input obtained at various stages of NSD

NSD stages [1]	Customer input	Modes of interaction
Initiation stage	Describe needs, problems and possible solutions	Face to face interviews
	Evaluate existing services by suggesting likes and dislikes	Focus group interview
Comprehension stage	Examine the overall sale-ability of a new service	Working as part of NSD team
	Jointly develop initial new service blue prints	Innovation retreat Working as part of NSD team
Corroboration stage	Observe the service delivery trial by the front-line service personnel	Innovation retreat
	Observe and participate in mock service delivery process by the key contact employees	Team meetings and discussion
	Participate in a simulated service delivery process as part of NSD team	Innovation retreat Working as part of NSD team
Execution stage	Adopt the service as a trial and provide feedback	Face to face Interviews
	Feedback about overall performance of the service along with desired improvements	Occasional meetings and feedback. Informal customer-manager mixer

After the completion of the research project, a total of 19 new service concepts were generated. These concepts were related to various financial services that the firm was planning to develop: group pension plans mutual fund investment and stocks and securities products, business insurance, money market products cash management systems, industrial asset management, direct equity investment. Due to confidentiality reasons we are unable to provide more details of these new services. We analysed the quality score ratings of these new services by collecting buying intent rating from a different set of customers (n = 61). We pre-screened these respondents to ensure that none of them was involved in the previous idea generation and NSD tasks. We sent a questionnaire to the respondents asking them to rate their willingness to buy/adopt the services and assess the strengths and weaknesses of each service concepts in the scale of 1–5. We received 42 completed questionnaires after two reminders. The majority of the respondents liked the service concepts and clearly indicated their willingness to buy the new services if offered by the company.

5   Implications of the Study

Scholars have argued that due to the fast changing marketing environment the traditional marketing methods are becoming less effective and the customers are demanding new modes of interaction [e.g., 37, 48]. Social media is one such mode of interaction and collaboration that has replaced many traditional mode of accessing and sharing information. Several recent studies have investigated such collaborations and interaction process over social media [e.g., 13, 29]. Yet, compared with the current literature, our research offers the first empirical evidence of the use of social media for customer interaction in NSD in the B2B context. In particular, we contribute to the emerging research on the importance of social media by studying the use of social media for customer interaction for NSD. The extant literature mostly investigates how to interact with the customers and obtain input from them for NSD. In contrast, we study how social media adds value to the interaction process. To that end, we delineate the actual process of obtaining customer input via the social media for NSD. Our second contribution is to empirically show that social media can be an additional mode of customer interaction and can complement the traditional interaction modes used by many firms and reported in the extant literature. Based on findings of this research we argue that various modes of customer interaction as reported in the literature still have a role to play, yet adding social media to the mix can provide better results. Thus, we extend the customer interaction literature by showing the importance of social media for the purpose of customer interaction.

Although the use of social media has grown rapidly and been the beneficiary of significant scholarly attention, we still know very little about how it is used for NSD. Particularly, the use of a microsite for customer interaction is relatively an un-researched area. As shown in this article, the social media tool of microsite provides greater customer value by facilitating connections among customers. Therefore, a company needs to invest in technologies and platforms that attract customers to their sites and encourage them to build their networks which in turn empowers them to communicate with each other and the managers. We agree with Perron and Kozinets [50] that these sites work best if there is a free flow of messages and communication as observed in this study. Our findings also support Rapp et al.'s [52] assertion that the use of social media is becoming rather important in the B2B context and collaborative communication is critical to the success of inter-firm relationship particularly for the B2B firms.

This research provides evidence that customer interaction in NSD is an iterative, interactive and experiential process. In addition, the intensity of interaction varies across various stages of NSD. Therefore, a variety of contact points are needed for a successful interaction strategy. For example, besides social media interactions, three other strategies are particularly critical to developing an effective interaction strategy: (1) conducting face-to-face interviews with the customers at various stages of NSD (2) increasing the amount of communication and informal interactions among the managers and customers, and (3) inducting innovative and expert customers into NSD teams. In summary, our research highlights the substantial benefits that accrue from harnessing customers' input and information by developing social media campaigns and combining the efforts both for social media and traditional marketing research. With the increased popularity of online and social media activities, the managers need to continuously develop strategies that harness the power of online social interactions and deciphering those interactions to develop new products or services.

6 Limitations and Future Research Directions

There are several limitations of this study that offer opportunities for further research. First, the exploratory nature of this study hinders the generalizability of the findings. Thus, a large scale quantitative empirical study is an avenue for further research. However, given the complex nature of this emerging research area, future research employing a pluralistic approach, integrating the use of interpretive and quantitative method is appropriate. Second, our data is based on B2B companies. Future research can replicate this study in other B2C service sectors and tangible products. Third, the single case design limits the generalizability of the result. Future research could use a multiple case research approach. Finally, this research relates to the financial services industry. A different pattern of findings may emerge in other service industries. Thus future research could investigate the process of customer interaction for developing several other types of services.

References

1. Alam, I.: Moving beyond the stage gate models for service innovation: The trend and the future. Int. J. Econ. Pract. Theory **4**(5), 637–646 (2014)
2. Alam, I.: Removing the fuzziness from the fuzzy front-end of service innovations through customer interactions. Ind. Mark. Manag. **35**(4), 468–480 (2006)
3. Alam, I.: An exploratory investigation of user involvement in new service development. J. Acad. Mark. Sci. **30**(3), 250–261 (2002)
4. Alam, I., Perry, C.: A customer-oriented new service development process. J. Serv. Mark. **16**(6), 515–534 (2002)
5. Baird, C.H., Parasnis, G.: From Social Media to Social CRM: What Customers Want. IBM Institute for Business Value, New York (2011)
6. Barras, R.: Towards a theory of innovation in services. Res. Pol. **15**(4), 161–173 (1986)
7. Bayus, B.: Crowdsourcing new product ideas over time: an analysis of the Dell IdeaStorm community. Manag. Sci. **59**(1), 226–244 (2012)
8. Biemans, W.G., Griffin, A., Moenaert, R.K.: New service development: how the field developed, its current status and recommendations for moving the field forward. J. Prod. Innov. Manag. **33**(4), 382–397 (2016)
9. Bosch-Sijtsema, P., Bosch, J.: User involvement throughout the innovation process in high-tech industries. J. Prod. Innov. Manag, **32**(5), 793–807 (2015)
10. Booz, Allen and Hamilton: New Product Management for the 1980s. Booz, Allen and Hamilton Inc., New York (1982)
11. Bowers, M.R.: Developing new services for hospitals: a suggested model. J. Health Care Mark. **7**(2), 35–44 (1987)
12. Bowers, M.R.: Developing new services: improving the process makes it better. J. Serv. Mark. **13**(1), 15–20 (1989)
13. Brodie, R.J., Ilic, A., Juric, B., Hollebeek, L.: Consumer engagement in a virtual brand community: an exploratory analysis. J. Bus. Res. **66**, 105–114 (2013)
14. Brown, S.L., Eisenhardt, K.M.: Product development: past research, present findings and future directions. Acad. Manag. Rev **20**(2), 343–378 (1995)
15. Bryman, A., Bell, E.: Business Research Methods, 4th edn. Oxford University Press, Oxford (2015)

16. Cadwallader, S., Burke, C., Bitner, M., Ostrom, A.L.: Frontline employee motivation to participate in service innovation implementation. J. Acad. Mark. Sci. **38**(2), 219–239 (2010)
17. Carlborg, P., Kindstrom, D., Kowalkowski, C.: The evolution of service innovation research: a critical review and synthesis. Serv. Ind. J. **34**(5), 373–398 (2014)
18. Chang, W., Taylor, S.A.: The effectiveness of customer participation in new product development: a meta analysis. J. Mark. **80**(1), 47–64 (2016)
19. Clifford, C.: Wal-Mart offers entrepreneurs a chance to compete for shelf space. Entrepreneur, 1 July 2013 (2013)
20. Cui, A.S., Wu, F.: Utilizing customer knowledge in innovation: antecedents and impact of customer involvement on new product performance. J. Acad. Mark. Sci. **44**(4), 516–538 (2016)
21. Cui, A.S., Wu, F.: The impact of customer involvement on new product development: contingent and substitutive effects. J. Prod. Innov. Manag. **34**(1), 60–80 (2017)
22. De Smet, D., Mention, A., Torkkeli, M.: Involving high net worth individuals (HNWI) for financial services innovation. J. Fin. Serv. Mark. **21**(3), 226–239 (2016)
23. Dubois, A., Gadde, L.E.: Systematic combining: an abductive approach to case research. J. Bus. Res. **55**(7), 553–560 (2002)
24. Fang, E.: Customer participation and the trade-off between new product innovativeness and speed to market. J. Mark. **72**(4), 90–104 (2008)
25. Fuchs, C., Shreier, M.: Customer empowerment in new product development. J. Prod. Innov. Manag. **28**(1), 17–32 (2011)
26. Geiger, S., Finch, J.: Making incremental innovation tradable in industrial service settings. J. Bus. Res. **69**, 2463–2470 (2016)
27. Ghose, A., Han, S.P.: An empirical analysis of user content generation and usage behavior on the mobile internet. Manag. Sci. **57**(9), 1671–1691 (2011)
28. Grimes, S.: The Rise and Stall of Social Media Listening. Information Week, 25 March 2013 (2013). (5)
29. Hajli, N., Shanmugamb, N., Papagiannidis, S., Zahay, D., Richard, M.: Branding co-creation with members of online brand communities. J. Bus. Res. **70**, 136–144 (2017)
30. Hauser, J., Tellis, G.J., Griffin, A.: Research on innovation: a review and agenda for marketing science. Mark. Sci. **25**(6), 687–717 (2006)
31. Henard, D.H., Szymanski, D.M.: Why some new products are more successful than others. J. Mark. Res. **38**, 362–375 (2001)
32. Hollebeek, L.D., Glynn, M.S., Brodie, R.J.: Consumer brand engagement in social media: conceptualization, scale development, and validation. J. Int. Mark. **28**(2), 149–165 (2014)
33. Homburg, C., Wilczek, H., Hahn, A.: Looking beyond the horizon: how to approach the customers' customers in business-to business markets. J. Mark. **78**(5), 58–77 (2014)
34. Hoyer, W.D., Chandy, R., Dorotic, M., Krafft, M., Singh, S.S.: Consumer cocreation in new product development. J. Serv. Res. **13**(3), 283–296 (2010)
35. Kindstrom, D., Kowalkowski, C.: Development of industrial service offerings: a process framework. J. Serv. Manag. **20**(2), 156–172 (2009)
36. Kumar, A., Bezawada, R., Rishika, R., Janakiraman, R., Kannan, P.K.: From social to sale: the effects of firm-generated content in social media on customer behavior. J. Mark. **80**(1), 7–25 (2016)
37. Lamberton, C., Stephen, A.T.: A thematic exploration of digital, social media, and mobile marketing: research evolution from 2000 to 2015 and an agenda for future inquiry. J. Mark. **80**(6), 146–170 (2016)
38. Lemon, K.N., Verhoef, P.: Understanding customer experience throughout the customer journey. J. Mark. **80**(6), 69–96 (2016)

39. Luo, L., Toubia, O.: Improving online idea generation platforms and customizing the task structure on the basis of consumers' domain-specific knowledge. J. Mark. **79**(5), 100–114 (2015)
40. Magnusson, P.R.: Exploring the contributions of involving ordinary users in ideation of technology-based services. J. Prod. Innov. Manag. **26**(5), 578–593 (2009)
41. McCracken, G.: The Long Interview (Qualitative Research Methods Series 13). Sage Publications, Newbury Park (1998)
42. Melton, H.L., Hartline, M.D.: Customer and frontline employee influence on new service development performance. J. Serv. Res. **13**, 411–425 (2010)
43. Moore, J.N., Raymond, M.A., Hopkins, C.D.: Social selling: a comparison of social media usage across process stage, markets, and sales job functions. J. Mark Theory Pract. **23**(1), 1–20 (2015)
44. Nambisan, S.: Designing virtual customer environments for NPD: toward a theory. Acad. Manag. Rev. **27**(3), 392–413 (2002)
45. Nambisan, S., Baron, R.A.: Interactions in virtual customer environments: implications for product support and customer relationship management. J. Int. Mark. **21**(2), 42–62 (2007)
46. Ordanini, A., Parasuraman, A.: Service innovation viewed through a service-dominant logic lens: a conceptual framework and empirical analysis. J. Serv. Res. **14**(1), 3–23 (2010)
47. Ottenbacher, M., Gnoth, J., Jones, P.: Identifying determinants of success in development of new high-contact services. Int. J. Serv. Ind. Manag. **17**(4), 344–363 (2006)
48. Park, E., Rishika, R., Janakiraman, R., Houston, M.B., Yoo, B.: Social dollars in online communities: the effect of product, user, and network characteristics. J. Mark. **82**(1), 93–114 (2018)
49. Patino, A., Pitta, D.A., Quinones, R.: Social media's emerging importance in market research. J. Consum. Mark. **29**(3), 233–237 (2012)
50. Perren, R., Kozinets, R.V.: Lateral exchange markets: how social platforms operate in a networked economy. J. Mark. **82**(1), 20–36 (2018)
51. Pettigrew, A.M.: Longitudinal field research on change: theory and practice. Org. Sci. **1**(3), 267–292 (1990)
52. Rapp, A., Beitelspacher, L.S., Grewal, D., Hughes, D.E.: Understanding social media effects across seller, retailer, and consumer interactions. J. Acad. Mark. Sci. **41**(5), 547–566 (2013)
53. Rosen, M.: Coming to terms with the field: understanding and doing organizational ethnography. J. Manag. Stud. **28**(1), 1–24 (1991)
54. Salunke, S., Weerawardena, J., McColl-Kennedy, J.R.: Towards a model of dynamic capabilities in innovation-based competitive strategy: insights from project-oriented service firms. Ind. Mark. Manag. **40**(8), 1251–1263 (2011)
55. Santos-Vijande, M., López-Sánchez, J., Rudd, J.: Frontline employees' collaboration in industrial service innovation: routes of co-creation's effects on new service performance. J. Acad. Mark. Sci. **44**(3), 350–375 (2016)
56. Scheuing, E.E., Johnson, E.M.: A proposed model for new service development. J. Serv. Mark. **3**(2), 25–35 (1989)
57. Schultz, D.E., Peltier, J.W.: Social media's slippery slope: challenges, opportunities and future research directions. J. Res. Int. Mark. **7**(2), 86–99 (2013)
58. Schweidel, D., Moe, W.A.: Listening in on social media: a joint model of sentiment and venue format choice. J. Mark. Res. **51**(August), 387–402 (2014)
59. Song, L.Z., Song, M., Di Benedetto, C.A.: A staged service innovation model. Decis. Sci. **40**(3), 571–599 (2009)
60. Srinivasan, S., Rutz, O.J., Pauwels, K.H.: Paths to and off purchase: quantifying the impact of traditional marketing and online consumer activity. J. Acad. Mark. Sci. **44**(4), 440–453 (2016)

61. Storey, C., Hull, F.M.: Service development success: a contingent approach by knowledge strategy. J. Ser. Manag. **21**(2), 140–161 (2010)
62. Toubia, O., Stephen, A.T.: Intrinsic vs. image-related utility in social media: why do people contribute content to Twitter? Mark. Sci. **32**(3), 368–392 (2013)
63. Tan, J.C.K., Lee, R.: An agency theory scale for financial services. J. Serv. Mark. **29**(5), 393–405 (2015)
64. Tyler, K., Stanley, E.: The role of trust in financial services business relationships. J. Serv. Mark. **21**(5), 334–344 (2007)
65. Zwass, V.: Co-creation: toward a taxonomy and an integrated research perspective. Int. J. Electron. Commer. **15**, 11–48 (2010)

Employee-Centric Service Innovation: A Viable Proxy for Customer-Intimacy for Product-Focused Enterprises

Michael Vössing[1]([envelope]), Jörg Siegel[1], Niels Feldmann[1], Thorsten Wuest[2], and Carina Benz[1]

[1] Karlsruhe Service Research Institute, Karlsruhe Institute of Technology, Kaiserstr. 12, 76131 Karlsruhe, Germany
{michael.voessing,niels.feldmann,carina.benz}@kit.edu
[2] Industrial and Management Systems Engineering, West Virginia University, Morgantown, WV 26506, USA
thwuest@mail.wvu.edu

Abstract. Servitization has received significant attention from scholars and practitioners over the last decade. However, despite substantial research contributions in the fields of new service development and service innovation, product-focused small and medium-sized enterprises struggle to develop sophisticated service offerings. This paper attempts to better understand this discrepancy and suggests ways to overcome it. We have conducted a case study with a medium-sized manufacturing company that currently undertakes first steps in the development of advanced services. In terms of a theoretical contribution, our study indicates a limited understanding of the perceived value of services during the fuzzy front end of service innovation. Therefore, companies need to adopt new ways to understand their customers (i.e. increase customer intimacy). However, they often struggle to directly involve customers in the co-development of advanced services. In terms of a practical contribution, our research suggests that employee-centric service innovation—the idea of utilizing front line employees as proxies of customers—is a viable strategy to mitigate the identified challenges.

Keywords: Front line employees · Service innovation · Servitization

1 Introduction

Due to constant product competition, manufacturing companies have to enhance their offerings continuously [14, 23]. As a result, they rarely focus on manufacturing alone, but typically offer a variety of supplementary services associated with their products [26]. Moreover, companies integrate products and services into product-service systems as means to lock out competitors, lock in customers, and increase the level of differentiation [26, 40]. This transformation is known as *servitization* and has been defined by Baines et al. [2, p. 1207] as "the innovation

© Springer Nature Switzerland AG 2018
G. Satzger et al. (Eds.): IESS 2018, LNBIP 331, pp. 88–100, 2018.
https://doi.org/10.1007/978-3-030-00713-3_7

of an organization's capabilities and processes to shift from selling products to selling integrated products and services that deliver value in use".

However, servitization is no guarantee for success. Research shows that even larger enterprises struggle to recoup the costs of service development [13,33] or achieve cost-effectiveness with their service offerings [9]. As a result, service development often does not materialize the expected competitive advantage [29]. To stay competitive, manufacturing companies need to move up the value chain and create increasingly sophisticated products and services. Therefore, the ability to incorporate and leverage emerging technologies—especially those associated with information and communications technology (ICT)—has become an essential enabler of servitization [26] and a crucial success factor for the development, integration, and delivery of sophisticated service offerings [1]. Companies currently benefit from two developments: Firstly, as ICT infrastructure is maturing and becoming commoditized, companies can develop services that rely on ICT with minimal upfront investment and scale them by utilizing variable cost structures [1]. Secondly, techniques for the automated collection and analysis of data are increasingly available. Companies can now leverage their data to better understand their customers, address complex needs on an individual level, and ultimately create more value through tailored and customized services [14].

This obligation to leverage technologies and data to develop advanced service offerings [9,35] adds an additional layer of complexity to the many documented challenges, risks, and barriers associated with servitization [21,28]. Schüritz et al. [35, p.4] specify this layer in their work and label the phenomenon as *datatization*—meaning the "innovation of an organization's capabilities and processes to change its value proposition by utilizing data analytics". Their work identifies challenges companies face when utilizing emerging technologies to adapt their value proposition (e.g., design of offering, design of revenue model, etc.). Thus, datatization can be considered a modern way of servitization that requires a transformation of the capabilities and processes of companies.

So far, no standard process on how to manage servitization and datatization has emerged [35]. Companies often struggle specifically in the early stages of the innovation process—which are known as the fuzzy front end. The term was initially coined by Smith and Reinertsen [37] to describe the early and intangible phase of product development in which critical properties of solutions are determined. The term was subsequently adopted to reference the early phases of service innovation—mainly idea generation, idea screening and concept development [6]. Research shows that this stage is crucial for the long-term success of service offerings [6,25]. However, few companies have adopted structured processes to manage this stage [42].

The study presented in this paper originates from our work with industry practitioners trying to find better ways to support service innovation in their organizations. Although the literature on servitization, service innovation, and service design is plentiful, there is a lack of alignment between the topics analyzed by scholars and those relevant for practitioners [4]. Neely [26] notes that while academic literature does highlight the importance of servitization, surprisingly little empirical evidence on the topic is available. Additionally, the topics have primarily been studied with large manufacturing companies. Nevertheless, product-service systems, servitization, and datatization are also intriguing for product-focused small and medium-sized enterprises (SME). Unfortunately, these companies often neither possess the required resources nor the experience to develop, deliver, or scale labor-intensive services—as they operate under stricter constraints [13,20]. In other words, the challenges outlined in this section are more severe for these companies and, thus, remain highly relevant.

In this study, we aim to contribute to the mitigation of the challenges of servitization, and more specifically of datatization, in the context of product-focused SME. According to Baines et al. [4], the challenges manufacturers face in the development of advanced services are rarely explored in literature and, so far, the capabilities companies need to succeed in the very early stages of innovation are little-researched [17]. Accordingly, we raise the following two research questions: (1) What capabilities are relevant in the fuzzy front end of service innovation for product-focused SME. (2) What are new and efficient ways for product-focused SME to overcome the challenges associated with these capabilities.

The methodology to answer these questions is twofold: To address the first question an empirical study is conducted. Building on the results of this study, the second question is approached through conceptual work. As there are currently few empirical studies available that address these topics, we decided to conduct an in-depth case study instead of utilizing a survey-based approach. The in-depth case study provides a unique, empirical perspective on the challenges manufacturers encounter in the fuzzy front end of service innovation and the capabilities they need to develop. Through a qualitative research approach—combining and contrasting perspectives from managers and employees with perspectives of multiple customers—two under-emphasized challenges faced by product-focused SME are outlined. Preliminary reasoning is provided that explains and emphasizes the importance of these challenges for servitization. Finally, the paper concludes by describing the concept of *employee-centric service innovation* and positioning it as a viable proxy for customer-intimacy.

The paper is structured as follows: Sect. 2 outlines the research methodology and presents how the empirical data was collected. Section 3 describes the results of the study—focusing on two capabilities essential in the fuzzy front end of service innovation. Finally, Sect. 4 outlines how companies can increase customer-intimacy by leveraging and empowering their front line employees.

2   Methodology and Data

This work aims to extend the understanding of capabilities SME require in the context of servitization and datatization—especially in the fuzzy front end of service innovation. To study this topic, an in-depth case-study is reasonable—especially as access to companies at the beginning of the transformation from selling products to offering advanced services is rare and empirical evidence, therefore, difficult to obtain [44]. To provide a comprehensive picture, incorporating an internal perspective (i.e., managers, customer service and technology partners) and an external perspective (i.e., customers) is essential. Overall, the applied methodology is based on grounded theory [15]. More specifically, a modified approach that adapts grounded theory for information systems research is applied [12].

A suitable company, at the beginning of the transformation from selling industrial products to offering advanced service was found in a market-leading manufacturing company involved in the sale and operation of auxiliary equipment in the plastics industry. The company's 230 employees manage a global distribution network that serves customers in multiple industries (e.g., health care and food packaging). While the company has traditionally focused on the sale of their approximately 450 distinct products, the company nowadays offers basic supplementary services (i.e., technical support, training). While these services have been well received by customers, limited resources prevent the company from offering these labor-intensive services at scale: "We have learned that we don't have the manpower [...] or required resources" (DP). Additionally, due to increasing competition, the currently offered services provide little long-term competitive advantage. As a result, the company has recently started working with technology providers to develop advanced services (i.e., predictive mainte-

Table 1. Internal research suspects

Interviewee	Role	Description	Data collection
FCR	Backoffice	Technical customers service	Interview
TSR	Backoffice	Technical customers service	Interview
FST	Technician	Technical field service	Interview
PSA	Technician	Product specialist	Interview
PSB	Technician	Product specialist	Interview
DP	Dispatcher	Dispatching technicians	Interview
FS	Management	Manages service team	Interview
SM	Management	Manages service division	Interview
MGMT	Management	Product management	Observation
TP	Technology	External partner	Observation

Table 2. External research suspects

Company	Industry	Revenue	Employees	Data collection
Alpha	Packaging	<$100M	<250	Interview
Beta	Plastic caps	<$1.000M	<10.000	Interview
Gamma	Logistics	<$500M	<10.000	Observation
Delta	Medical	<$1.000M	<10.000	Observation
Epsilon	Logistics	<$20.00M	<100.000	Observation
Zeta	Plastic hoses	<$100M	<250	Observation

nance) that can replace the current labor-intensive services and overall improve the availability of their customers' machinery.

Empirical data has been collected over a four-month period in the summer of 2017. The study consists of semi-structured interviews—varying in duration between thirty minutes and two hours—with eight employees of the company and two customers (see Tables 1 and 2). Additionally, participant observations—spanning several days in total—were conducted with two employees as well as multiple customers to supplement and verify the data obtained from the interviews through a close observation of work responsibilities and environments in the industry. Studies on the capabilities needed for servitization generally present results on a high level of abstraction. To derive provisional codes for our analysis, we compiled a list of 30 capabilities through a structured literature review based on Webster and Watson [43]. The analysis of the data focused on systematically comparing the recorded practices, challenges, and concerns with the collected servitization capabilities. These findings were further examined and contrasted with the prevailing understanding in the servitization literature. While a number of documented capabilities were confirmed in the interviews, through the analysis of the data two capabilities emerged as essential for product-focused SME. Primarily, as the employees and customers agreed on their importance while presenting conflicting perspectives on the reasons for their importance.

3   Essential Capabilities of Product-Focused Enterprises in the Fuzzy Front End of Service Innovation

As outlined by Brax [7], industrial services are difficult to manage as they require organizational settings different from those needed for the production of physical goods. Therefore, providing industrial services is not just a matter of changing the offering, but requires organizations to "re-focus [their] attention" [7, p.152] and adopt new practices and technologies [3]. Competing through services fundamentally requires companies to change their strategy for market leadership. While product-focused companies typically focus on *product leadership* or *operational excellence*, the service economy requires companies to focus on *customer intimacy*. Customer intimacy is typically defined as a company's ability to acquire and "combine detailed customer knowledge with operational flexibility" to quickly respond to customer needs and tailor offerings accordingly [38,39]. In practice, however, manufacturing companies often struggle with this paradigm shift and need to progressively develop "the capability to design and deliver services rather than product[s]" [26, p. 114]. In this context, we have identified two capabilities crucial for the servitization of product-focused SME—especially in the fuzzy front end of service innovation: (a) managing shifting expectations of value-in-use, and (b) collaboration and learning in service systems.

3.1   Managing Shifting Expectations of Value-in-Use

Traditionally, manufacturing companies rely on their customers to integrate purchased products into their work environment and maximize value-in-use by themselves [5]. Driven by the servitization of these companies, today the focus is shifting from the delivery of products to the facilitation of value creation [18,26,41]. This transformation is particularly challenging for previously product-focused SME [22]. Primarily, because customer engagement [21,27] requires resources, capabilities, and activities to be integrated and coordinated [18].

While our case study has confirmed these challenges from the perspective of the service provider, it has also revealed that customers face similar challenges. Customer Beta expected implementation efforts of servitization to be high due to operational barriers and the necessity of change management when "starting from scratch". Customer Alpha further highlighted the importance of understanding the scope of service offerings: "I think it's a big undertaking. [...] To say 'yeah let's go do this', I have to understand where the transformation starts and stops". The requirement of new forms of collaboration and the intangible nature of services further complicate the evaluation of service offerings. In particular, customers were missing prior experiences to asses the impact of utilizing advanced services to reduce downtime: "We don't say 'that machine was down for two hours and that costs us X dollar'. Nobody tries to do that." (Beta). Further, customers emphasized different objectives. For example, Alpha and Delta focus foremost on product quality, while Epsilon focuses on productivity. Accustomed to optimize value-in-use of products by themselves, customer transferred this mindset to service offerings. This behaviour led to misaligned expectations of how services would create value. Further, advanced services irritated customers as they blur the boundaries between supporting and replacing their core competencies: "I'm not sure how we feel about giving all those [responsibilities] up" (Alpha). Thus, developing new capabilities to systematically identify and manage these implicit expectations is essential for overcoming hesitant customer behavior. Especially at the early stages of the service innovation process, uncertainties regarding value-in-use explain this 'dilemma of closeness' [32].

3.2 Collaboration and Learning in Service Systems

Servitization requires three new forms of collaboration: (a) between internal units [17,19], (b) with partner networks [14,18] and (c) with customers in long-term relationships [9,26]. Whereas different variants of collaboration are discussed in literature [4,35], the impact of cross-sector partnerships (e.g., with new technology partners) has not yet been analyzed. Close collaboration with customers facilitates a learning relationship beneficial for the development of new services [32]. However, it requires a shift in the mindset and culture of product-focused SME [2,28,36]. Today, companies no longer only sell products, but enter into long-term relationships with their customers through the services they sell. The servitization of manufacturing companies, therefore, fundamentally "changes the nature of what is being sold" [26, p. 115].

The case study indicates that companies need to adopt a service system view on collaboration. In particular, the case company and the TP intensified their cooperation gradually to meet additional challenges arising from co-developing services with customers. However, we observed multiple challenges: Due to limited resources, the high efforts required to build customer intimacy, and the difficulty of convincing customers to "[allow] their data to leave their facility" (SM) only a small number of customers was involved in the co-creation. Additionally, in many cases the opinions and expectations of the customers contradicted each other (e.g., which machines should be connected first). While managers valued the "hands-on feedback" (MGMT) from customers, they quickly became concerned about the narrow selection of partners and feared "losing sight of the bigger picture" (MGMT). Primarily, as only long-term customers were selected and volunteered to participated in the development of these services. Including more customers in co-development was expected to decrease the willingness of these participants to share sensitive data, because "many [...] customers are competitors" (SM). Unfortunately these 'handpicked' customers ultimately provided neither relevant insights (i.e., as they already expressed above-average satisfaction with the products) nor were they representative of the entire customer base. As including more customers was not a sustainable solution, new approaches to collect data on the concerns and needs of customers were required.

4   Towards Employee-Centric Service Innovation

To meet the challenges outline in Sect. 3, companies need to increase customer intimacy. While this insight is not new, the problem remains unsolved for SME as scholars have primarily studied large corporations. However, SME differ from these cases by their limited resources and varying capabilities. Therefore, how SME can increase customer intimacy remains unanswered. Building on the data obtained in the case study, the following conceptual work outlines an efficient way for product-focused SME to overcome these challenges. The section outlines (a) the challenges of customer-centric service innovation, and (b) the advantages of employee-centric service innovation.

4.1   Challenges of Customer-Centric Service Innovation

Manufacturing companies primarily have two responsibilities: producing existing products reliably, and driving innovation for new products and services. Patrício et al. [31, p. 9] outline that focusing on what is perceived valuable by the customer is essential. As customers "know their own contexts best", they propose that companies should integrate customers through active participation in their innovation processes (i.e., co-development).

However, our study shows that SME struggle to find practical ways to acquire and integrate customer knowledge through direct involvement of their customers. Engaging a representative set of customers and managing the company's resources to actively participate in the co-development with customers often is challenging. More specifically, attaining customer-intimacy traditionally means instilling top-down a customer-centric mindset in employees. However, for

SME this often leads to an innovation fallacy: Approaches such as design thinking and co-development address these issues, however, are time-consuming and challenging for product-focused SME. Focusing on these approaches, therefore, might even be counterproductive for these companies, as co-developing solutions with few customers might not address the needs of their heterogeneous customer base: "[...] our equipment is used everywhere" (SM).

While customer intimacy is, in fact, essential for servitization, our research shows that leveraging the potential of the company's front-line employees, subsequently called *employee-centric service innovation*, is a viable proxy for an efficient representation of customers' perspectives in a service innovation endeavor. Companies can benefit by moving from a top-down and outward-focused approach to customer intimacy (i.e., becoming more customer-centric by getting closer to customers) towards a bottom-up and inward-focused approach (i.e., becoming more customer-centric by leveraging internal expertise). The idea of utilizing front-line employees to facilitate service innovation has been mentioned before [34]. However, the concept is typically discussed in the context of professional service firms [11]. However, product-centric companies also have employees with naturally high levels of customer-intimacy (i.e., service technicians). Unfortunately, these employees do not necessarily collaborate with employees that develop new services and could benefit from their knowledge (e.g., service product owner). Moreover, their job descriptions rarely emphasize knowledge sharing beyond their own work environment. As our interview partners put it, the information their front-line employees capture "rather serves for [service] documentation purposes" (FS). Hence, product specialists have to "pick up on [machine failure] trends" themselves.

4.2   Advantages of Employee-Centric Service Innovation

Ostrom et al. [30, pp. 133–135] note that rapidly changing contexts in which services are delivered require companies to increase "employee engagement to improve service outcomes" by "incorporating the voice of the employee in service innovation". According to Benkenstein et al. [5], service work in the context of servitization is insufficiently understood and needs further research. Only few authors examined the potential of front line employees (i.e., customer service and sales) on the service innovation process [8,10,16,24]. Notably, Feldmann et al. [11] explore multiple instruments (e.g., enterprise crowd funding) for engaging front line employees in the service innovation process—primarily focusing on professional service firms and knowledge intensive domains.

Our study shows that front line employees of product-focused SME are exceptionally customer-centric, since customers have "[problems] all across the board" (FCR) which require employees to constantly "read between the lines" (FCR). Therefore, knowledge about customers often is acquired as a side-product of customer support for example when "customers don't understand how to configure a machine properly" (FST) or are running a machine "completely different than it was intended to" (PSA). Customers already engage with front line employees through co-creation where it meets their needs. This requires service employees to "[continuously] learn" (FCR) and understand the "huge variety of different use-cases [...] for the same piece of equipment" (PSA). Employees acquire this knowledge implicitly through their work, but only leverage it themselves. Therefore, most of this knowledge is neither captured nor used in the service innovation process. The vast majority of the knowledge is lost without structured processes.

There are two ways front line employees can be leveraged for service innovation: (a) customer knowledge assimilation, and (b) customer co-development. Customer knowledge assimilation is about absorbing the knowledge of customers. In this case, front line employees act as proxies for customers in internal service innovation projects by contributing their in-depth understanding of customers. The second level is the facilitation of co-development with customers. This, more active, role leverages their personal relationships with customers to facilitate, by proxy, the engagement of customers in the service innovation process. Both activities benefit from the interpersonal skills of front line employees, the trust they have earned as problem solvers for customers, and their in-depth knowledge of a wide range of customer concerns acquired through their daily job responsibilities. Unquestionably, this knowledge is difficult to obtain from customers through other means.

Managing job enlargement of front line employees is essential for their integration into the service innovation process. While our study outlines the high levels of expertise and interpersonal skills front line employees possess, the study also outlines how incorporating front line employees would add significant complexity (i.e. recording data, managing concerns, managing contacts). Collecting data through smart devices can help bridge this knowledge gap, but without human interpretation few insights are generated as "someone has to understand the data that is coming in" (FCR). While the high number of customer encounters constitutes a unique resource for service innovation, leveraging these employees requires managers to find ways to relieve them of other tasks. However, the same technology that can be used to develop new services for customers (i.e. Internet of Things) can be used to relieve front line employees of repetitive tasks. Managers need to allow employees to focus on their core competency to ultimately enable them to additionally support the companies co-creation efforts. Managers need to take their employees 'job-to-be-done' into account and align them with the design of new service offerings. New tools that facilitate the co-creation and enable new forms of communication are required to validate insights from customer experiments and facilitate a targeted knowledge generation.

5 Conclusion and Outlook

This paper reports on our work to understand the practical challenges product-focused SME face in the fuzzy front end of service innovation. While emerging technologies do enable the development of increasingly complicated services, managing servitization and datatization requires companies to overcome a number of challenges. Through an in-depth case study, this work contributes to the understanding of these capabilities. While most literature analyzes servitization from the perspective of service providers, this study additionally explores the customer perspective [4].

The study shows that developing complex services requires SME to gain a deeper understanding of their customers—especially as expectations and perceptions of value-in-use differ significantly from product-focused contexts. Customers tend to apply and transfer their mindset of 'buying a product'—which requires maximizing value-in-use on their own—to service offerings. The study shows that failing to manage these concerns can lead to a discrepancy of expected risk and effort. Therefore, companies need to adjust how they learn from their customers in these service systems. The study provides support that especially front line employees can play a key role in acquiring a deeper understanding of customers required for service innovation. This led to the hypothesis that 'employee-centric service innovation is a viable proxy for customer-intimacy for product-focused small and medium-sized enterprises'. Contrary to top-down approaches to customer intimacy, often positioned in the literature, our findings indicate that managers should focus on learning from their customer-centric front line employees and incorporating them into the service innovation process. Our observations lead to the following conclusions: While improving customer-intimacy certainly is an essential requirement for servitization—and is already outlined by servitization literature as such—the concept of *employee-centric innovation* is a viable means for product-focused SME to improve customer-intimacy. Companies can benefit tremendously from the experience their front line employees have gained through customer support and the close relationships they have built with decision makers of customers.

These results align with earlier research. They confirm Brax [7] hypothesis that industrial services require an overhaul of organizational culture. They further contribute to the research priorities positioned by Ostrom et al. [30] by outlining—based on empirical evidence obtained through an illustrative case study—a concrete example in which incorporating employees in the service innovation process is not only beneficial but preferable to other service innovation approaches. Our study further builds on the work of Feldmann et al. [11], by transferring the proposed concepts to SME—specifically embedding the proposed ideas in the manufacturing domain.

Overall, the identified capabilities are not meant to be exhaustive, nor can the drawn conclusions be applied to all companies. However, the study provides empirical support that companies—so far product-focused SME—should utilize their front line employees in their service innovation processes. Learning how to leverage the knowledge of front line employees will likely increase in importance given the growing share of services in our economy. While this research emphazises the value of front line employees, it also outlines multiple challenges (e.g., managing job enlargement) of incorporating them into the service innovation process. Hence, more research is needed to derive and evaluate practical frameworks for leveraging front line employees for service innovation.

References

1. Ardolino, M., Rapaccini, M., Saccani, N., Gaiardelli, P., Crespi, G., Ruggeri, C.: The role of digital technologies for the service transformation of industrial companies. Int. J. Prod. Res. **7543**, 1–17 (2017)
2. Baines, T.S., Lightfoot, H.W., Kay, J.M.: Servitized manufacture: practical challenges of delivering integrated products and services. Proc. Inst. Mech. Eng. Part B: J. Eng. Manuf. **223**(9), 1207–1215 (2009)
3. Baines, T., et al.: Towards an operations strategy for product-centric servitization. Int. J. Oper. Prod. Manag. **29**(5), 494–519 (2009)
4. Baines, T., Ziaee Bigdeli, A., Bustinza, O.F., Shi, V.G., Baldwin, J., Ridgway, K.: Servitization: revisiting the state-of-the-art and research priorities. Int. J. Oper. Prod. Manag. **37**(2), 256–278 (2017)
5. Benkenstein, M., et al.: Topics for service management research - a European perspective. J. Serv. Manag. Res. **1**(1), 4–21 (2017)
6. Boukis, A., Kaminakis, K.: Exploring the fuzzy front-end of the new service development process - a conceptual framework. Procedia Soc. Behav. Sci. **148**, 348–353 (2014)
7. Brax, S.: A manufacturer becoming service provider - challenges and a paradox. Manag. Serv. Qual. Int. J. **15**(2), 142–155 (2005)
8. Cadwallader, S., Jarvis, C.B., Bitner, M.J., Ostrom, A.L.: Frontline employee motivation to participate in service innovation implementation. J. Acad. Mark. Sci. **38**(2), 219–239 (2010)
9. Coreynen, W., Matthyssens, P., Van Bockhaven, W.: Boosting servitization through digitization: pathways and dynamic resource configurations for manufacturers. Ind. Mark. Manag. **60**, 42–53 (2017)
10. Dubruc, N., Peillon, S., Farah, A.: The impact of servitization on corporate culture. Procedia CIRP **16**, 289–294 (2014)
11. Feldmann, N., Fromm, H., Satzger, G., Schüritz, R.: Using employees' collective intelligence for service innovation: theory and instruments. In: Maglio, P.P., Kieliszewski, C.A., Spohrer, J.C. (eds.) Handbook of Service Science, vol. 2. Springer (2018, forthcoming)
12. Fernández, W.D.: The grounded theory method and case study data in IS research: issues and design. In: Information Systems Foundations: Constructing and Criticising Workshop, pp. 43–59 (2004)
13. Gebauer, H., Fleisch, E., Friedli, T.: Overcoming the service paradox in manufacturing companies. Eur. Manag. J. **23**(1), 14–26 (2005)
14. Gebauer, H., Ren, G., Valtakoski, A., Reynoso, J.: Service-driven manufacturing. J. Serv. Manag. **23**(1), 120–136 (2012)
15. Glaser, B.G., Strauss, A.L.: The Discovery of Grounded Theory: Strategies for Qualitative Research, vol. 1. Aldine Transaction, London (1967)
16. Hsieh, J.K.: The effect of frontline employee co-creation on service innovation: comparison of manufacturing and service industries. Procedia Soc. Behav. Sci. **224**, 292–300 (2016). August 2015
17. Karlsson, A., Larsson, L., Öhrwall Rönnbäck, A.: Product-service system innovation capabilities: linkages between the fuzzy front end and subsequent development phases. Int. J. Prod. Res. **7543**(November), 1–15 (2017)
18. Kindström, D., Kowalkowski, C.: Service innovation in product-centric firms: a multidimensional business model perspective. J. Bus. Ind. Mark. **29**(2), 96–111 (2014)

19. Kindström, D., Kowalkowski, C., Sandberg, E.: Enabling service innovation: a dynamic capabilities approach. J. Bus. Res. **66**(8), 1063–1073 (2013)
20. Kowalkowski, C., Witell, L., Gustafsson, A.: Any way goes: identifying value constellations for service infusion in SMEs. Ind. Mark. Manag. **42**(1), 18–30 (2013)
21. Lightfoot, H., Baines, T., Smart, P.: The servitization of manufacturing. Int. J. Oper. Prod. Manag. **33**, 1408–1434 (2013)
22. Martinez, V., Neely, A., Velu, C., Leinster-Evans, S., Bisessar, D.: Exploring the journey to services. Int. J. Prod. Econ. **192**(March), 66–80 (2017)
23. Meier, H., Völker, O., Funke, B.: Industrial product-service systems (IPS2): paradigm shift by mutually determined products and services. Int. J. Adv. Manuf. Technol. **52**(9–12), 1175–1191 (2011)
24. Melton, H.L., Hartline, M.D.: Customer and frontline employee influence on new service development performance. J. Serv. Res. **13**(4), 411–425 (2010)
25. Meuris, D., Herzog, M., Bender, B., Sadek, T.: IT support in the fuzzy front end of Industrial Product Service design. Procedia CIRP **16**, 379–384 (2014)
26. Neely, A.: Exploring the financial consequences of the servitization of manufacturing. Oper. Manag. Res. **1**(2), 103–118 (2009)
27. Ng, I.C., Maull, R., Yip, N.: Outcome-based contracts as a driver for systems thinking and service-dominant logic in service science: evidence from the defence industry. Eur. Manag. J. **27**(6), 377–387 (2009)
28. Oliva, R., Kallenberg, R.: Managing the transition from products to services. Int. J. Serv. Ind. Manag. **14**(2), 160–172 (2003)
29. Opresnik, D., Taisch, M.: The value of big data in servitization. Int. J. Prod. Econ. **165**, 174–184 (2015)
30. Ostrom, A.L., Parasuraman, A., Bowen, D.E., Patrício, L., Voss, C.A.: Service research priorities in a rapidly changing context. J. Serv. Res. **18**(2), 127–159 (2015)
31. Patrício, L., Gustafsson, A., Fisk, R.: Upframing service design and innovation for research impact. J. Serv. Res. **21**(1), 3–16 (2018)
32. Raja, J.Z., Frandsen, T., Mouritsen, J.: Exploring the managerial dilemmas encountered by advanced analytical equipment providers in developing service-led growth strategies. Int. J. Prod. Econ. **192**, 120–132 (2017). December 2016
33. Reinartz, W., Ulaga, W.: How to sell services more profitably. Harv. Bus. Rev. **86**(5), 90–96 (2008)
34. Schneider, B., Bowen, D.E.: Winning the service game. In: Maglio, P., Kieliszewski, C., Spohrer, J. (eds.) Handbook of service science. SSRI, pp. 31–59. Springer, Boston (2010)
35. Schüritz, R., Seebacher, S., Satzger, G., Schwarz, L.: Datatization as the next frontier of servitization: understanding the challenges for transforming organizations. In: International Conference on Information Systems (2017)
36. Smith, L., Maull, R., Ng, I.C.L.: Servitization and operations management: a service dominant-logic approach. Int. J. Oper. Prod. Manag. **34**(2), 242–269 (2014)
37. Smith, P.G., Reinertsen, D.G.: Developing products in half the time: new rules, new tools. Van Nostrand Reinhold, New York (1991)
38. Treacy, M., Wiersema, F.: Customer intimacy and other value disciplines. Harv. Bus. Rev. **71**, 84–93 (1993)
39. Treacy, M., Wiersema, F.: The Discipline of Market Leaders: Choose Your Customers, Narrow Your Focus Dominate Your Market. Addison-Wesley, Boston (1997)
40. Vandermerwe, S., Rada, J.: Servitization of business: adding value by adding services. Eur. Manag. J. **6**(4), 314–324 (1988)

41. Vargo, S.L., Maglio, P.P., Akaka, M.A.: On value and value co-creation: a service systems and service logic perspective. Eur. Manag. J. **26**(3), 145–152 (2008)
42. Wagner, L., Baureis, D., Warschat, J.: How to develop product-service systems in the fuzzy front end of innovation. Int. J. Technol. Intell. Plann. **8**(4), 333 (2012)
43. Webster, J., Watson, R.: Analyzing the past to prepare for the future: writing a literature review. MIS Q. **26**(2), 13–23 (2002)
44. Yin, R.K.: Case Study Research and Applications: Design and Methods. Sage publications, Thousand Oaks (2017)

Towards Managing Smart Service Innovation: A Literature Review

Caroline Götz[✉], Sophie Hohler, and Carina Benz

Karlsruhe Service Research Institute (KSRI),
Karlsruhe Institute of Technology (KIT), Karlsruhe, Germany
{caroline.goetz,sophie.hohler}@student.kit.edu,
carina.benz@kit.edu

Abstract. Smart services are increasingly gaining in popularity amongst diverse industries. Their special character–combining physical components, smart components, and connectivity components supported by embedded ICT and big data analytics–allows for an entirely new approach of service offering. Adopting smart services within their solution portfolio, confronts companies with challenges related to servitization and digital transformation that are not only impacting their operation, but also their innovation. Based on a literature review, this study investigates the current state of research on innovation management for smart services. Findings are conceptualized within six categories: Topics, Resources, Knowledge & Information, Processes, Principles, and Methods & Tools.

Consequently, this study consolidates existing knowledge on challenges, changes and approaches for smart service innovation in a structured manner and identifies the need as well as potential for further research in this field.

Keywords: Smart service · Innovation management · Service Design

1 Introduction

Digital transformation is challenging diverse industries–forcing companies to redefine the way they think of their products, as well as their entire business. Two megatrends are omnipresent and particularly fueling the urge to transform [1–3]: servitization and digitalization. Not only are companies increasingly trying to enhance their traditional products and business models with additional or supplementary services, most of them also feel the need to adopt the use of digital technologies within their organizations and offerings in order to not be left behind [4].

As a result, smart services emerge in the intersection of both challenges by (1) expanding the service portfolios of companies and (2) following the technological trend of digitalization. While services reflect "the application of specialized competences (knowledge and skills) through deeds, processes, and performances for the benefit of another entity" [5] the term 'smart' emphasizes the use of intelligent digital technologies in service provision [6, 7]. MICHELIN Fleet Solutions, a service offered by the tire manufacturing company Michelin, uses sensors that measure miles travelled and thus allows truck fleet managers to rent tires as a service instead of buying them.

© Springer Nature Switzerland AG 2018
G. Satzger et al. (Eds.): IESS 2018, LNBIP 331, pp. 101–111, 2018.
https://doi.org/10.1007/978-3-030-00713-3_8

CLAAS, an agricultural machinery manufacturer, uses digital technologies within their machines to—for example—precisely determine the amount of fertilizer needed in a specific field section based on the planting density.

Smart services, thus, reflect a special type of services that are composed of three core elements: physical components, smart components, and connectivity components [9]. Their data-driven character is facilitated by sensors, cyber-physical or embedded systems, and cloud computing [6] which allows for sensing conditions and surroundings and giving in-time feedback [11].

Smart services differ from traditional (e-)services in multiple ways: (1) Smart services rely on embedded information and communications technology (ICT) that allows for data transmission and information generation [13]. (2) Smart services integrate and are enabled by big data analytics [14]. (3) Smart services are completely or at least partly automated and they are perfectly aligned with human interaction. Such automated service actions are only possible by the integration of smart components like cognitive systems [14]. (4) From a customer perspective, smart services allow for greater customization of services by reacting on environmental-conditions or customer-requests (e.g. smart services adapt based on users' location data).

The new smart level of services not only results in various new opportunities, but also in challenges—especially for the innovation management in companies. Innovation and implementation of additional smart features for services need a completely new technology infrastructure and skillset including software development, systems engineering, data analytics and online security expertise [16]. Furthermore, the additional technology perspective within smart service systems are related to extended opportunities for both value creation solutions and value co-creation between different companies. Based on the integrated smart features in the service system, completely new service offerings are made possible and nearly every existing industry might get chances to design new smart services. At the same time, the value chain is getting more interfaces for cooperation. Thus, more actors and industries could be involved in the value creation process of smart services [15].

Despite the popularity of smart services in industry, they are barely represented in academia, especially in non-technical contexts. Regarding the nature of smart services, being different from products and mere services e.g. in terms of value proposition or skills and departments necessary for their development and operation, both the potential and the need can be perceived to further explore the challenges and changes evoking by smart services.

This study especially focuses on how smart services are innovated and developed. While innovation is a key competence to stay competitive, it is explored, whether and how innovation management must adapt to the specifics of smart services.

In context of this paper **smart service innovation** is understood as the development and market introduction of a new, redesigned or substantially improved solution [13] consisting of physical, smart and connective elements. The meaning of **innovation management** is understood as "[s]haping frame conditions in order to enable the emergence and successful implementation of new ideas" as well as the "[a]ctive search, development and implementation of innovation ideas". [18] It describes the initiation, planning, execution and control activities necessary to cause innovations. By connecting these definitions, smart service innovation management can be defined as activities shaping frame conditions and support actively the emergence, development and successful implementations [18] of new, redesigned or substantially improved goods and services [19] consisting of physical, smart and connective elements.

In order to start this endeavor, we conduct a literature review, focused on pre-existing knowledge on how to innovate smart services. Our research, therefore, contributes to the current body of knowledge by providing an overview of existing theoretical knowledge on smart service innovation management, and by identifying gaps for further research. The remainder of this study is structured in the following manner: Within the next section, the design of the literature review is described. Chapter three summarizes and conceptualizes the findings, and chapter 4 discusses them. Finally, a short conclusion and implications for future work are given.

2 Design of the Literature Review

The literature review follows the approach of Webster and Watson [20]: Relevant papers are identified by a database research and investigated by authors cited forward as well as backward and findings in the chosen papers are processed in a matrix by authors and paper contents.

Starting with the database search, the two following combinations of relevant keywords are applied in the four scientific databases Academic Research, Google Scholar, JSTOR and Science Direct: First "Smart AND Service AND Innovation" and second "(Digital OR IoT) AND Service AND Development".

Papers that were published between January 2007 and November 2017 are selected if they fulfill at least one of the following selection criteria: Either a direct relation to innovation processes for smart services can be seen or an indirect relation to the innovation process for smart services can be assumed. More precisely, four guiding questions are used to decide if a paper is relevant: (1) WHAT topics are relevant for SSI, (2) WHO should participate in the SSIM process, including what knowledge does this person has to have, (3) WHICH assets are needed for SSI (e.g. resources or processes), (4) HOW is SSI be done, meaning methods and tools are needed. The questions are oriented on the 8W model [42]. The 8W model is used to understand and describe processes in companies. Since the model is very comprehensible and straight forward, it serves as a good starting point to structure content of unknown and not yet completely defined processes—which is the case for the process of SSIM. A paper has to provide answers to at least one of those questions. In the first iteration, the titles and abstracts of the papers are screened. In the second iteration, the papers that are selected in the first iteration are completely scanned to decide if a paper is relevant. Next, backward search (BS) and forward search (FS) are conducted to complete final papers selection.

This literature basis is analyzed regarding approaches and tendencies in the field of SSIM. The findings are clustered in defined categories for each paper and are connected in concept clusters. By mapping papers and relevant findings, a comprehensive overview of current research tendencies is generated.

3   Findings

In total, 3.304 papers have been found with the data base search, including possible overlaps. In the first step the paper titles and abstracts were screened, which resulted in 30 potentially relevant papers. Those papers were screened considering the complete paper in a second iteration. Finally, 14 papers were identified as relevant after those two steps. In the BS, six more papers could be found. In the following FS, ten additional papers were selected.

Combining all literature research steps, in total 27 relevant papers are identified. The distribution of papers across databases and search process steps is presented in Fig. 1.

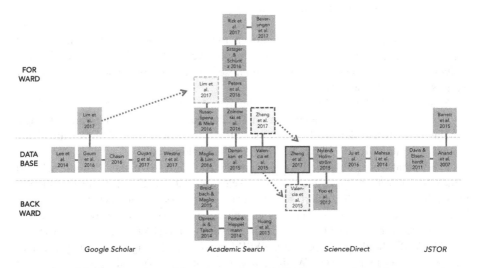

Fig. 1. Overview of the papers selected including research overlap

To structure the relevant papers and to give an overview over the content of the papers, a combined top down – bottom up approach is used. In the first step the relevant papers were structured bottom down in the four clusters defined by the four guiding questions explained in section two. A paper that provides answers to more than one of the four questions is assigned to all relevant clusters. In the second step, clusters are divided bottom up to sharpen the categories. Ultimately, the combination of (1) the questions from the 8 W model used in the top down analysis and (2) the further findings based upon the bottom up subdivision of the clusters, results in the following six categories: *Topics* from the question 'WHAT', *Knowledge & Information* from the question 'WHO', *Resources and Processes*, both evolving from 'WHICH' as well as *Principles* and *Methods & Tools* as categories from the 'HOW' cluster. A description and differentiation of the categories, and their connection to the four guiding questions is shown in Fig. 2.

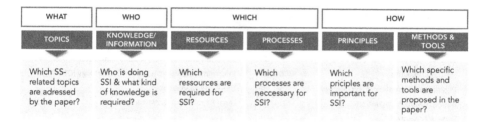

Fig. 2. Paper content categories.

The main findings within each category are described in the following.

Topics. A huge variety of topics are discussed and covered in the papers. This also represents the complexity of SSIM. Topics dealing with general *design principles and approaches* (e.g. Valencia et al. [12]) including data driven innovation processes [10, 21], the importance of *big data analytics* (e.g. Ju et al. [22]) and new ways of *stakeholder interaction & communication* (e.g. Mehrsai et al. [23]) appear most frequently. Moreover, *information & communication technology (ICT)* [24–26], *digital transformation & technology* [8, 17, 27], and changing *service ecosystems* [22, 28, 29] are discussed. Some authors also focus on *business model innovation* [28, 30] complete *life-cycle management* [8], *knowledge integration & structure* [31, 32], new approaches for *idea generation* [10, 33], as well as the need for *legal regulations & warranty issues* [9, 15].

Resources. SSIM requires an increasing set of resources. First of all, the complexity of smart services require *strong R&D capabilities* [34] including researchers from diversified disciplines [25]. To handle the huge amount of knowledge from different industries and areas, a *knowledge based structure* of the resources is necessary [32]. Moreover, *systems engineers* are required to cope with the increasing complexity of the systems [9]. Additionally, several authors raise demand for an improved *collaboration and ICT infrastructure*, to ensure seamless interaction between the stakeholders [31]. With reference to this, both, an excellent ICT infrastructure ensuring efficient communication [23, 28, 35], and approaches such as an *open ecosystem* [22] and digital platforms (e.g. for co-creation) are discussed in the papers [36]. Besides, smart services require a new *technology infrastructure* [9], for example, as already mentioned, a state of the art ICT infrastructure including all time connectivity, this time to ensure a flawless use of smart services for users. Beverungen et al. [29] also mention smart products as resources to develop smart services. Ultimately, data related resources, such as a *quality scale for smart service systems* [14] to guarantee basic quality standards for *data analytics,* are seen as relevant [22] (Fig. 3).

TOPICS

Design Approaches & Principles	[10][12][21][33][31][37][38]	Knowledge Integration & Structure	[31][32]
Idea Generation	[10][33]	Stakeholder Interaction & Communication	[23][28][34][35]
Big Data Analytics	[14][15][22][39][41]	Information & Communication Technology ICT	[24][25][36]
Digital Transformation & Technology	[8][17][27]	Life-Cycle Management	[8]
Business Model Innovation	[28][30]	Legal Regulations & Warranty Issues	[9][15]
Service Ecosystem	[22][28][29]		

RESOURCES

Big Data (Analytics)	[22][37][30]	Strong R&D Capabilities	[25][34]
Systems Engineers	[9]	Collaboration & ICT Infrastructure	[23][28][31][35]
Technology Infrastructure / Platforms	[9][13][17][29][36]	Knowledge Based Structure	[32]
Open Ecosystem	[22]	Quality Scale SSS	[14]

KNOWLEDGE / INFORMATION

Big Data Analytics	[9][25][37]	Continuous Learning	[27]
New Technologies	[8]	Heterogeneous Knowledge	[17]
Data / Information Exchange btw. Stakeholders	[28]	Data Privacy & Security / Legal Policies	[8][9][15]

PROCESSES

Service Development & Improvement	[23][31][35][37]	Business Model Generation	[17][30]
Idea Generation	[10][33][38]	Alignment of IT Architecture & Infrastructure	[39]
Technology Choice	[9][30]	Value Co-creation	[15][31][34]
Big Data Scanning & Analytics	[25][30]		

PRINCIPLES

Human Centered	[10][12][15][21][25][38][39][40]	Value Co-creation	[31][36]
Dynamic Innovation	[17][27]	Intra-Organizational Collaboration	[9][13][23]
Data as a Service	[30]	Total Life-Cycle Management	[8]
Information Layer	[41]	Smart Products as a Foundation	[29]

METHODS / TOOLS

Innovation Approaches	[10][27][29][38][39][40]	Collaboration & Communication	[23][25][31][35]
Business Model Innovation	[21][22][28][30][37]	Need Identification	[33]
Big Data Integration	[14][15][37][41]	Resources & Infrastructure	[9][24]

Fig. 3. Concept clusters of current literature related to SSIM.

Knowledge & Information. Papers dealing with topics related to knowledge or information, put a clear focus on knowledge around data. For instance, Demirkan et al. [15] raise the need for knowledge in the area of *data privacy and data security*, incl. existing regulations and standards and Thomas [8] states that new *legal policies* are required. Furthermore, several authors highlight the need of expert knowledge in *data analytics* [9, 25, 37]. Stakeholders involved in the service creation are in need of an easy and safe way to *exchange data and information* as well [28]. From a more general perspective, *heterogeneous knowledge* [17] e.g. on *new technologies* [8] is required. Finally, *continuous learning* mechanisms are needed to keep up with a fast-changing environment [27].

Processes. New processes have to be developed in the areas of *idea generation* (e.g. Lee et al. [38]), *service development and continuous improvement* (e.g. Mehrsai et al. [23]). In this context, Lim et al. [37] propose nine concrete process steps "From Data to value". Moreover, it should be specifically concentrated on the process to define the functions needed for smart services [9] and to *choose* the right *technology* by using *data analytics* [30]. Schuritz et al. [30] also state, data analytics should be integrated as an additional specific part in the *business model generation*. Yoo et al. [17] focus on creating business models around platforms, including license agreements for technology use. Furthermore, *value co-creation* processes [31, 34]—also across industry boarders [15]—have to be defined. Since service innovation is more and more distributed [17], and many players are involved, *IT architecture and the service infrastructure have to be aligned* [39].

Principles. Several underlying principles can be identified in the papers. Many of them deal with demonstrating needs for new services from a *human centered* perspective [40]. Some of them combine the user perspective with market gaps [38] and new technologies as enablers [10]. Beverungen et al. [29] name *smart products as a foundation* for developing smart services, since they serve as boundary objects between the physical world and the service layer. Moreover, *dynamic innovation* is necessary. Specifically, innovation needs to be continuous, relentless and fast [17] and balanced between structure and flexibility [27]. Schuritz et al. [30] name *data as a service* as new business principle, while Opresnik et al. [41] introduce the *information layer* as a new principle. Additionally, *value co-creation* and *intra-organizational collaboration* are important SSIM principles [13, 26]. Ultimately, it is important to consider the *complete lifecycle* for SSIM [8].

Methods & Tools. More than two thirds of the papers propose methods and tools for SSIM. However, many of them are rather concepts, ideas and needs for future methods and are not yet tested or verified. Many authors propose methods that serve as *innovation approaches* such as the "integrated market pull technology push morphology" by Geum et al. [10]. Methods and tools in the area of *business model innovation* are proposed by Schuritz et al. [30] ("5 patterns of data-infusion") and by Ju et al. [22] ("IoT Business Model Canvas"), amongst others. Chasin's "Social Representations Model" [33] describes an approach to *identify future needs*, by analyzing social media. Several authors propose the *integration of big data*, e.g. Opresnik et al. [41] in form of a big data strategy. Furthermore, several *collaboration and communication* tools are explained—such as a VR toolset [35] and Collaborative Innovation Centers [31]—to enable seamless interaction between stakeholders. Finally, it is dealt with the development of *resources and infrastructure*. While Breidbach et al. [24] pledge to reallocate the existing resources to develop smart services, Porter et al. [9] propose to establish a completely new technology stack.

Reflecting the results for each category by looking at the leading questions 'WHAT', 'WHO', 'WHICH' and 'HOW' makes different levels of detail visible and demonstrates the current status in research on SSIM. Most findings are referring to the questions on 'WHAT' *Topics* and 'HOW' in relation to *Methods & Tools*. But, whereas the described *Topics* are already representing a good overview on the most relevant components that should be included in SSIM, the granularity of the identified *Methods & Tools* only give a broad answer to how SSIM can be realized and mainly refer to already existing approaches from (non-smart) service innovation processes. In comparison to this, less results are found for the questions on 'HOW' in relation to *Principles,* 'WHO' in relation to *Knowledge & Information* and 'WHICH' in relation to *Resources & Processes*. Apparently, even the relevant topics seem to be obvious, the translation on what is specifically needed is not clearly exposed so far. In addition, for each category focal points can be defined by the number of sources found: Regarding to the category *Topics*, Design Principles and Approaches as well as Big Data Analytics are connected with SSIM most frequently. In terms of *Resources* needed in the SSIM process, Technology as well as ICT- and Collaborative Infrastructures are the focused. The most relevant *Information & Knowledge* sources seem to be Big Data Analytic skills and experiences combined with Data Security and Legal Policies. Besides, Service Design and Improvement is the main component when looking on the category of *Processes*. In relation to the *Principles,* this should be especially combined with a Human Centered approach. Finally, the *Methods & Tools* focus on innovation approaches first, followed by Business Model innovation and Big Data Integration.

4   Discussion

Our literature review revealed that only very few papers which are directly related to "smart service" or "smart service innovation" still exist. First relevant publications are from 2012, whereas nearly all relevant papers were published within the last five years. The number of published papers relating to SSIM are slightly increasing in the last years, but nevertheless it still does not seem to be a focus topic. Even though the need and potential for smart services is mentioned, a clear research trend for new approaches to tackle the challenges in SSIM cannot be observed, so far.

Considering key topics identified through our research approach, the central topics mentioned in context of SSI are Big Data Analytics, ICT infrastructure and Service Design. Overarching frameworks for re-designing innovation processes and activities to account for the new nature of smart services, are still not existing. Current concepts indicate that scholars are rather concentrating on showing ways of how smart services can be understood. They consequently remain on a general level. This can also be seen for methods and tools mentioned in context of SSIM. Although, methods and tools to support the smart service innovation process are discussed, these tools rather stem from to service development in general rather than introducing new dedicated solutions for SSIM. In general, the review of literature reveals that previous studies relating to SSIM mainly focus on single aspects within the smart service innovation process, but no comprehensive approach or framework is found. Accordingly, research needs to account for an overall framework, procedure or tools for the management of smart service innovations.

5 Conclusion and Future Research

Smart services constitute a particular new type of service combining physical, smart and connectivity components. For corporate innovation management it is necessary to account for the specifics of smart services. The goal of this study, therefore, was to generate an understanding of current research on smart service innovation management. Based on a literature review, we identified 27 papers with relevant contribution to the research area. These papers' content was structured along six major categories (Topics, Resources, Knowledge & Information, Processes, Principles and Methods & Tools) in order to summarize pre-existing knowledge on smart service innovation management Our analysis reveals, a lack of practical knowledge on the transformation of innovation activities and processes in organizations necessary to account for the new smart and connective nature of smart services. The relevant challenge will be to define how the three different levels of smart services – physical, smart and connectivity - can be aligned and managed within the smart service innovation process.

Finally, this study not only contributes by summarizing the status quo perspective on innovation management for smart service, but also by identifying starting points for future research in the field of SSIM. However, the literature review is only a first step towards holistically understanding the impact, smart services will have on innovation management. On basis of the overview given within this paper, future research will need to create a more comprehensive understanding of the innovation management of smart services. In particular, it is suggested to conduct several case studies to fully understand the challenges and changes, of the introduction of smart services into a company's solution portfolio.

References

1. Oliva, R., Kallenberg, R.: Managing the transition from products to services. Int. J. Serv. Ind. Manag. **14**, 160–172 (2003)
2. Nudurupati, S.S., Lascelles, D., Wright, G., Yip, N.: Eight challenges of servitisation for the configuration, measurement and management of organisations. J. Serv. Theory Pract. **26**, 745–763 (2016)
3. Bilgeri, D., Wortmann, F., Fleisch, E.: How digital transformation affects large manufacturing companies' organization. In: Thirty Eighth International Conference on Information Systems, South Korea (2017)
4. Gimpel, H., Röglinger, M.: Digital Transformation: Changes and Chances—Insights Based on an Empirical Study. Project Group Business and Information Systems Engineering (BISE), Fraunhofer Institute for Applied Information (FIT), Augsburg/Bayreuth (2015)

5. Vargo, S.L., Lusch, R.F.: Evolving to a new dominant logic for marketing. J. Mark. **68**, 1–17 (2004)
6. Kagermann, H.: Change through digitization—value creation in the age of industry 4.0. In: Albach, H., Meffert, H., Pinkwart, A., Reichwald, R. (eds.) Management of Permanent Change, pp. 23–45. Springer, Wiesbaden (2015). https://doi.org/10.1007/978-3-658-05014-6_2
7. Allmendinger, G., Lombreglia, R.: Four strategies for the age of smart services. Harv. Bus. Rev. **83**, 131–145 (2005)
8. Thomas, O., Hrsg, M.F.: Smart Service Engineering (2017)
9. Porter, M.E., Heppelmann, J.E.: How smart, connected products are transforming competition. Harv. Bus. Rev. **92**, 64–88 (2014)
10. Geum, Y., Jeon, H., Lee, H.: Developing new smart services using integrated morphological analysis: integration of the market-pull and technology-push approach. Serv. Bus. **10**, 531–555 (2016)
11. Sawatani, Y., Spohrer, J., Kwan, S., Takenaka, T.: Serviceology for Smart Service System. Springer, Tokyo (2017). https://doi.org/10.1007/978-4-431-56074-6
12. Valencia, A., Mugge, R., Schoormans, J.P.L., Schifferstein, H.N.J.: The design of smart product-service systems (PSSs): an exploration of design characteristics. Int. J. Des. **9**, 13–28 (2015)
13. Barrett, M., Davidson, E., Vargo, S.L.: Service innovation in the digital age: key contributions and future directions. MIS Q. **39**, 135–154 (2015)
14. Maglio, P.P., Lim, C.: Innovation and big data in smart service systems. J. Innov. Manag. **1**, 1–11 (2016)
15. Demirkan, H., Bess, C., Spohrer, J., Rayes, A., Don, C., Cisco, A.: Communications of the association for information systems innovations with smart service systems: analytics, big data, cognitive assistance, and the internet of everything. Commun. Assoc. Inf. Syst. **37**, 35 (2015)
16. Porter, M.E., Heppelmann, J.E.: How smart, connected products are transforming IT. Harv. Bus. Rev. **92**(11), 64–88 (2014)
17. Yoo, Y., Boland, R.J., Lyytinen, K., Majchrzak, A.: Organizing for innovation in the digitized world. Org. Sci. **23**, 1398–1408 (2012)
18. Hengsberger, A.: Was Ist Innovationsmanagement? (2016). www.lead-innovation.com, http://www.lead-innovation.com/blog/was-ist-innovationsmanagement
19. Businessdictionary.com: Product innovation. http://www.businessdictionary.com/definition/product-innovation.html
20. Webster, J., Watson, R.T.: Analyzing the past to prepare for the future: writing a literature review reproduced with permission of the copyright owner. Further reproduction prohibited without permission. MIS Q. **26**, xiii–xxiii (2002)
21. Zolnowski, A., Christiansen, T., Gudat, J.: Business model transformation patterns of data-driven innovations. In: 24th European Conference Information Systems ECIS 2016, pp. 0–16 (2016)
22. Ju, J., Kim, M.S., Ahn, J.H.: Prototyping business models for IoT service. Procedia Comput. Sci. **91**, 882–890 (2016)
23. Mehrsai, A., Henriksen, B., Røstad, C.C., Hribernik, K.A., Thoben, K.D.: Make-to-XGrade for the design and manufacturing of flexible, adaptive, and reactive products. Procedia CIRP. **21**, 199–205 (2014)
24. Breidbach, C., Roelens, B., Lemey, E., Poels, G.: A service science perspective on the role of ICT in service innovation, pp. 1–9 (2011)
25. Huang, M.H., Rust, R.T.: IT-related service: a multidisciplinary perspective. J. Serv. Res. **16**, 251–258 (2013)

26. Weiß, P., Kölmel, B., Bulander, R.: Digital service innovation and smart technologies: developing digital strategies based on industry 4.0 and product service systems for the renewal energy sector. In: Proceedings of the 26th Annual RESER Conference, Naples, Italy, pp. 274–291 (2016)
27. Nylén, D., Holmström, J.: Digital innovation strategy: a framework for diagnosing and improving digital product and service innovation. Bus. Horiz. **58**, 57–67 (2015)
28. Zheng, M., Ming, X., Wang, L., Yin, D., Zhang, X.: Status review and future perspectives on the framework of smart product service ecosystem. Procedia CIRP **64**, 181–186 (2017)
29. Beverungen, D., Müller, O., Matzner, M., Mendling, J., Vom Brocke, J.: Conceptualizing smart service systems. Electron. Mark. 1–12 (2017)
30. Schuritz, R., Satzger, G.: Patterns of data-infused business model innovation. In: Proceedings of CBI, 18th IEEE Conference Business Informatics, vol. 1, pp. 133–142 (2016)
31. Ouyang, Q., et al.: Collaborative Innovation Centers (CICs): toward smart service system design. In: Sawatani, Y., Spohrer, J., Kwan, S., Takenaka, T. (eds.) Serviceology for Smart Service System, pp. 385–391. Springer, Tokyo (2017). https://doi.org/10.1007/978-4-431-56074-6_42
32. Anand, A.N., Gardner, H.K., Morris, T., Anand, N., Gardner, H.K.: Knowledge-based innovation: emergenece and embedding of new practice areas in management consulting firms. Acad. Manag. J. **50**, 406–428 (2014)
33. Chasin, F.: Business analysis of digital discourse for new service development: a theoretical perspective and a method for uncovering the structure of social representations for improved service development. In: Proceedings of 49th Annual Hawaii International Conference System Science (HICCS 2016), pp. 1567–1576 (2016)
34. Davis, J.P., Eisenhardt, K.M.: Rotating leadership and collaborative innovation: recombination processes in symbiotic relationships. Adm. Sci. Q. **56**, 159–201 (2011)
35. Westner, P., Hermann, S.: VR|ServE: a software toolset for service engineering using virtual reality. In: Sawatani, Y., Spohrer, J., Kwan, S., Takenaka, T. (eds.) Serviceology for Smart Service System, pp. 237–244. Springer, Tokyo (2017). https://doi.org/10.1007/978-4-431-56074-6_26
36. Russo-Spena, T., Mele, C.: What's Ahead in Service Research (2016)
37. Lim, C., Kim, K.H., Kim, M.J., Heo, J.Y., Kim, K.J., Maglio, P.P.: From data to value: a nine-factor framework for data-based value creation in information-intensive services. Int. J. Inf. Manag. **39**, 121–135 (2018)
38. Lee, J., Kao, H.A.: Dominant innovation design for smart products-service systems (PSS): strategies and case studies. In: Annual SRII Global Conference SRII, pp. 305–310 (2014)
39. Rizk, A., Bergvall-Kåreborn, B., Elragal, A.: Digital Service Innovation Enabled by Big Data Analytics—A Review and the Way Forward, pp. 1247–1256 (2017)
40. Peters, C., Maglio, P., Badinelli, R., Harmon, R.R., Maull, R.: Emerging digital frontiers for service innovation. Commun. Assoc. Inf. Syst. **39**, 8 (2016)
41. Opresnik, D., Taisch, M.: The value of big data in servitization. Int. J. Prod. Econ. **165**, 174–184 (2015)
42. Weiß, P., et al.: Methoden und Instrumente zur Messung und Verbesserung der Produktivität industrieller Dienstleistungen in KMU, pp. 1–51 (2014)

Open Innovation in Ecosystems – A Service Science Perspective on Open Innovation

Carina Benz[(✉)] and Stefan Seebacher

Karlsruhe Institute of Technology, Karlsruhe, Germany
{carina.benz,stefan.seebacher}@kit.edu

Abstract. Fostered by technological developments, a growing tendency towards interconnectedness of people, solutions, and organizations can be observed. In close alignment to this trend, the notion of 'ecosystems' is becoming popular to describe a system of complex relationships between diverse actors in analogy to natural ecosystems. With growing prominence of ecosystems, a stronger need for collaboration and co-creation beyond traditional supply chain networks is arising—including co-innovation amongst a diverse set of loosely coupled partners. Research in service science and on open innovation capture the topics of collaboration in systems and joint innovation, but still, open innovation research uses to focus on mainly firm-centric aspects of distributed innovation. The service ecosystems view, adopted by service science research, provides a means for studying how the co-creation of value is performed in systems of weakly tied actors.

The purpose of this paper is to broaden the perspective of research in open innovation and to advance the understanding of open innovation in ecosystems by combining insights from service science research with the concept of open innovation. Consequently, this paper stems from four propositions that address gaps in knowledge related to the understanding of co-innovation in dynamic ecosystems of multiple actors and, thus, proposes directions for future research.

Keywords: Open innovation · Innovation ecosystems · Service ecosystems

1 Introduction

Today's world is more connected than ever: social networks connect people all over the globe [1], companies connect with their customers in co-creation initiatives [2], and the paradigm of Internet of Things (IoT) connects devices [3]. The notion of people, organizations or devices being part of a network of interdependent relationships is described as interconnectedness [4]. Not least because of currently emerging technologies, a growing tendency towards interconnectedness can be observed: Blockchain technology—from a technical point a "distributed database, which is shared among and agreed upon a peer-to-peer network" [5]—is suggesting a novel way of providing trust amongst diverse parties in (business) networks. As a technical infrastructure, blockchain technology enables the formation of service systems, allowing for trusted collaboration, and resource integration in a decentralized setting, without the need for an intermediary. Since all participants of the network are thoroughly connected, sharing a

© Springer Nature Switzerland AG 2018
G. Satzger et al. (Eds.): IESS 2018, LNBIP 331, pp. 112–124, 2018.
https://doi.org/10.1007/978-3-030-00713-3_9

common platform, language and understanding, the exchange of information between the parties is facilitated.

Similar trends can be observed for the current rise of the Internet of Things. As products and services become smarter and connected, they require a rethinking of traditional assumptions on business, processes and structures [6]. Products become components of broader systems. Take popular smart home application systems: Using geo-fencing technology on the user's smartphone, the smart home system is informed that he is leaving his workplace heading home and smart thermostats are beginning to heat up the temperature. When the smartphone logs into his apartment's WLAN, the smart lighting system automatically turns on the light and a verbal command send to the home assistant ensures that a smart TV is turned on.

These above-mentioned technologies not only connect people, organizations and devices, they have the potential to radically change the organization structure of companies and industries. At the heart of this development is the idea of ecosystems —"complex relations built into the web of dependencies among [...] different 'species'" [7]—and one finds ecosystems appearing in discussion and commentary on an enormous range of topics. With the emergence of ecosystems, a stronger need for collaboration and co-creation beyond traditional supply chain networks will become obvious [6]. As Moore describes it: "In a business ecosystem, companies co-evolve capabilities around a new innovation: they work cooperatively and competitively to support new products, satisfy customer needs, and eventually incorporate the next round of innovations" [8]. Research-sides, collaboration in systems and joint innovation are topics captured by service science [9], as well as research on open innovation (OI) [10].

Service science as the study of service systems in which value is co-created within complex constellations of integrated resources [9] has increasingly embraced a systemic perspective within the last years with a growing focus on value co-creation in ecosystems [11, 12]. This service ecosystems view constitutes a shift from a focus on dyadic exchange to various forms of interaction and resource integration including the highlighting of largely loosely coupled relationships [11]. Distributed innovation processes with purposively managed knowledge flows across organizational boundary [13], known as open innovation, are gaining rising popularity among scholars and practitioners. As the original concept of OI is firm-centric [14], research in this field has mainly been focusing on how a single organization can best leverage internal and external knowledge flows to accelerate its innovation performance. With the trend towards ecosystems, which is particularly fueled through technological advances, conventional conceptualizations of dyadic and unidirectional innovation collaborations no longer adequately describe the dynamic multidirectional processes needed to innovate within ecosystems.

This paper shall, therefore, provide an understanding of the current research in open innovation with a focus on (eco)systemically oriented concepts. On this basis, we propose a research agenda that spans across levels of analysis for open innovation in ecosystems. To do this, we specifically draw on knowledge on service ecosystems, to develop a deeper understanding of how service ecosystem actors collaboratively co-create value in dynamic and loosely coupled relationships.

The remainder of this paper proceeds as follows: Sect. 2 provides the theoretical foundations on service (eco-)systems and innovation in such settings. Sect. 3 gives an overview of research in the field of open innovation with a focus on current studies stressing an ecosystem-view on open innovation. In Sect. 3, it is discussed, how the two research streams can cross-fertilize each other and research propositions for the study of open innovation in ecosystems are discussed. We conclude our research by summing up our contribution, critically examining limitations and giving an outlook on future work.

2 Service (Eco-)Systems: A Concept from Service Science

Service science is the study of service systems including the co-creation of value within complex constellations of integrated resources, and therefore inherently embraces system thinking. In this context, service systems are defined as value co-creation configurations within constellations of integrated resources [9]. They are composed of organizations, people, and technology, and shared information that interact to create value [15]. Service systems scale up from systems composed of multiple types of knowledge workers, to enterprise-, industrial-, national- and global service systems [16].

The notion of the service system is in line with the service-dominant logic (SD-logic) [17, 18], which provides a theoretical framework for research in service science. SD-logic postulates that value is always co-created, meaning that value is created in a mutually beneficial way through resource integration and relationships within and among service systems. Special focus is placed on resource integration as it is a central element to value co-creation [11]: Organizations (firms, customers, etc.) exist to integrate and transform e.g. specialized competences into complex services [18].

Starting with a rather narrow view on value co-creation between firms and customers, there has been a pivotal shift in the last years, which is triggered by the rising scientific importance of the SD-logic, to take a broader perspective and to allow for a more holistic and dynamic approach on value creation, e.g. regarding a more inclusive configuration of actors [19]. From the former perception of dyadic, bi-directional exchange between firms and customers, SD-logic has evolved to consider interactive processes among various entities, including different types of stakeholders (e.g. governmental actors). In this context, Vargo emphasizes actor-to-actor approaches regarding value co-creation relationships, instead of distinguishing between B2B or B2C relationships [20]. This zooming out has also resulted in a major turn towards an (eco-)system orientation. Service ecosystems, can be defined as "relatively self-contained, self-adjusting system[s] of resource-integrating actors [which] are connected by shared institutional arrangements and mutual value creation through service exchange" [19]. A service ecosystem, thus, consists of mainly loosely coupled, value-proposing social and economic actors, that co-produce service offerings, engaging in mutual service provisioning and co-creation of value [21].

This new view on services accounts for the complex and dynamic nature of social and economic systems consisting of a diverse set of stakeholder (e.g. firms, customers, government entities), in which interaction and resource integration occurs [11, 21].

With the concept of institutions, SD-logic also emphasizes, how social contexts influence and are influenced by service ecosystem actors in the value co-creation process [11]. Institutions can, therefore, be interpreted as social norms and common rule systems, serving as a means for coordinating and governing interactions [11, 19]. As coordination mechanisms, they facilitate some forms of co-creation, e.g. tasks and relationships which are in need of a formalized setting, while potentially constraining other co-creation activities [22].

Traditionally, service research can be divided into streams of different view points on service innovation [23]. While research following the assimilation perspective believes that service and goods innovation follow the same rules, researchers taking the demarcation perspective assume substantial differences between service and goods innovation, therefore calling for new models and theories for service innovation [23]. The service ecosystems perspective on innovation is in line with the third perspective, the synthesis view [22]. With its systemic perspective on value creation between actors, the idea of service ecosystems zooms out from a notion of 'products vs. services' and 'producers vs. consumers' to value being co-created by actors [22]. Innovation in service ecosystems, consequently, is not considered as the introduction of new products or services, but rather addresses transformations in the value co-creation process [24]. Institutions are the core object of change for innovation form a service ecosystem perspective [12]. In this sense, innovation in service ecosystems is characterized as "innovation as the institutionalized changes in service ecosystem structures that stem from either a new configuration of resources or a new set of schemas (social norms and rules) and result in new practices that are valuable to the actors in a specific context" [25]. Koskela-Huotari et al. [22] emphasized that this process of re-configuration depends on successfully balancing the breaking of old and making of new rules of resource integration, while maintaining some of the prevailing institutions and institutional structures – a concept referred to as "institutionalization" [26].

3   Open Innovation: Managing Purposive Knowledge In- and Outflows

3.1   The Concept of Open Innovation

Open innovation is a phenomenon gaining increasing importance in research as well as practice since the publication of Chesbrough's seminal book in 2003 [10]. Open innovation refers to a "distributed innovation process that relies on purposively managed knowledge flows across organizational boundaries" [27]. In accordance to the open innovation paradigm, firms' organizational boundaries are permeable allowing for knowledge in- and outflows. This is often illustrated with the picture of a porous innovation funnel [28]. These knowledge flows can be classified in three core process archetypes: the outside-in process, the inside-out process, and the coupled process [29]. The outside-in process describes the enrichment of a company's knowledge base through the integration of knowledge from customers, suppliers, and other external knowledge sources to increase the company's innovativeness. Innovation contest, crowdsourcing, or innovation communities are common methods firms use to acquire and integrate knowledge from external actors [30, 31]. The inside-out process refers to the idea of bringing un- or under-utilized internal ideas to the market by selling IP rights, licensing, or through spin-offs. When inside-out and outside-in are combined, one is talking about the coupled process which occurs, when complementary partners are collaborating.

Open innovation has become a paradigm for managing innovation. It assumes that in order to preserve innovativeness, organizations can and should use internal as well as external ideas, and intern as well as external paths to bring ideas to the market [32]. The reasons, firms engage in OI practices are manifold. Chesbrough [10] argues that due to the increasing mobility of knowledge workers and rising potential to manufacture high technology products, also in emerging countries, product lifecycles get shorter, development costs become higher and companies can no longer rely solely on internal knowledge for innovation. The potential success of open innovation practices, in turn, have been demonstrated in various qualitative studies on individual cases, but also in quantitative studies [33–36].

Considering research on open innovation, Randhawa et al. [37] identify three focus areas in open innovation research: (1) firm-centric aspects, (2) management of open innovation networks, and (3) the role of users and communities. Research on the firm-centric aspects of OI is centering around topics such as open innovation strategies [38–40], challenges and organizational competencies for OI [41, 42], or open innovation business models [43]. From a theoretical point, absorptive capacity [44], and dynamic capabilities [45] are building the basis for firm-centric studies. Within the second focus area, regarding the management of open innovation networks, studies broach the issue of inter-organizational innovation [46], living labs as innovation networks [47], or the general role of intermediaries in OI networks [48, 49]. Tools for user-integration [31], user-types (e.g. lead user method) [50] and communities [51] are topics that fall within the third focus area.

While most attention has been paid to firm-centric aspects [7], the role of users and communities, as well as the management of networks, has received relatively little attention from OI researchers. Furthermore, the fast majority of the studies on open innovation focus on inside-out, rather than outside-in or coupled processes [32, 52].

Consequently, the following chapter will specifically focus on open innovation in ecosystems.

3.2   Open Innovation in Ecosystems

Innovation ecosystems can exist on different levels regarding their geographical boundaries or facilitating infrastructure [53]. Corporate innovation ecosystems typically consist of suppliers, users, and partners. District- or city-based, regional, and national innovation ecosystems (e.g. hubs or clusters) are distinguished based on their geographical boundaries. Digital ecosystems, which are often facilitates by online platforms, allow actors to build synergistic relationships–often centered around a core offering (e.g. Google's or Apple's developer ecosystems) [53]. While prior research on innovation systems [54], innovation networks [55], or R&D alliances [56] exists, these concepts hardly embrace the dynamic, and loosely coupled nature of (service) ecosystems.

The creation open innovation ecosystems is subject to different case studies that investigate how firms manage to establish such an ecosystem [57–59]. Traitler et al. [60] identify 10 major topics from leadership and internal experts to passion, that determine the success of OI ecosystems provide specific recommendations for the reinvention of R&D in open innovation ecosystems. Furthermore, research on innovation ecosystems concentrates on strategies to align internal innovation strategies with the innovation ecosystem [61–63]. Gawer et al. [64] demonstrate the impact of industry platforms and ecosystems on product innovation. The concept is addressed by Riedl et al. [65], as well, as they highlight the role of central platforms bringing together multiple actors that follow the OI paradigm within an ecosystem. They furthermore propose a framework that shows individual actors' capabilities with regard to services innovation, and strategies for ecosystems to exploit them in order to advance service innovation.

A promising current approach to emphasize the importance of broadening the perspective of open innovation from a firm to an ecosystem view is the idea of Open Innovation 2.0 (OI2.0) [66, 67]. While in his seminar book, Chesbrough [10] presented case studies on the collaboration between two organizations, he now acknowledges that the concept of open innovation is being used to "orchestrate a large number of players across multiple phases of the innovation process" [13]. Curley et al. [68] define Open Innovation 2.0 as new paradigm based on principles of integrated cross-organizational innovation collaboration based on co-created shared value and mission, cultivated in ecosystems of diverse actors (quadruple helix of government/public, academia, industry, citizens). Settings where innovation in the notion of Open Innovation 2.0 is performed, are starting to be observable in real-world: Avalog, a banking software provider, allows partners, including Fintechs to integrate their banking software solutions into the Avalog platform which is offered to financial institutions within the Avalog community. Philips opened up their research lab in Eindhoven and generated a research campus where they co-innovate with researchers from more than 140 other firms [69]. With a move from bilateral cooperation to innovation ecosystems, new approaches to understand, design and manage dynamic innovation in ecosystems, become necessary [13].

4 Open Innovation 2.0: Rethinking Innovation from a Service (Eco)System Perspective

Open innovation and service science are first brought together in Chesbrough's [70] work on 'Open Services Innovation' that demonstrates how companies can shift their perspective from product- to service-centricity while relying on the principles of open innovation. With its latest conversion towards an ecosystem view, service science has gone through a transition, which the concept of open innovation is most likely to equally go through within the next years. This becomes particularly obvious, when considering similarities between service ecosystems and the vision of open innovation 2.0: The idea OI2.0 aligns closely with SD-logic's system view, as both center on dynamic interactions among multiple stakeholders. SD-logic suggests that the reason why different actors interact, is their goal to create value for themselves and for others in the service provision process. As compared to OI2.0, service science applies a broader view on reasons for firms, customers, and other stakeholders to collaborate than just for inno-vation activities. Nevertheless, both research streams agree, that all activities performed, are accomplished in a mutually beneficial manner—or as Curley [69] states in a "Win more—Win more" situation. Drawing on the integration of social and economic actors, SD-logic suggests that customers, employees, but also other stakeholders (e.g. gov-ernments) are actors and therefore resource integrators [11], which complements the view of Open Innovation 2.0's quadruple helix system of innovation actors.

Contrary to the attractiveness open innovation has had on research and industry, conceptual frameworks and robust empirical investigations—especially in the field of open innovation in ecosystems—are still under represented [37]. Building on the perception, that service science and the ecosystem-perspective of open innovation have substantial potential for cross-fertilization, the proposition for future research on open innovation ecosystems (Table 1) is organized along the elements of the service sci-ence's value co-creation framework [71].

Table 1. Emerging topics and questions for research on open innovation ecosystems

Value co-creation framework	Topics addressed in service science	Overarching research propositions for open innovation in ecosystems
Actors	• Actors Engagement [72–74] • Inter-actor relationships [75, 76]	Investigate how actors with different roles (organizations, employees, public institutions etc.) can be engaged to participate in innovative activities within OI ecosystems
involved in Resource Integration and Exchange	• Actor interaction [77] • Co-creation practices [78]	Investigate how innovative co-creation practices proceed within OI ecosystems and identify ways to further facilitate co-creation in ecosystems
enabled and constrained by	• Governance and orchestration [79] • Value capturing and sharing (i.e. business models) [80–82]	Investigate how value in ecosystemic innovations can be captured and shared among different actors
Institutions & Institutional Arrangements	• Emergence and formation of service ecosystems [11, 83] • Evolution of service ecosystems [84, 85]	Investigate how and why OI ecosystems emerge and evolve and to what extent digital technologies influence the emergence of OI ecosystems
establishing nested and interlocking Service Ecosystems *of*		

According to this framework, service ecosystems consist of different actors that are involved in resource integration and service exchange which is enabled and constrained by institutions and institutional arrangements. Consequently, four foundational concepts—actors, resource integration and service exchange, institutions and institutional arrangements, and service ecosystems—can be distinguished, and build the foundation for further conceptual exploration. Topics of relevance and emerging themes are extracted from ongoing research endeavors in the field of service science with focus on service ecosystems. We therefore, screened the databases Web of Science (WoS) and Elsevier Scopus for articles with the keyword 'service ecosystem' in either topic (WoS) or title, abstract or keywords (Scopus). Papers were further selected based on whether they contribute by addressing general aspects of service ecosystems in a theoretical, conceptual or empirical manner rather than describing ecosystems in certain domains (e.g. health care) or regions. Subsequently, key topics were extracted and categorized according to the value co-creation framework (Table 1). Based on the four foundational concepts of the value co-creation framework, we identify overarching research propositions and theoretical perspectives that have the potential to further enhance the understanding of open innovation in ecosystems.

The first category refers to the concept of actors in services science, which posits that all social and economic actors are service-providing and value co-creating entities of varying sizes ranging from individuals to firms [21]. Service scholars strongly emphasize actor engagement as prerequisite for inter-actor value creation [72–74] and aim to analyze the nature of relationships between actors [75, 76]. While in open innovation research firm-centric aspects and co-innovation with customers have been in the focus of attention, the ecosystem-view of service science proposes to generalize from firms and firm-customer innovation as focal point to a wider circle of parties involved (e.g. governments, public organizations etc.). How these diverse actors can be engaged in innovative practices in open innovation ecosystems, needs to be investigated in future studies.

Service innovation performance is contingent on the way in which resources are integrated and service exchange is performed. Key topics identified within this category are interactions between actors [77] and co-creation practices [78]. Co-creation practices encompass "activities, where actors engage collaboratively in activities through interactions within a specific social context", [78]. In the case of service ecosystems, interactions in environments of dynamic nature are the reasons, co-creation practices are being re-examined through the ecosystem lens. This will also be the case for understanding interactions in open innovation ecosystems. Therefore, we suggest for future research to investigate, how innovative co-creation practices proceed within OI ecosystems. In a second step, identifying and testing how these co-creation activities can be further facilitated to enhance the ecosystem output (e.g. by shaping or altering environmental settings) can be addressed in future studies.

The third category covers the service science concept of institutions and institutional arrangements. Service scholars aim to understand governance and orchestration mechanisms [79] that allow service ecosystems to function, as well as the processes in which they are created and agreed upon. In addition, capturing and sharing jointly created value are topics addressed in service ecosystem research [80–82]. As shared value as well as a shared mission represent foundational elements of innovation in ecosystems, these concepts extend OI research beyond the topic of IP right management. Research is needed to thoroughly explore how value in ecosystemic innovations can be captured and shared among different actors.

Emergence and evolution of service ecosystems, are overarching topics in current service science research endeavors. Research on open innovation, until lately, has mainly investigated innovation settings, in which individuals (e.g. Wikipedia) collaborate or in which firms as initiators ask for participation of individuals (e.g. innovation contests). Future research needs to cast light on how and why OI ecosystems, consisting of different actors form and evolve. This includes the identification of initiators, potential roles of hub-actors that facilitate OI ecosystem emergence, and the motivation of actors to participate in ecosystem innovation. In addition, future research could investigate the impact of actors being tied together in digitally or technologically induced ecosystems (e.g. on industry platforms, or in IoT or blockchain networks) on innovation activities amongst them [64, 86].

5 Conclusion

The service ecosystems view suggests that service systems are made up of dynamic connections with permeable system boundaries. Because of this, exchange flows take place both within and among service systems, ultimately interweaving a multitude of service systems into a larger ecosystem. We argue, that the concept of service ecosystems, which evolved in service science research, is closely aligned with the ecosystem view on open innovation.

This paper, therefore, provides a general understanding of the concept of service ecosystems, and presents an overview of current research in open innovation with a special focus on (eco)systemically-oriented concepts. Furthermore, we identify research gaps for open innovation in ecosystems and formulate research propositions that span across the concept of value co-creation in service ecosystems including the focal points *actors*, *resource integration and exchange*, *institutions and institutional arrangements*, and *service ecosystems*. Our paper contributes to the current body of knowledge with its interdisciplinary approach on several levels of analysis. First, we provide a unifying lens on both concepts, SD-logic and open innovation arguing for further cross-fertilization between these areas of research. In particular, we argue that the service ecosystem view on value co-creation and the according process of zooming out from dyadic to systemic views on value-creation will be a necessary step for the field of open innovation. Consequently, research directions and emerging topics in the context of service ecosystems can guide research endeavors in understanding open innovation in ecosystems. Second, we systematically identify current research topics related to the concept of service ecosystems. We demonstrate that related concepts are relevant, yet not extensively adapted in the open innovation literature. Our third and final contribution stems from four propositions that address gaps in knowledge related to the understanding of how co-innovation can take place in dynamic ecosystems of multiple actors. These proposition, consequently, provide opportunities for future research. While the research propositions posed in this paper are by no means exhaustive, this paper not only aims to highlight the potential of reciprocal knowledge exchange and adaption between the two research streams but emphasizes the relevance of an ecosystem view on open innovation.

References

1. Hanna, R., Rohm, A., Crittenden, V.L.: We're all connected: the power of the social media ecosystem. Bus. Horiz. **54**, 265–273 (2011)
2. Wong, T.Y.T., Peko, G., Sundaram, D., Piramuthu, S.: Mobile environments and innovation co-creation processes and ecosystems. Inf. Manag. **53**, 336–344 (2016)
3. Lee, I., Lee, K.: The internet of things (IoT): applications, investments, and challenges for enterprises. Bus. Horiz. **58**, 431–440 (2015)
4. Ritter, T.: A framework for analyzing interconnectedness of relationships. Ind. Mark. Manag. **29**, 317–326 (2000)
5. Seebacher, Stefan, Schüritz, Ronny: Blockchain technology as an enabler of service systems: a structured literature review. In: Za, Stefano, Drăgoicea, Monica, Cavallari, Maurizio (eds.) IESS 2017. LNBIP, vol. 279, pp. 12–23. Springer, Cham (2017)
6. Porter, M.E., Heppelmann, J.E.: How smart, connected products are transforming companies. Harv. Bus. Rev. **93**, 96–115 (2015)
7. Remneland Wikhamn, B., Wikhamn, W.: Structuring of the open innovation field. J. Technol. Manag. Innov. **8**, 173–185 (2013)
8. Moore, J.F.: Predators and prey: a new ecology of competition harvard business review. Harv. Bus. Rev. **71**, 75–86 (1993)
9. Maglio, P.P., Spohrer, J.: Fundamentals of service science. J. Acad. Mark. Sci. **36**, 18–20 (2008)
10. Chesbrough, H.: Open Innovation: The New Imperative for Creating and Profiting from Technology. Harvard Business School Press, Boston (2003)
11. Vargo, S.L., Akaka, M.A.: Value cocreation and service systems (Re)formation: a service ecosystems view. Serv. Sci. **4**, 207–217 (2012)
12. Vargo, S.L., Wieland, H., Akaka, M.A.: Innovation through institutionalization: a service ecosystems perspective. Ind. Mark. Manag. **44**, 63–72 (2015)
13. Chesbrough, H.: The future of open innovation. Res. Manag. **60**, 29–35 (2017)
14. Bogers, M., et al.: The open innovation research landscape: established perspectives and emerging themes across different. Ind. Innov. 1–49 (2017, in press)
15. Vargo, S.L., Maglio, P.P., Akaka, M.A.: On value and value co-creation: a service systems and service logic perspective. Eur. Manag. J. **26**, 145–152 (2008)
16. Maglio, P.P., Srinivasan, S., Kreulen, J.T., Spohrer, J.: Service systems, service scientists, SSME, and innovation. Commun. ACM. **49**, 81–85 (2006)
17. Vargo, S.L., Lusch, R.F.: Service-dominant logic: continuing the evolution. J. Acad. Mark. Sci. **36**, 1–10 (2008)
18. Lusch, R.F., Vargo, S.L.: Service-dominant logic: reactions, reflections and refinements. Mark. Theory **6**, 281–288 (2006)
19. Vargo, S.L., Lusch, R.F.: Institutions and axioms: an extension and update of service-dominant logic. J. Acad. Mark. Sci. **44**, 5–23 (2016)
20. Vargo, S.L.: Toward a transcending conceptualization of relationship: a service-dominant logic perspective. J. Bus. Ind. Mark. **24**, 373–379 (2009)
21. Vargo, S.L., Lusch, R.F.: It's all B2B…and beyond: toward a systems perspective of the market. Ind. Mark. Manag. **40**, 181–187 (2010)

22. Koskela-Huotari, K., Edvardsson, B., Jonas, J.M., Sörhammar, D., Witell, L.: Innovation in service ecosystems: breaking, making, and maintaining institutionalized rules of resource integration. J. Bus. Res. **69**, 2964–2971 (2016)
23. Coombs, R., Miles, I.: Innovation, measurement and services: the new problematique. In: Metcalfe, J.S., Miles, I. (eds.) Innovation Systems in the Service Economy. Measurement and Case Study Analysis, pp. 85–103. Kluwer, Boston (2000)
24. Koskela-Huotari, K., Siltaloppi, J., Vargo, S.L.: Designing institutional complexity to enable innovation in service ecosystems. In: 49th Hawaii International Conference on System Sciences, pp. 1596–1605 (2016)
25. Aal, K., Di Pietro, L., Edvardsson, B., Francesca Renzi, M., Guglielmetti Mugion, R.: Innovation in service ecosystems: an empirical study of the integration of values, brands, service systems and experience rooms. J. Serv. Manag. **27**, 619–651 (2016)
26. Archpru, M., Vargo, S.L., Wieland, H.: Extending the context of innovation: the co-creation and institutionalization of technology and markets. In: Russo-Spena, T., Nuutinen, C., Nuutinen, M. (eds.) Innovating in Practice, pp. 43–57. Springer, Cham (2017)
27. West, J., Salter, A., Vanhaverbeke, W., Chesbrough, H.: Open innovation: the next decade. Res. Policy **43**, 805–811 (2014)
28. Cooper, R.G.: Stage-gate systems: a new tool for managing new products. Bus. Horiz. **33**, 44–54 (1990)
29. Gassmann, O., Enkel, E.: Towards a theory of open innovation: three core process archetypes. In: R&D Management Conference, Lisbon (2004)
30. Haller, J., Bullinger, A., Möslein, K.: Innovation contests: an IT-based tool for innovation management. Bus. Inf. Syst. Eng. **3**, 103–106 (2011)
31. Möslein, K., Neyer, A.-K.: Open Innovation: Grundlagen, Herausforderungen, Spannungsfelder. In: Zerfaß, A., Möslein, K. (eds.) Kommunikation als Erfolgsfaktor im Innovationsmanagement. Gabler, Wiesbaden (2009)
32. Chesbrough, H.: Open innovation: where we've been and where we're going. Res. Manag. **55**, 20–27 (2012)
33. Chesbrough, H.: The era of open innovation. MIT Sloan Manag. Rev. **44**, 35–42 (2003)
34. Chesbrough, H.: The governance and performance of Xerox's technology spin-off companies. Res. Policy **32**, 403–421 (2003)
35. Laursen, K., Salter, A.: Open for innovation: the role of openness in explaining innovation performance among U.K. manufacturing firms. Strateg. Manag. J. **27**, 131–150 (2006)
36. Chesbrough, H., Brunswicker, S.: A fad or a phenomenon?: The adoption of open innovation practices in large firms. Res. Manag. **57**, 16–25 (2015)
37. Randhawa, K., Wilden, R., Hohberger, J.: A bibliometric review of open innovation: setting a research Agenda. J. Prod. Innov. Manag. (2016)
38. Chesbrough, H., Appleyard, M.M.: Open innovation and strategy. Calif. Manag. Rev. **50**, 57–76 (2007)
39. Bogers, M., West, J.: Managing distributed innovation: strategic utilization of open and user innovation. Creat. Innov. Manag. **21**, 61–75 (2012)
40. Igartua, J.I., Garrigós, J.A., Hervas-Oliver, J.L.: How innovation management techniques support an open innovation strategy. Res. Technol. Manag. **53**, 41–52 (2010)
41. van de Vrande, V., de Jong, J.P.J., Vanhaverbeke, W., de Rochemont, M.: Open innovation in SMEs: trends, motives and management challenges. Technovation. **29**, 423–437 (2009)
42. Lichtenthaler, U., Lichtenthaler, E.: A capability-based framework for open innovation: complementing absorptive capacity. J. Manag. Stud. **46**, 1315–1339 (2009)
43. Chesbrough, H.: Open Business Models: How to Thrive in the New Innovation Landscape. Harvard Business School Press, Boston (2010)

44. Cohen, W.M., Levinthal, D.A.: Absorptive capacity: a new perspective on learning and innovation. Adm. Sci. Q. **35**, 128–152 (1990)
45. Teece, D.J., Pisano, G., Shuen, A.: Dynamic capabilities and strategic management. Strateg. Manag. J. **18**, 509–533 (1997)
46. Mention, A.-L.: Co-operation and co-opetition as open innovation practices in the service sector: which influence on innovation novelty? Technovation **31**, 44–53 (2011)
47. Leminen, S., Westerlund, M., Nyström, A.-G.: Living labs as open-innovation networks. Technol. Innov. Manag. Rev. **2**, 6–11 (2012)
48. Sieg, J.H., Wallin, M.W., Von Krogh, G.: Managerial challenges in open innovation: a study of innovation intermediation in the chemical industry. R&D Manag. **40**, 281–291 (2010)
49. Katzy, B., Turgut, E., Holzmann, T., Sailer, K.: Innovation intermediaries: a process view on open innovation coordination. Technol. Anal. Strateg. Manag. **25**, 295–309 (2013)
50. Mahr, D., Lievens, A.: Virtual lead user communities: drivers of knowledge creation for innovation. Res. Policy **41**, 167–177 (2012)
51. West, J., Lakhani, K.R.: Getting clear about communities in open innovation. Ind. Innov. **15**, 223–231 (2008)
52. Huizingh, E.K.R.E.: Open innovation: state of the art and future perspectives. Technovation. **31**, 2–9 (2011)
53. Oh, D.-S., Phillips, F., Park, S., Lee, E.: Innovation ecosystems: a critical examination. Technovation. **54**, 1–6 (2016)
54. Nelson, R.R.: National innovation systems: a comparative analysis. Oxford University Press, New York (1993)
55. Lyytinen, K., Yoo, Y., Boland, R.J.: Digital product innovation within four classes of innovation networks. Inf. Syst. J. **26**, 47–75 (2015)
56. Oxley, J.E., Sampson, R.C.: The scope and governance of international and alliances. Strateg. Manag. J. **15**, 723–749 (2004)
57. Chesbrough, H., Kim, S., Agogino, A.: Chez Panisse: building an open innovation ecosystem. Calif. Manag. Rev. **56**, 144–172 (2014)
58. Rohrbeck, R., Hölzle, K., Gemünden, H.G.: Opening up for competitive advantage–how Deutsche Telekom creates an open innovation ecosystem. R&D Manag. **39**, 420–430 (2009)
59. Isckia, T., Lescop, D.: Open Innovation within business ecosystems: a Tale from Amazon.com. Commun. Strateg. **2**, 37–54 (2009)
60. Traitler, H., Watzke, H.J., Saguy, I.S.: Reinventing R&D in an open innovation ecosystem. J. Food Sci. **76** (2011)
61. Adner, R.: Match your innovation strategy to your innovation ecosystem. Harv. Bus. Rev. **84**, 98–107 (2006)
62. Holgersson, M., Granstrand, O., Bogers, M.: The evolution of intellectual property strategy in innovation ecosystems: uncovering complementary and substitute appropriability regimes. Long Range Plann. Corrected proof (2017)
63. Bosch-Sijtsema, P., Bosch, J.: Aligning innovation ecosystem strategies with internal R&D. In: Proceedings of the 2014 IEEE ICMIT, Singapore, pp. 424–430 (2014)
64. Gawer, A., Cusumano, M.A.: Industry platforms and ecosystem innovation. J. Prod. Innov. Manag. **31**, 417–433 (2014)
65. Riedl, C., Böhmann, T., Leimester, J.M., Krcmar, H.: A framework for analysing service ecosystem capabilities to innovate. In: Proceedings of the 17th European Conference on Information Systems, Verona, Italy (2009)
66. Curley, M., Salmelin, B.: Open Innovation 2.0: The New Mode of Digital Innovation for Prosperity and Sustainability. Springer, Berlin (2017)
67. European Union: Open Innovation 2.0 Yearbook 2016. Belgium, Brussels (2016)

68. Curley, M., Salmelin, B.: Open Innovation 2.0: A New Paradigm. In: OISPG White Paper (2013)
69. Curley, M.: Twelve principles for open innovation 2.0: evolve governance structures, practices and metrics to accelerate innovation in an era of digital connectivity. Nature **533**, 314–316 (2016)
70. Chesbrough, H.: Open services innovation: rethinking your business to grow and compete in a New Era (2011)
71. Vargo, S.L., Lusch, R.F.: Service-dominant logic 2025. Int. J. Res. Mark. **34**, 46–67 (2017)
72. Hollebeek, L.D., Andreassen, T.W., Smith, D.L.G., Gronquist, D., Karahasanovic, A., Marquez, A.: Epilogue-service innovation actor engagement: an integrative model. J. Serv. Mark. **32**, 95–100 (2018)
73. Finsterwalder, J.: A 360-degree view of actor engagement in service co-creation. J. Retail. Consum. Serv. **40**, 276–278 (2017)
74. Alexander, M.J., Jaakkola, E., Hollebeek, L.D.: Zooming out: actor engagement beyond the dyadic. J. Serv. Manag. **29**, 333–351 (2018)
75. Selander, L., Henfridsson, O., Svahn, F.: Transforming ecosystem relationships in digital innovation. In: Proceedings of the Thirty First International Conference on Information Systems. St. Louis (2010)
76. Ketonen-Oksi, S.: Creating a shared narrative: the use of causal layered analysis to explore value co- creation in a novel service ecosystem. J. Futur. Res. **6** (2018)
77. Soe-Tsyr, D.Y., Shiou-Tian, H.: Enhancing service system design: an entity interaction pattern approach. Inf. Syst. Front. **19**, 481–507 (2017)
78. Frow, P., Mccoll-Kennedy, J.R., Payne, A.: Co-creation practices: their role in shaping a health care ecosystem. Ind. Mark. Manag. **56**, 24–39 (2016)
79. Still, K., Huhtamäki, J., Russell, M.G., Rubens, N.: Title Insights for orchestrating innovation ecosystems: the case of EIT ICT Labs and data-driven network visualisations Insights for Orchestrating Innovation Ecosystems: Case EIT ICT Labs and Data-driven Network Visualizations. Int. J. Technol. Manag. **66**, 243–265 (2014)
80. Ritala, P., Agouridas, V., Assimakopoulos, D., Gies, O.: Value creation and capture mechanisms in innovation ecosystems: a comparative case study. Int. J. Technol. Manag. **63**, 244–267 (2013)
81. Haile, N., Altmann, J.: Value creation in software service platforms. Future Gener. Comput. Syst. **55**, 495–509 (2016)
82. Wieland, H., Hartmann, N.N., Vargo, S.L.: Business models as service strategy. J. Acad. Mark. Sci. **45**, 925–943 (2017)
83. Fujita, S., Vaughan, C., Vargo, S.L.: Service ecosystem emergence from primitive actors in service dominant logic: an exploratory simulation study. In: Proceedings of the 51st Hawaii International Conference on System Sciences, Waikoloa, Hawaii, pp. 1601–1610 (2018)
84. Banoun, A., Dufour, L., Andiappan, M.: Evolution of a service ecosystem: longitudinal evidence from multiple shared services centers based on the economies of worth framework. J. Bus. Res. **69**, 2990–2998 (2016)
85. Di Pietro, L., Edvardsson, B., Reynoso, J., Renzi, M.F., Toni, M., Guglielmetti Mugion, R.: A scaling up framework for innovative service ecosystems: lessons from Eataly and KidZania. J. Serv. Manag. **29**, 146–175 (2018)
86. Lusch, R.F., Nambisan, S.: Service innovation: a service-dominant logic perspective. MIS Q. **39**, 155–175 (2015)

Smart Service Processes

Crowdsensing-Based Road Condition Monitoring Service: An Assessment of Its Managerial Implications to Road Authorities

Kevin Laubis[✉], Florian Knöll, Verena Zeidler, and Viliam Simko

FZI Research Center for Information Technology, Karlsruhe, Germany
{laubis,knoell,zeidler,simko}@fzi.de

Abstract. The ubiquity of smart devices in vehicles, such as smartphones allows for a crowdsensing-based information gathering of the vehicle's environment. For example, accelerometers can reveal insights into road condition. From a road authorities' perspective, knowing the road condition is essential for scheduling maintenance actions in an efficient and sustainable manner. In Germany, expensive laser-based road inspections are scheduled every four years. In future, they could be extended or completely replaced with a crowd-based monitoring service. This paper determines whether the lower accuracy of crowdsensing-based measurements is redeemed by its potential of near-real time data updates. Partially observable Markov decision processes are applied for determining maintenance policies that minimize roads' life-cycle costs. Our results show that substituting laser-based road condition inspections by a crowdsensing-based monitoring service can decrease total costs by 5.9 % while an approach, which combines both monitoring approaches, reduces the costs by 6.98 %.

Keywords: Crowdsensing · Internet of Things
Predictive maintenance · Road condition
Partially observable Markov decision processes
Reinforcement learning

1 Introduction

The road network is regarded as a valuable asset for road authorities. Next to the decisions "when", "where" and "how" to construct new road links for expanding the network, also the existing infrastructure has to be maintained. Besides considering maintenance costs, road authorities have to consider road user costs related to the road condition [15].

A sustainable road maintenance requires knowledge of the actual roads' conditions for coming up with efficient decisions about maintenance actions. Next to this, knowing the current road condition is of interest for road user as well since they can avoid vehicles wear by taking a detour on a smoother road [11].

© Springer Nature Switzerland AG 2018
G. Satzger et al. (Eds.): IESS 2018, LNBIP 331, pp. 127–137, 2018.
https://doi.org/10.1007/978-3-030-00713-3_10

Road conditions can be inspected with specialized equipment, such as cameras and lasers attached to special-purpose vehicles. These can provide highly accurate measurements and serve as a sound information basis for maintenance decisions. However, due to limited resources, inspections are performed irregularly at long time intervals or are not performed at all. In the case of Germany, the federal road network is monitored at coarse-granular four year intervals. Furthermore, roads with multiple lanes are not covered fully. Just the right lane is considered. For roads managed by counties and municipalities, there is not even a uniform procedure for road condition monitoring. The assessment is sometimes carried out with pen and paper leading to rather subjective results.

The growing number of smart devices such as smartphones, mobile navigation systems, wearables, etc. represents a vast potential for a crowdsensing-based improvement of the current road condition monitoring, while serving as a road condition information service for road users and authorities. Collecting and analyzing such data beside the position and speed from multiple vehicles is called an Extended Floating Car Data (XFCD) approach [9]. In a former study of ours, we describe how data from accelerometer, gyroscope and GPS sensors from heterogeneous cars and smartphones can be analyzed automatically for predicting the road's condition [12]. Thus, heterogeneous cars and sensor types can easily participate without any manual calibration effort. The main potential of such a crowdsensing approach is the spatio-temporal coverage. Next to this, cars are driving on the roads anyway. Gathering and analyzing their data in a centralized backend system would constitute only a small additional effort. Even though the measurements performed by individual cars are not as accurate as those from official inspection drives are, they are performed by many more cars and at a much higher rate—nearly at real-time. Figure 1 outlines the objective of our research. It depicts the usage of conventional road condition inspections by a road maintenance decision support system (DSS) and it indicates inspections based on smart technologies as providing a substitutional or supplemental potential. Based on our former findings focusing on the technological perspective of a crowdsensing-based road condition monitoring, the focus of this paper is on the economic benefits of analysing and exploiting XFCD for road authorities.

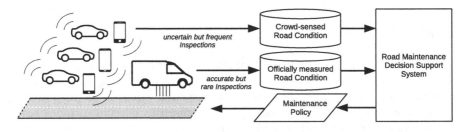

Fig. 1. Outline of research objective: a road condition monitoring service for supporting maintenance decisions based on official measurements with specially equipped inspection vehicles and a crowdsensing-based approach as substitute or supplement.

Thus, we quantify the monetary effect of this crowd-based approach on the performance of maintenance policies. Accordingly, the research questions answered in this paper are as follows:

1. Given crowdsensed road condition inspections, what is the effect of (a) uncertainty and its (b) potential of being performed more frequently on an optimal maintenance policy that currently relies on accurate laser-based measurements?
2. To what extent can maintenance and road user costs be reduced by determining an optimal maintenance policy making use of both, crowdsensing and laser-based road condition inspections?

Knowing these effects, from a managerial perspective, road authorities are able to decide whether crowdsensed inspections alone or combined with current inspection cycles are beneficial.

The paper is structured as follows: In the next section, related work is provided. This is followed by a foundation section about partially observable Markov decision processes (POMDP) and related Markov models. Section 4 describes our approach of parameterizing the POMDP models for answering the research questions. Section 5 concludes the paper by summarizing and exposing possible future research directions.

Table 1. Differentiation of Markov models by the controllability of transitions due to actions and by observability of actual states.

	Control over transitions	States fully observable
Markov chain	×	✓
Markov decision process	✓	✓
Hidden Markov model	×	×
Partially observable Markov decision process	✓	×

2   Related Work

Smart devices have recently become a great source of sensor data from accelerometers, gyroscopes and GPS units. This made the crowdsensing-based road condition monitoring an interesting research field [2,6,8,13,14,22]. In former work, we present an approach that allows for a seamless integration of new participants with different smart devices and vehicle types in such crowdsensing-based systems for predicting the road condition without the need for manual calibration [12].

A widely used metric for quantifying the road condition that can be measured with smartphone equipped vehicles is the international roughness index (IRI) developed in the course of the road roughness experiment by the World Bank [18]. The IRI is defined as the ratio of the accumulated movement of a suspension system and the hereby driven distance. Common units are m/km and $inches/mile$.

Extensive work has been done in the modeling of road deterioration and maintenance task scheduling through Markov decision processes (MDP). Gao and Zhang developed a MDP model for a road maintenance optimization problem including road user costs and compute an optimal maintenance policy [7]. Smilowitz and Madanat extended MDP approaches by considering uncertainty within the inspection methods [20]. They call this approach latent MDP. Similar approaches are also known as POMDP. The proposed approach considers discrete condition states. Schöbi and Chatzi describe the implementation of a continuous state POMDP that likewise considers inspection uncertainty within maintenance planning [19]. The results are compared to a discrete model.

Even though there has been research done in both fields—crowdsensing-based road condition monitoring and optimal scheduling of road maintenance tasks based on road conditions—to our best knowledge a combination of both fields has not been investigated. This is where our paper contributes.

3 Partially Observable Markov Decision Processes

Markov chains have been proved useful for modeling of stochastic processes [17]. The key characteristic of Markovian chains is the Markov property, which describes the fact that Markov chains are "memoryless", i. e., the conditional probability distribution of a system's state at time step $t + 1$ only depends on the present state at time t. This property allows for computing probabilities for state transitions and thus related costs. The basic model can be extended by including uncertainty about the system's state, what is called a hidden Markov model, or influencing the state transition probabilities by the agent's actions (MDP). Furthermore, applying a POMDP both can be modeled, uncertainty and control over state transitions. Table 1 provides an overview of the characteristics of the mentioned Markov models. A verbal differentiation is given below.

Hidden Markov models account for the fact that the information about a system's state is often imperfect, e. g. due to measurement errors [1]. This can be modeled by assuming an underlying Markov chain, which is unobservable, i. e., its states are hidden. However, observations are made which relate to the real states according to a predefined probability distribution. Using Bayesian updating, a probability distribution of the possible sequences of state transitions for a given sequence of actions and observations can be computed to define the most probable hidden state and reduce the uncertainty. This probability distribution is called the belief state and contains all information about the past.

A *Markov decision process* involves the possibility to perform certain actions, which are associated with different state transition probability matrices [16]. Assuming perfect information, policies for performing optimal actions, i. e., maximizing the expected rewards, can be determined by value iteration or policy iteration algorithms.

Partially observable Markov decision processes combine both approaches [3, 10]. Thus, it can be modeled that states are not completely observable and actions can be taken to control the state transitions. A formal description of a POMDP is provided in Definition 1.

Definition 1. (Partially observable Markov decision processes). *A POMDP can be formally described as a tuple* $(S, A, T, R, \Omega, O, d)$ *where:*

- S *is a finite set of states that are not completely observable.*
- A *is a finite set of actions that can be taken to control the state transitions.*
- T *is a probabilistic state-transition function* $S \times A \mapsto \Pi(S)$.
- R *is an immediate reward function* $S \times A \mapsto \mathbb{R}$.
- Ω *is a finite set of observations.*
- O *is a probabilistic observation function* $S \times A \mapsto \Pi(\Omega)$.
- $d \in [0, 1]$ *is a discount factor.*

The question that shall be answered by a POMDP is which action is optimal at a certain stage of the process given the belief state and the current observation. Therefore, the expected cumulated reward associated to each action is iteratively determined. Information of the history of observations and actions is inherently contained in the belief state, which is updated in each iteration of the process. The computational effort for solving a POMDP increases exponentially with the number of states, actions and observations.

For a discrete state-space POMDP, there exist several algorithms that provide exact or approximate solutions, e. g., Enumeration [21], Two-Pass [21], Witness [3], Linear Support [5] and Incremental Pruning [4,23]. All of them are based on an iterative forward-backward approach. The belief state is calculated for a given time horizon and subsequently the optimal value function—that maximizes the expected reward associated to actions and states—is computed recursively.

4   Assessment

For assessing the relevance of crowdsensing-based road condition inspections to road authorities, we compare four different scenarios. The scenarios differ in the type and frequency of inspection (laser-based and crowd-based) performed. Laser-based inspections can be performed at the beginning of each year or at four year intervals as it is the case for the German federal road network. Crowd-based road condition measurements are gathered within one year and aggregated at the end of each year for having a single robust crowd-based inspection. Combining these inspection types, we came up with the following four scenarios: *scenario 1*—laser-based inspection every year, *scenario 2*—laser-based inspection every fourth year, *scenario 3*—crowd-based inspection every year and *scenario 4*—laser-based inspection every fourth year and crowd-based inspections every other year. For each scenario we apply a POMDP for finding an optimal maintenance policy. Given these policies, a road authority knows in which year and at which condition a certain maintenance decision should be performed for minimizing the expected total costs (maintenance costs and road user costs) for a time horizon of 100 years. Its parameterization is described in the following section.

Table 2. Road condition states S are defined by whole numbers of the RQI [7]. Next to the RQI and the corresponding IRI ranges, each state is provided with representative RQI and IRI values, which are considered within the analysis.

	IRI range (m/km)	RQI range	Representative IRI (m/km)	Representative RQI
s_1	0.327–1.169	$4 \leq RQI \leq 5$	0.683	4.5
s_2	1.170–2.529	$3 \leq RQI < 4$	1.784	3.5
s_3	2.530–4.409	$2 \leq RQI < 3$	3.405	2.5
s_4	4.410–6.808	$1 \leq RQI < 2$	5.544	1.5
s_5	6.809–9.726	$0 \leq RQI < 1$	8.202	0.5

4.1 Parameterization

Using the POMDP Definition 1, the states S reflect road conditions, the actions A reflect different maintenance tasks. Transition probabilities T describe the effects of maintenance tasks and road deterioration. The observation probabilities O indicate the accuracy of the inspection methods. The immediate rewards R consist of both cost components, maintenance and road user costs. The states, actions, transition probabilities and rewards are defined in correspondence to the work [7]. The observation probabilities for the crowd-based inspections are defined based on our findings in [12]. All these parameters are defined and described below.

For defining the *road condition states* $S = \{s_1, s_2, s_3, s_4, s_5\}$ the ride quality index (RQI) is derived from the IRI by using Eq. 1 and splitting the resulting range into whole numbers, which facilitates classifying IRI values into discrete quality states.

$$RQI = 6.122 - 1.963 \cdot \sqrt{IRI} \qquad (1)$$

The states are shown in Table 2. A low IRI and a high RQI indicate a good road condition. Thus, s_1 stands for good condition, whereas s_5 stands for complete infrastructure failure.

There are three *maintenance actions* $A = \{a_1, a_2, a_3\}$ that can be scheduled by road authorities. These are *reconstruction* a_1, *resurfacing* a_2 and just doing minor tasks or *nothing* a_3. Reconstructing a road segment is considered as resetting it in the best possible state s_1. A resurfacing action on an asphalt road section improves the condition according to Eq. 2 [15].

$$IRI_t - IRI_{t+1} = \frac{0.66 \cdot IRI_t}{7.15 \cdot IRI_t + 18.3} \theta \qquad (2)$$

Thus, the IRI after resurfacing IRI_{t+1} depends on the IRI before the maintenance task IRI_t and on the resurfacing thickness θ in millimeters. In this analysis, a thickness of 40 mm is chosen to keep comparability with the study from Gao and Zhang [7]. It is obvious that performing no maintenance action has no direct effect on the road condition.

Table 3. Transition probabilities T composed of rehabilitation due to maintenance action A performed at the beginning of the year and annual deterioration (rounded) [7].

	a_1 (reconstruction)					a_2 (resurfacing)					a_3 (nothing)				
	s_1	s_2	s_3	s_4	s_5	s_1	s_2	s_3	s_4	s_5	s_1	s_2	s_3	s_4	s_5
s_1	0.74	0.26	0	0	0	0.74	0.26	0	0	0	0.74	0.26	0	0	0
s_2	0.74	0.26	0	0	0	0.74	0.26	0	0	0	0	0.83	0.18	0	0
s_3	0.74	0.26	0	0	0	0.28	0.61	0.11	0	0	0	0	0.86	0.14	0
s_4	0.74	0.26	0	0	0	0	0.19	0.70	0.11	0	0	0	0	0.87	0.13
s_5	0.74	0.26	0	0	0	0	0	0.09	0.79	0.11	0	0	0	0	1

Transition probabilities T are determined by the maintenance actions performed and the road deterioration. The annual deterioration of an asphalt road segment can be describe with Eq. 3 [7,15].

$$IRI_{t+1} = (IRI_t + \alpha)e^{\beta} \tag{3}$$

In this equation, $\alpha = 0.2$ is defined as $\epsilon(1 - e^{-\beta})$, where ϵ depends on road type and traffic and $\beta = 0.0153$. This parameterization is analogous to [7]. The actions, which are described above, are performed at the beginning of each year. If a maintenance action is performed (a_1 or a_2), first, the condition is improved as described above and second, the annual deterioration is considered based on the maintained road. This combination of maintenance and deterioration probabilities results in the combined transition probabilities, which are shown in the columns headed with a_1 and a_2 in Table 3. If no maintenance action is performed at the beginning of a year, the road is expected to deteriorate according to the probabilities provided in the right columns a_3 of Table 3. These deterioration probabilities are directly derived from Eq. 3.

A finite set of possible *observations* Ω is defined accordingly to the set of states $\Omega = S$. *Observation probabilities* O depend on the inspection method (laser- and crowd-based). For the laser inspections we assume a 100 % accuracy. This means that given a RQI value observation, it is assumed that it reflects the actual road condition. The observation probabilities for this certain inspection method can be modeled as an identity matrix, which is depicted in the first laser-based columns of Table 4. For the crowdsensed inspections, we consider an accuracy in observing the actual state of 85%. This accuracy results from our former study in which we built supervised models for predicting the IRI with smartphone equipped vehicles [12]. Since we drove on a 2.28 km road link of a district road in Germany, which was in an overall good condition, we extrapolated the hereby determined accuracy from the states s_1 and s_2 to the states s_3, s_4 and s_5. These empirically determined observation probabilities are provided in the crowd-based columns of Table 4. In two of our considered, scenarios the laser-based inspections are not performed on an annual basis, but on four year intervals. Thus, we extend the state set by the cross product with the four

Table 4. Observation probabilities O for laser-based inspections are defined as an identity matrix since certainty is assumed. For crowd-based inspections, findings from a former study of ours are considered [12].

	Laser-based					Crowd-based				
	o_1	o_2	o_3	o_4	o_5	o_1	o_2	o_3	o_4	o_5
s_1	1	0	0	0	0	0.850	0.150	0	0	0
s_2	0	1	0	0	0	0.075	0.850	0.075	0	0
s_3	0	0	1	0	0	0	0.075	0.850	0.075	0
s_4	0	0	0	1	0	0	0	0.075	0.850	0.075
s_5	0	0	0	0	1	0	0	0	0.150	0.850

Table 5. Annual maintenance costs and road user costs in Euro/year for one road segment depending on segment state and maintenance action performed at the beginning of the year. Rewards R are defined as the sum of maintenance costs and road user costs. Costs are from [7] and are translated from Hong Kong Dollar into Euro at the closing exchange rate 0.1162 on 02.11.2016.

	a_1 (reconstruction)		a_2 (resurfacing)		a_3 (nothing)	
	MC	RUC	MC	RUC	MC	RUC
	(€/year)	(€/year)	(€/year)	(€/year)	(€/year)	(€/year)
s_1	36.249,91	158.037,99	13.942,30	87.739,89	0	82.891,46
s_2	36.249,91	158.037,99	13.942,30	87.739,89	0	193.477,04
s_3	36.249,91	158.037,99	13.942,30	157.234,91	0	357.251,23
s_4	36.249,91	158.037,99	13.942,30	324.099,64	0	573.788,56
s_5	36.249,91	158.037,99	13.942,30	555.065,01	0	823.950,81

different treated years $Y = \{y_1, y_2, y_3, y_4\}$. This results in 20 states $s_{i,j}$ with $i \in \{1, \ldots, 5\}, j \in \{1, \ldots, 4\}$. Likewise, we extend the transition and observation probabilities in a way that a state $S_{i,j}$ will always result in a state $s_{i,k}$ with one year offset, i. e. $k = (j + 1)$ *modulo* 4. However, in our description we keep the way of addressing states without mentioning a special year.

Below describe both cost components considered in this study, road maintenance and road user costs. The *maintenance costs* depend on the maintenance action A performed at the beginning of a year. A reconstruction a_1 is expected to cost €45.32 /m^2 and a resurfacing a_2 €17.43 /m^2 [7]. Assuming a 200 m asphalt road segment having a width of 4 m, the aggregated maintenance costs per year are given in Table 5. The annual *road user costs* are also provided in this table. They are composed of vehicle operating costs and costs caused due to additional travel time. The vehicle operating costs are composed of expenses for fuel, tire wear and vehicle repair. These costs linearly depend on the road's roughness [15]. The costs for travel delay depend on maintenance actions, since work zones can cause traffic jams and thus forces road users to stop or drive slower. The

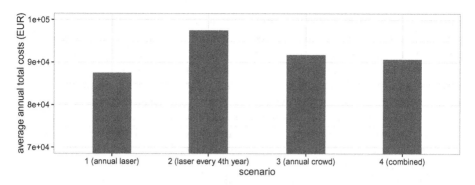

Fig. 2. Average annual total costs of investigated inspection scenarios (with and without utilizing crowd-based road condition monitoring service).

determination of the road user cost aggregates is provided by Gao and Zhang [7], while considering the annual average daily traffic, the road's capacity of three lanes, the duration and effects on speed of work zones and the wear of two different car types (passenger cars and trucks). In this research in progress we first consider the effects for the middle lane. While considering multiple years in our analysis we apply a discount factor $d = 0.95$ to future rewards.

4.2   Results

The scenario specific policies lead to the following expected total costs (maintenance and road user costs, discounted for a 100 years horizon and for a 200 m asphalt road segment): *scenario 1 — €1,739,908, scenario 2 — €1,938,836, scenario 3 — €1,824,365* and *scenario 4 — €1,803,470.* Figure 2 provides the corresponding average annual total costs for each scenario and segment. It can be seen that performing laser-based inspection every year results in the lowest annual total costs (€87,513.53) compared to the other scenarios. Due to limited resources, however, this frequency cannot be applied to the German federal road network. Performing the laser-based inspections just every fourth year—as it is done in Germany—causes the highest annual total costs (€97,519.17). Relying on inspections of a crowd-based monitoring serve instead reduces the costs to €91,761.53. A further cost reduction to €90,710.55 can be achieved by complementing four year laser-based inspections with annual crowd-based inspections.

5   Conclusion and Outlook

For assessing the implications of a crowdsensing-based road condition monitoring service to road authorities, we applied POMDP for determining maintenance policies for different scenarios. The scenarios differ in the inspections' accuracy and frequency. For considering inaccuracy in the scenarios that model crowdsensing inspections, we made use of observation probabilities which we derived from our former study [12]. The effect of this uncertainty and the potential of performing crowdsensing-based inspections more frequently is addressed by the first research question. The results show that having annually, but less accurate crowd-based inspections reduces maintenance and road user costs by 5.9% compared to performing laser-inspection at four year intervals. This finding shows that from a managerial perspective the inaccuracy of a crowdsensing-based monitoring is overcompensated by its potential of providing measurements more frequently. The second research question is answered by supplementing laser-based measurement with annual crowd-based inspection. It is shown that the combination of both inspection types reduces the total costs by 6.98%. With respect to a DSS for road maintenance, the use of a crowdsensing-based approach can be recommended as a substitute and as well as a supplement to the nowadays expensive laser-based monitoring. Thus, analysing and exploiting XFCD should be considered within road monitoring systems.

There are limitations that can be addressed in future work. We empirically determined the observation probabilities on road segments, which were in condition states s_1 and s_2. We were required to extrapolate the determined uncertainty to the other states. This can be addressed by an empirical determination of the probabilities for the residual states. Furthermore, except to the differentiation in inspection methods, up to now we consider a fixed set of parameter values. Even though they are well chosen, an analysis of the results' sensitivities to parameter changes such as different road types, lane types and the amount of traffic can broaden the current findings. Next to these limitations, future work can consider costs for the different inspection methods and thus allow for an optimal scheduling of the measurements.

References

1. Baum, L.E., Petrie, T.: Statistical inference for probabilistic functions of finite state Markov chains. Ann. Math. Stat. **37**(6), 1554–1563 (1966)
2. Bhoraskar, R., Vankadhara, N., Raman, B., Kulkarni, P.: Wolverine: traffic and road condition estimation using smartphone sensors. In: 2012 4th International Conference on Communication Systems and Networks, COMSNETS 2012. IEEE, Bangalore (2012)
3. Cassandra, A., Kaelbling, L.P., Littman, M.L.: Acting optimally in partially observable stochastic domains. In: Proceedings of the Twelfth AAAI National Conference on Artificial Intelligence, AAAI 1994, pp. 1023–1028. AAAI, Seattle (1994)
4. Cassandra, A., Littman, M.L., Zhang, N.: Incremental pruning: a simple, fast, exact method for partially observable Markov decision processes. In: Proceedings of the Thirteenth Conference on Uncertainty in Artificial Intelligence, UAI 1997, pp. 54–61. Morgan Kaufmann Publishers Inc., San Francisco (1997)
5. Cheng, H.T.: Algorithms for partially observable Markov decision processes. Ph.D. thesis, University of British Columbia (1988)
6. Eriksson, J., Girod, L., Hull, B., Newton, R., Madden, S., Balakrishnan, H.: The pothole patrol: using a mobile sensor network for road surface monitoring. In: Proceedings of the 6th International Conference on Mobile systems, Applications, and Services - MobiSys 2008, pp. 29–39. ACM, New York (2008)

7. Gao, H., Zhang, X.: A Markov-based road maintenance optimization model considering user costs. Comput.-Aided Civ. Infrastruct. Eng. **28**(6), 451–464 (2013)
8. Hara, T., Springer, T., Muthmann, K., Schill, A.: Towards a reusable infrastructure for crowdsourcing. In: 2014 Proceedings of the 7th International Conference on Utility and Cloud Computing, pp. 618–623. IEEE (2014)
9. Irschik, D., Stork, W.: Road surface classification for extended floating car data. In: Proceedings of the 2014 IEEE International Conference on Vehicular Electronics and Safety (ICVES), pp. 78–83. IEEE (2014)
10. Kaelbling, L.P., Littman, M.L., Cassandra, A.: Planning and acting in partially observable stochastic domains. Artif. Intell. **101**(1), 99–134 (1998)
11. Laubis, K., Simko, V., Schuller, A.: Crowd sensing of road conditions and its monetary implications on vehicle navigation. In: 2016 International IEEE Conferences on Ubiquitous Intelligence & Computing. Advanced and Trusted Computing, Scalable Computing and Communications, Cloud and Big Data Computing, Internet of People, and Smart World Congress, pp. 833–840. IEEE, Toulouse (2016)
12. Laubis, K., Simko, V., Schuller, A.: Road condition measurement and assessment: a crowd based sensing approach. In: Thirty Seventh International Conference on Information Systems. AIS, Dublin (2016)
13. Masino, J., Pinay, J., Reischl, M., Gauterin, F.: Road surface prediction from acoustical measurements in the tire cavity using support vector machine. Appl. Acoust. **125**, 41–48 (2017)
14. Mohan, P., Padmanabhan, V.N., Ramjee, R.: Nericell: rich monitoring of road and traffic conditions using mobile smartphones. In: Proceedings of the 6th ACM Conference on Embedded Network Sensor Systems, SenSys 2008, pp. 323–336. ACM, New York (2008)
15. Ouyang, Y., Madanat, S.: Optimal scheduling of rehabilitation activities for multiple pavement facilities: exact and approximate solutions. Transp. Res. Part A Policy Pract. **38**(5), 347–365 (2004)
16. Puterman, M.L.: Markov Decision Processes: Discrete Stochastic Dynamic Programming. Wiley-Interscience, Hoboken (2005)
17. Revuz, D.: Markov Chains, 2nd edn. North-Holland, Amsterdam (1984)
18. Sayers, M.W., Gillespie, T.D., Queiroz, C.A.V.: The international road roughness experiment: establishing correlation and a calibration standard for measurements. Technical report 45, The World Bank, Washington (1986)
19. Schöbi, R., Chatzi, E.N.: Maintenance planning using continuous-state partially observable Markov decision processes and non-linear action models. Struct. Infrastruct. Eng. **12**(8), 977–994 (2016)
20. Smilowitz, K., Madanat, S.: Optimal inspection and maintenance policies for infrastructure networks. Comput.-Aided Civ. Infrastruct. Eng. **15**(1), 5–13 (2000)
21. Sondik, E.J.: The optimal control of partially observable Markov decision processes. Ph.D. thesis, Stanford University (1971)
22. Yi, C.W., Chuang, Y.T., Nian, C.S.: Toward crowdsourcing-based road pavement monitoring by mobile sensing technologies. IEEE Trans. Intell. Transp. Syst. **16**(4), 1905–1917 (2015)
23. Zhang, N., Liu, W.: Planning in stochastic domains: problem characteristics and approximation. Technical report HKUST-CS96-31, The Hong Kong University of Science and Technology, Hong Kong (1996)

Digitalization of Field Service Planning: The Role of Organizational Knowledge and Decision Support Systems

Michael Vössing[(✉)], Clemens Wolff, and Volkmar Reinerth

Karlsruhe Service Research Institute, Karlsruhe Institute of Technology,
Kaiserstr. 12, 76131 Karlsruhe, Germany
{michael.voessing,clemens.wolff}@kit.edu

Abstract. Increasing competition and decreasing margins force manufacturers of industrial machinery to augment their products with complementary services—several dependent on field service planning. Hence, the efficient delivery of field services has become a main competitive differentiator and is driving companies to digitalize their processes and utilize decision support systems. Based on a qualitative interview study of maintenance providers, the paper provides empirical insight on the role of organizational knowledge and decision support systems in this digital transformation. The study shows that while maintenance providers have digitized their field service processes, many are only beginning the process of digitalization. Today, employees heavily rely on tacit and embedded knowledge in their decision-making. Therefore, knowledge-driven decision support systems—which have not yet been adopted in the industry—have been identified as an important cornerstone of the coming digitalization of field service planning.

Keywords: Decision support systems · Digitalization
Field service planning

1 Introduction

Since the invention of the programmable logic controller in the 1960s, machines have supported human work in a variety of industries [17]. In an increasingly digital world, technological advances such as cyber-physical systems, the Internet of Things, and constantly improving analytical capabilities (e.g. machine learning) are once again transforming how humans and machines collaborate. Finding new ways to leverage the individual strengths of humans and machines is a challenge faced in many industries—with an enormous influence on established work structures, business models, and processes. However, even though the emerging technologies have matured in recent years, only few industries have incorporated these technologies into their processes. In many, transforming established processes and overcoming organizational habits have proven to be more challenging than companies have expected. According to Ross [28], many companies

© Springer Nature Switzerland AG 2018
G. Satzger et al. (Eds.): IESS 2018, LNBIP 331, pp. 138–150, 2018.
https://doi.org/10.1007/978-3-030-00713-3_11

have underestimated the challenges associated with digitalization by overemphasizing the technological aspect of their digitization initiatives—an important prerequisite of digital transformation. Digitalization, therefore, requires "fundamental changes in how people work". Employees play a crucial role in the digital transformation of organizational processes due to their high familiarity with the processes on an operational level. Nowadays, companies are exploring human-centered ways to accelerate this digitalization of work systems and design services based on their employees' requirements.

Ostrom et al. [21] note that business-to-business services are currently underrepresented in service research. As a contribution to this field, this study examines how providers of maintenance services leverage machines—in the form of decision support systems (DSS)—for field service planning. In recent years, increasing competition and decreasing margins have emphasized the importance of continually reshaping and improving the organizational, controlling and manufacturing aspects of products and services. As a result, there is an increasing awareness among maintenance providers that decision-making processes associated with service delivery need to be improved—making it a suitable use-case for this research. Many companies augment their products with complementary services (e.g. repair, maintenance, and overhaul) to offer their customers integrated solutions. Efficient field service planning—a subset of workforce management focused on the allocation of service requests to a company's field service workforce—is essential for the delivery of these services. Research shows that providing good after-sales product support is an important competitive advantage [15] and that improving field service planning processes, therefore, is an important source of profit and future growth for maintenance providers [5]. However, field service planning practices have not kept up with technological advanced and changing work environments.

To illustrate how dispatchers, team leaders, and service managers utilize machines in their decision-making, this study explores field service planning processes through a study of seven maintenance providers. The research focuses on the adoption and usage of DSS as well as the role of organizational knowledge. The structured analysis shows that while maintenance providers do utilize communication-driven, document-driven, and data-driven DSS in their decision-making, companies rarely leverage knowledge-driven DSS. Additionally, employees emphasize the prevailing importance of tacit and embedded knowledge in their decision-making. Based on these insights, the study shows that to further digitalize their processes companies should focus on developing knowledge-driven DSS.

The paper is structured as follows: In Sect. 2, relevant literature on organizational knowledge and DSS is summarized to provide the theoretical foundation for the study. Section 3 describes the research methodology—focusing on the interview design, data collection, and analysis of the empirical data. Section 4 presents the results of the analysis by outlining the role of organizational knowledge and DSS in the context of field service planning. In Sect. 5, the results of the analysis are discussed and put in the context of digitization and digitalization. Finally, Sect. 6 summarizes the results, provides managerial implications, and acknowledges limitations.

2   Theoretical Foundations and Related Work

This chapter puts the research in context to relevant literature. First, literature on organizational knowledge is presented to illustrate the relevance of different types of knowledge in decision-making. Second, five distinct types of DSS are presented, and their importance in information systems is outlined. Third, field service planning is introduced and relevant studies are outlined.

2.1   Organizational Knowledge

The concept of knowledge has been a significant focus of research for many years. However, as an increasing number of researchers have adopted a "knowledge-based" view of companies, the concept has gained a new strategic importance for the development of information systems [2]. Today, the importance of organizational knowledge for product innovation and organizational innovation is undisputed. Researchers view organizational knowledge as an invaluable source of competitive advantage [20] and an important synergistic advantage [2]. Nonaka [20] describes knowledge as a "multifaceted concept with multilayered meanings" that is traditionally defined as "justified true belief".

This study adopts Horvath's [11] definition of three distinct types of knowledge: explicit, tacit, and embedded knowledge. Explicit knowledge refers to formalized and codified knowledge that can be expressed in words and numbers [24]. It is easy to identify, store, and retrieve and is labeled by Brown and Duguid [2] as "know-what". It is typically disseminated within companies through official documents such as drawings, operating procedures, and manuals, and is transmittable in formal, systematic language. Tacit knowledge is defined by Howells [12] as "non-codified, disembodied knowhow that is acquired via the informal take-up of learned behavior and procedures". Tacit knowledge is intuitive, difficult to define, and largely experience-based knowledge of individual employees. Brown and Duguid [2] label this type of knowledge as "know-how". According to Grant and Gregory [7], it is often encountered in situations where the actions of employees are embedded in the context of individual perspective. Typically, it is prevalent when an action requires simultaneous information processing or the recognition of patterns in diffuse and subsidiary clues. The acquisition of tacit knowledge has largely been attributed to learning by doing (e.g. the personal assimilation of knowledge). When tacit knowledge is commonplace, the overall knowledge of companies does not increase over time but becomes increasingly codified in rules of thumb and routines [7]. Embedded knowledge, while similar to tacit knowledge, specifically encompasses knowledge that is not individually acquired by people but embedded in products, processes, organizational culture, and routines.

2.2   Decision Support Systems

Decision support systems (DSS) play an important role in organizational decision-making. Power et al. [26] define a DSS as an ancillary or auxiliary "information system that supports decision-making activities [and] enhances a person or group's ability to make decisions". These computer-based systems improve decision-making by facilitating communication, structuring knowledge, analyzing situations, or recommending actions. Burstein and Holsapple [3] emphasize that these systems "relax cognitive, temporal, and economic limits of decision makers" and amplify "decision makers' capacities for processing knowledge". DSS compensate weaknesses in human decision making by providing decisional guidance [33] and increasing the effectiveness and efficiency of decision-making [32]. According to Hoch and Schkade [10], they enable human decision makers to capitalize on their strengths while compensating for their weaknesses. A large body of research literature recognizes the use of computer-based DSS as an important driver of organizational change alongside human change agents [33].

A number of academic disciplines provide an essential foundation for the development of DSS. Today companies leverage multiple types of either off-the-shelf or custom-designed DSS. Power and Sharda [25] classification of DSS is adopted in this research. Their work defines five distinct types of DSS: Communication-driven DSS that facilitate and enhance collaboration and enable shared decision making. Data-driven DSS that emphasizes access to and manipulation of internal data. Document-driven DSS that provide document (e.g. procedures, corporate documents) retrieval and analysis. Knowledge-driven DSS that use artificial intelligence and statistical inference to recommend actions. Model-driven DSS that emphasize access to and manipulation of optimization models to provide decision support. This distinction provides a valuable basis for further research.

2.3   Field Service Planning

The term field service refers to offerings in which spatially distributed employees (e.g. field service technicians) must be allocated to spatially distributed service demand. Field service planning is encountered in a variety of industries such as telecommunication [4], in-home health care services [19], military aircraft operation [29], or industrial maintenance [23]. Traditionally, these businesses are conducted in a transactional fashion. Customers signal service demand to the provider, who then assigns and delivers the required service at a suitable time. This scheduling and dispatching of delivery resources to service demand is referred to as field service planning [34]. The process is traditionally performed by human dispatchers who follow simple dispatching strategies and rely on their personal experience [9]. However, DSS are increasingly adopted to increase the efficiency of service delivery and develop new value propositions for customers.

Overall, humans play a crucial role in the delivery of these services. So far, few studies have been conducted from a human-centered perspective to better understand the needs and requirements of employees: A study by Yamauchi et al. [36] explores patterns of information usage by service technicians during problem-solving. The study outlines implications for the general use of technology to improve the organization of work. A similar study conducted by Makri and Neely [16] identifies barriers and facilitators of incident reporting in the context of servitized manufacturers through a series of semi-structured interviews. In the context of healthcare services, Wong and Blandford [35] study how dispatchers make decisions in the process of scheduling emergency ambulance to support the development of novel DSS. In a related study of medical dispatchers, Møller et al. [18] study barriers in the process of handling emergency calls and conclude that continuous professional development and DSS are potent levers to improve medical dispatching. However, the challenges and requirements of many key stakeholders in the field service planning process—namely dispatchers, service technicians, and customers—have not yet been studied similarly.

3   Research Methodology

This work aims at extending the body of knowledge in information systems by understanding how organizations leverage DSS for field service planning, and how their use of organizational knowledge is related to the development of these systems. The research explores these phenomena through an empirical investigation. Data is collected via in-depth interviews, and qualitative analysis is used to derive an understanding of the explored concepts [27]. The following chapter outlines the overall research methodology. More specifically, the interview design and data collection (see Sect. 3.1) and the analysis of the data (see Sect. 3.2) are highlighted.

3.1   Interview Design and Data Collection

As outlined in Sect. 1, field service planning is essential in many industries. Therefore, the study must be sufficiently broad to encompass any variations across these industries and sufficiently rigorous to ensure the results are reliable. An exploratory in-depth interview study is used to broaden the understanding of the presented concepts in the context of field service planning. The intention is to learn about the companies individual challenges, through a rigorous analysis of interviewees' circumstances, experiences, perspectives, and backgrounds [27]. In this context, the flexibility of exploratory research is advantageous as it allows the direction of the research to change in response to new insights [31]. Collected data is explored to identify themes and issues to concentrate on in following interviews. Therefore, explanations emerge as a result of the research process [31]. As the purpose of the study is exploratory, the research utilizes comparison focused sampling, a non-probability sampling technique, in which cases are selected to compare, contrast, and learn about factors that explain the similarities and differences [31]. The number of interview candidates to draw upon is limited in the context of maintenance providers, as few companies have developed extensive field service networks relevant for the study. Further, access to interviewees is limited, because the efficient delivery of field service is an important competitive advantage in the industry.

The selection process led to a sample of seven companies (cf. Table 1) that offer maintenance, repair, and overhaul services. To avoid contextual bias, the study covers perspectives from technical service providers (e.g. facility management and cleaning services, maintenance and assembly services), producers of industrial machinery (e.g. machine tools or reserve vending), as well as producers of household appliances. The cases draw on expert interviews to gather insights on the companies' use of DSS and organizational knowledge.

Table 1. Overview of cases

Case	Interviewee role	Industry	Employees	Revenue
Alpha	Dispatching, management	Maintenance services	10k–50k	€0.5B–€1B
Beta	Dispatching, management	Maintenance services	<10k	€0.2B–€0.5B
Gamma	Dispatching, team lead	Industrial machinery	<10k	€0.5B–€1B
Delta	Dispatching, team lead	Domestic appliances	10k–50k	€1B–€5B
Epsilon	Dispatching, technician	Robotics	>50k	€5B–€50B
Zeta	Dispatching, team lead	Industrial machinery	10k–50k	€1B–€5B
Eta	Dispatching, technician	Domestic appliances	>50k	€5B–€50B

3.2   Interview Analysis

To derive a sound understanding of the role of knowledge and the adoption of decision support tools from the non-standardized interviews, a qualitative analysis was used [30]. Meaningful analysis of the data collected in the non-standardized interviews requires the data to be condensed (summarized) and grouped (categorized) before the data can be interpreted through the use of conceptualization [31]. The data extracted from the interviews was analyzed through a theoretical coding approach—sometimes referred to as selective coding [6,30]. The analysis primarily relies on predefined central categories derived from the academic literature on DSS and organizational knowledge presented in Sect. 3. However, the analysis was open to emergent concepts and ideas that could have led to new patterns of association. Codes resulting from an initial open coding phase are aggregated into these core categories, which are presented in Sect. 4 and serve as a basis for the discussion in Sect. 5. The transcripts were separately coded by two researchers and the presented results (see Tables 2 and 3) are based on a critical comparison of the individual codes.

4   Results

Section 4.1 present the results of the study, starting with an overview of the adoption of DSS across companies, which is followed by a section outlining the role of organizational knowledge in field service planning. The section is structured to first reflect on the usage and adoption of DSS and the importance of different types of knowledge in the process of field service planning. Based on the analysis of the process of field service planning in seven organizations. Section 4.2 synthesizes the data and presents the results following the two central topics: decision support systems and organizational knowledge. Each subsection consists of a condensed synthesis of the insights that were identified during the in-depth interviews. The insights are enriched with quotes from the interviews.

4.1   Decision Support Systems in Field Service Planning

The study shows that companies leverage DSS to support a variety of activities (see Table 2). Companies utilize these tools to increase the responsiveness of their processes, prefer the use of commercial products over in-house developments, and utilize a number of specialized products to support individual tasks.

Communication-driven DSS are the most prevalent and are leveraged across companies to increase communication among stakeholders—most commonly in-between field employees and dispatchers. "Technicians have mobile devices and received their assigned tasks daily" (Gamma). Tools that improve communication are omnipresent in field service planning to simplify the flow of information relevant for a variety of processes. "Technicians use mobile clients [...] to write and share their reports" (Epsilon). Likewise, document-driven DSS are common-place to simplify the retrieval and access to documents relevant for the creation of work orders (e.g. contracts, service-level agreements). "We can see the location of the technician [...] and the location of the customer" (Zeta). "We have a support software where product information is available" (Delta). Overall, the majority of DSS simplify preliminary activities in the planning process. Additionally, data-driven DSS simplify access to internal data and provide basic analysis capabilities—for example for the calculation of deterministic maintenance cycles. "Our [maintenance cycles] are calculated from historical data" (Beta).

However, "planning an optimal route [remains] a manual process" (Zeta) in many companies. Interviewees that previously have used model-driven DSS—systems that could support route planning through the use of optimization models—attribute their dissatisfaction with model-driven DSS to the fact that the complexity of their processes exceeds the capabilities of the available solutions. "We tried a semi-automatic dispatching tool, but it could not handle our situation" (Gamma). Finally, interviewees across different cases suggest that planning is a process that needs to be managed by human dispatchers: "Humans do this work [...] no [available] software can do [planning] right" (Alpha). The adoption of DSS is similar across the companies (see Table 2). Interestingly knowledge-driven DSS—systems that use artificial intelligence and statistical inference to recommend actions—are rarely utilized for field service planning.

4.2   Organizational Knowledge in Field Service Planning

The study shows that companies, despite the availability of decision support systems, still heavily rely on tacit and embedded knowledge in their decision-making (see Table 3). Across companies, interviewees state that knowledge pri-

Table 2. Adoption of decision support systems for field service planning

DSS	Alpha	Beta	Gamma	Delta	Epsilon	Zeta	Eta
Data-driven	●	●	●	●	○	●	●
Knowledge-driven	○	●	○	○	○	○	○
Model-driven	○	●	○	●	●	○	●
Document-driven	●	●	●	●	●	●	●
Communication-driven	●	○	●	●	●	●	●

● : Adopted   ○ : Not Adopted

marily resides in the heads of dispatchers and back office employees. The concentration of organizational knowledge in "communities-of-practice" makes knowledge essential for field service planning difficult to contextualize and document. Interviewees emphasize that knowledge of many aspects of field service planning is primarily accumulated through operational exposure [7]. "I prioritize customers based on my past experiences with them. [...] Intuition and experience define what we do" (Gamma). The description of these behavioral patterns corresponds to the definitions of tacit and embedded knowledge outlined in Sect. 2. Knowledge is rarely explicitly recorded. Interviewees outline that knowledge is often deeply embedded in situation-specific contexts (i.e. the current backlog of work orders or utilization of technicians) and require fast decision-making (i.e. to respond to unexpected machinery breakdowns)—which is typically for tacit knowledge.

Table 3. Role of knowledge in field service planning

Knowledge	Alpha	Beta	Gamma	Delta	Epsilon	Zeta	Eta
Explicit	●	○	○	○	○	●	○
Tacit	●	●	●	●	●	●	●
Embedded	○	●	●	●	●	●	●

● : Utilized   ○ : Not Utilized

Dispatchers refer to their experience and skills as 'feel', and emphasize its role in recognizing patterns in diffuse situations—for example when selecting suitable employees for work orders, when estimating traffic conditions, or when estimating the duration of work orders. "Successfully estimating a customer's problem depends on the experience of the dispatcher. [...] Experience defines a dispatcher" (Alpha). Dispatchers leverage their accumulated knowledge specifically for work order specification, work order prioritization, and disruption recovery. These routines and mostly repetitive tasks characterize operational field service planning but are difficult to transmit to new employees. Additionally, these skills can not be learned by mastering their fragments, but only by discovering how to coordinate all aspects of field service planning simultaneously. As a result, many companies train their employees on the job [7].

5 Discussion

The study reveals multiple insights: First, companies utilize a variety of data-driven, document-driven, and communication-driven DSS to support human decision-making in field service planning. Second, employees emphasize the prevailing importance of their accumulated knowledge (i.e. tacit knowledge) and established routines (i.e. embedded knowledge) and rely heavily on both in their decision-making. Third, model-driven DSS are rarely used and knowledge-driven DSS have not yet been adopted, even though interviewees show a willingness to adopt DSS and also explicitly express the demand for DSS that further simplify their work. Based on these insights, the following conclusions can be drawn.

DSS are an essential aspect of field service planning. However, knowledge-intensive planning activities (i.e. work order specification and quantification, work order prioritization and planning, and disruption recovery) are rarely supported by DSS, even though interviewees do express a demand for improved support systems. As available support systems currently cannot manage the complexity and uncertainty of field service planning, employees have to rely on tacit and embedded knowledge in their decision-making. However, companies can already envision how knowledge-driven support systems could leverage emerging technologies to better support their decision-making. "Knowing the risk that some event might overturn the schedule—something that certainly could be estimated with big data and statistics—that would simplify the planning process significantly" (Beta). They further envision support systems that recommend actions, automate repetitive tasks, and simplify complex aspects of their work. "Rescheduling takes too much time. If a system could automatically suggest a solution [...], this would be much better than what we can do today" (Zeta).

These findings are especially interesting in the context of digitization and digitalization. While these conceptual terms are often used interchangeably, digitization refers to the transformation of information from an analog to a digital medium meanwhile digitalization refers to the adoption of digital or computer technologies in companies [1]. Digitization drives the standardization of business processes and is an essential cornerstone of efficiency, operational excellence, and predictability in many industries [28]. Digitalization, however, is the adoption of digital technologies to capture value—sometimes on the foundation of digitized processes [14]. Ross [28] outlines that failing to distinguish digitization from digitalization is a fatal mistake for companies. According to Hess [8], today, many organizations are in the process of digitization, but not necessarily in the process of digitalization. The conducted interviews show that maintenance providers have successfully digitized their field service process and use data-, document-, and communication-driven DSS to facilitate the access, transfer, and use of information. However, the study shows that many companies are only beginning their digitalization. Field service planning still relies heavily on tacit knowledge of human dispatchers to capitalize on the available digital information. The available systems do not utilize digital technologies in the decision-making on a higher level yet. The results of the study suggest that digitalization is the key to reducing the reliance on tacit and embedded knowledge in field service planning.

6 Conclusion and Outlook

Advances in information and communication technology (ICT) have fundamentally changed the context in which services are delivered [22]. In the context of field service planning, cyber-physical systems, and improved analytical capabilities have paved the way for the transformation of established processes. Reducing the dependence of processes on tacit and embedded knowledge is a complex challenge—but a valuable opportunity for the digitalization of field service planning. The conducted study provides empirical insights into the way organizational knowledge and decision support systems (DSS) affect field service planning. Notably, the study identifies that maintenance providers are heavily depended on tacit and embedded knowledge. This study shows that the development of knowledge-driven DSS can be a viable strategy for the digitalization of field service planning. This paper provides a consolidated overview of the adoption of DSS across multiple industries, identifies a shared demand for knowledge-driven DSS, and thereby lays the foundation for future research. Leveraging emerging technologies to better manage the use of tacit and embedded knowledge provides an important opportunity to advance field service planning and accelerate the digitalization of maintenance providers.

6.1 Implications

The research has multiple implications: First, managers need to recognize the importance of organizational knowledge. Through the interviews, tacit and embedded knowledge has been identified as a fundamental aspect of field service planning. Second, DSS are often limited to the retrieval and processing of information, which cannot replace the tacit knowledge of field service planners. Consequently, DSS are not able to support human decision-making on a advanced level. However, emerging technologies—such as cyber-physical systems and advanced analytical capabilities—provide novel ways to streamline processes, simplify planning, and reduce the dependency on tacit knowledge. Third, managers need to reassess how digital technologies can enhance their capabilities and create new customer value [28]. Therefore, managers need to investigate DSS designed for their environments instead of purchasing generic DSS.

Fundamentally, reorganizing established work structures requires companies to empower their stakeholders. Therefore, researchers and practitioners need to collectively explore how established processes can leverage emerging technologies and how novel knowledge-driven DSS can be developed. In the context of field service planning, machines can increasingly predict performance problems as well as maintenance demands, and automatically communicate these insights to support systems. Lee et al. [13] define this continuous assessment of the current or past condition of a machine and the automatic reaction to the output of this assessment as "self-awareness". Knowledge-driven DSS can use this information to automatically reorganize schedules to prevent machines from malfunctioning.

6.2 Limitations and Future Research

The research poses limitations that need to be addressed in future research. First, additional maintenance providers (e.g. viticulture machinery) and companies not involved in maintenance but dependent on field service planning (e.g. elderly care) should be included to gain additional insights. Second, even though all cases support the drawn conclusion and indicate a qualitative saturation, no quantitative verification of the results has been pursued yet. Future research should, therefore, be used to confirm the conclusions. Third, while the research has shown the need for knowledge-driven DSS, further research is needed to understand why companies have not yet adopted these systems and how emerging technologies can be leveraged in their design. Accelerating the digitalization of field service planning requires further systematic research. However, the study outlines the wide-spread demand for knowledge-driven DSS and therefore not only lies the foundation for further research, but also guides practitioners in the digital transformation of their field service processes.

References

1. Brennen, J.S., Kreiss, D.: Digitalization. In: The International Encyclopedia of Communication Theory and Philosophy, pp. 1–11 (2016)
2. Brown, J.S., Duguid, P.: Organizing knowledge. Calif. Manag. Rev. **40**(3), 90–111 (1998)
3. Burstein, F., Holsapple, C.: Handbook on Decision Support Systems 1. Springer, Heidelberg (2008). https://www.springer.com/de/book/9783540487128
4. Cordeau, J.F., Laporte, G., Pasin, F., Ropke, S.: Scheduling technicians and tasks in a telecommunications company. J. Sched. **13**(4), 393–409 (2010)
5. Finke, G.R., Hertz, P.: Uncertainties in after-sales field service networks. In: The 2nd International Research Symposium in Service Management, pp. 186–194 (2011)
6. Glaser, B.G., Strauss, A.L.: Discovery of Grounded Theory: Strategies for Qualitative Research. Routledge (2017)
7. Grant, E., Gregory, M.: Tacit knowledge, the life cycle and international manufacturing transfer. Technol. Anal. Strat. Manag. **9**(2), 149–162 (1997)
8. Hess, T.: Digitalisierung. In: Enzyklopädie der Wirtschaftsinformatik. Lehrstuhl für Wirtschaftsinformatik, Universität Potsdam (2016)
9. Hill, A.V.: An experimental comparison of dispatching rules for field service support. Decis. Sci. **23**(1), 235–249 (1992)
10. Hoch, S.J., Schkade, D.A.: A psychological approach to decision support systems. Manag. Sci. **42**(1), 51–64 (1996)

11. Horvath, J.: Working with tacit knowledge. In: The Knowledge Management Yearbook 2000–2001, pp. 34–51 (2000)
12. Howells, J.: Tacit knowledge, innovation and technology transfer. Technol. Anal. Strat. Manag. **8**(2), 91–106 (1996)
13. Lee, J., Kao, H.A., Yang, S.: Service innovation and smart analytics for industry 4.0 and big data environment. Procedia CIRP **16**, 3–8 (2014)
14. Legner, C., et al.: Digitalization: opportunity and challenge for the business and information systems engineering community. Bus. Inf. Syst. Eng. **59**(4), 301–308 (2017)
15. Lele, M.M., Karmarkar, U.S.: Good product support is smart marketing. Harv. Bus. Rev. **61**(6), 124–132 (1983)
16. Makri, C., Neely, A.: Barriers and Facilitators to Incident Reporting in Servitized Manufacturers (2017)
17. Mikell, G.: Automation, Production Systems, and Computer-Integrated Manufacturing. Pearson Education, London (2016)
18. Møller, T.P., Jensen, H.G., Viereck, S., Lippert, F.K., Østergaard, D.: 37 emergency medical dispatchers' perception of barriers in handling emergency calls. A qualitative study. BMJ Open **7**, A15 (2017)
19. Nickel, S., Schröder, M., Steeg, J.: Mid-term and short-term planning support for home health care services. Eur. J. Oper. Res. **219**(3), 574–587 (2012)
20. Nonaka, I.: A dynamic theory of organizational knowledge creation. Organ. Sci. **5**(1), 14–37 (1994)
21. Ostrom, A.L., et al.: Moving forward and making a difference: research priorities for the science of service. J. Serv. Res. **13**(1), 4–36 (2010)
22. Ostrom, A.L., Parasuraman, A., Bowen, D.E., Patrício, L., Voss, C.A.: Service research priorities in a rapidly changing context. J. Serv. Res. **18**(2), 127–159 (2015)
23. Paz, N.M., Leigh, W.: Maintenance scheduling: issues, results and research needs. Int. J. Oper. Prod. Manag. **14**(8), 47–69 (1994)
24. Polanyi, M.: The Tacit Dimension, 1st edn. Garden City, New York (1967)
25. Power, D.J., Sharda, R.: Decision Support Systems, pp. 1539–1548. Springer, Heidelberg (2009)
26. Power, D.J., Sharda, R., Burstein, F.: Decision support systems. Wiley Encycl. Manag. **93**(12), 629–633 (2015)
27. Ritchie, J., Lewis, J., O'Connor, W.: Qualitative Research Practice - A Guide for Social Science Students and Researchers. SAGE, London (2003)
28. Ross, J.: Don't Confuse Digital With Digitization (2017). https://sloanreview.mit.edu/article/dont-confuse-digital-with-digitization/
29. Safaei, N., Banjevic, D., Jardine, A.K.S.: Workforce-constrained maintenance scheduling for military aircraft fleet: a case study. Ann. Oper. Res. **186**(1), 295–316 (2011)
30. Saldaña, J.: The Coding Manual for Qualitative Researchers. Sage Publications Ltd (2009)
31. Saunders, M., Lewis, P., Thornhill, A.: Research Methods For Business Students, vol. 53. Pearson Education Limited, London (2016)
32. Sharda, R., Barr, S.H., McDonnell, J.C.: Decision support system effectiveness: a review and an empirical test. Manag. Sci. **34**(2), 139–159 (1988)
33. Silver, M.S.: Decision support systems: directed and nondirected change. Inf. Syst. Res. **1**(1), 47–70 (1990)

34. Vössing, M.: Towards managing complexity and uncertainty in field service technician planning. In: IEEE 19th Conference on Business Informatics, pp. 312–319 (2017)
35. Wong, B.L.W., Blandford, A.: Analysing ambulance dispatcher decision making: trialing emergent themes analysis. Ergonomics **41**(12–13), 1698–1718 (1998)
36. Yamauchi, Y., Whalen, J., Bobrow, D.G.: Information use of service technicians in difficult cases. In: Proceedings of the SIGCHI Conference on Human Factors in Computing Systems, vol. 5, pp. 81–88 (2003)

Co-creation in Action: An Acid Test of Smart Service Systems Viability

Francesco Polese[1], Luca Carrubbo[2(✉)], Francesco Caputo[3],
and Antonietta Megaro[1]

[1] Department of Business Sciences - Management & Innovation Systems,
University of Salerno, Fisciano, SA, Italy
{fpolese,amegaro}@unisa.it
[2] Department of Medicine, Surgery and Dentistry "Salernitana Medical School",
University of Salerno, Fisciano, SA, Italy
lcarrubbo@unisa.it
[3] Department of Pharmacy, University of Salerno, Fisciano, SA, Italy
fcaputo@unisa.it

Abstract. *Purpose* - The topic of this paper is about Smart Service Systems (SSS), as recent evolution of service systems. The aim is to focus on the connection between structural/system traits of SSS and the value co-creation therein, and how the achievement of a co-creative experience is effectively supportive of the SSS viability.

Design/methodology/approach - This conceptual paper deepens two scientific propositions: (i) Are we able to distinguish between 'structural' and 'system' features of SSS as drivers for value co-creation processes? (ii) Are we able to define effective value co-creation processes the blue-print for SSS viability? The lens used is, in particular, the Viable Systems Approach (VSA).

Findings - Findings confirm a link among traits of SSS and value co-creation processes, regarding (i) multi-part specialized contribution, (ii) unopportunistic behaviours and (iii) co-working and enjoying for results (win-win logic). The smartness of SSS is considered as the basis of their viability.

Research limitations/implications - Limits of this work depends on the theoretical nature of reflections introduced up-to-know. This framework will stimulate future studies by deepening case studies or empirical surveys.

Practical implications - Applying a smart context to modern service 'events', we can enjoy important changes in our life. These assumptions find practical evidence on stage of studies on Service worldwide.

Originality/value - Advances in Literature on Service Science (SS) lead us to better understand the essence, organization and design of SSS. Studies and reflections on System Thinking support observers to highlight static preconditions and dynamic determinants featuring with value co-creation processes.

Keywords: Smart service systems · Value co-creation
Structure/system dichotomy · System viability

© Springer Nature Switzerland AG 2018
G. Satzger et al. (Eds.): IESS 2018, LNBIP 331, pp. 151–164, 2018.
https://doi.org/10.1007/978-3-030-00713-3_12

1 Research Purpose and Scientific Propositions

The topic of Smart Service Systems (SSS), seen in Service Science (SS) studies as the recent evolution of service systems, is interesting to deal with. Worldwide practical use of SSS as Service ground is evident [1], as well as they affect multi-domain emphasis from many theoretical foundations of systems view.

In many field of interest, SSS are able to totally reword all devices definitions, such as smart-phone, smart-grids, smart-box, and so on. We can find SSS in any sector as Healthcare, Tourism, Energy, Education, Retail, Logistics or ICT [2]. The attention paid by every player in markets has increased the relevance of SSS in the economy as a whole. Progresses in technologies (not only in computer science) bridge this evolution showing us what is really going on.

Advances in literature on SS lead us to better understand a number of insights in Service on the essence, organization and design of SSS. In the same way, Systems' studies and System Thinking reflections appear helpful to support observers particularly in defining structural and system traits [3–5], in order to highlight static preconditions and dynamic determinants featuring with value co-creation processes. SSS are scalable, functional, user-friendly, updated and able to learn and upgrade the level of performances if needed. SSS are able to efficiently adapt themselves to external changes by re-configuring the organization structure and modifying actions, which also fits with value co-creation.

This paper aims to focus on the existing connection between structural and system traits of SSS and inner value co-creation, and how reaching a co-creative experience is effectively supportive of the viability of SSS in terms of survival over time (being 'viable').

The two main scientific propositions are there as follow.

(1) RQ1: Are we able to distinguish between 'structural' and 'system' features of SSS (according to the SS Paradigm) as drivers for value co-creation processes?
(2) RQ2: Are we able to define an investigated effective value co-creation processes as the acid test for SSS viability and their ability to survive in the long run?

The scientific lens used for this kind of interpretation is represented by System studies [3–7] and Viable Systems Approach (VSA) in particular [8–13]. In detail, the main contributions from VSA to this study are the Foundational Concepts (FCs) recently outpointed, while historical assumptions borrowed from System studies are system openness, system interactions, system hierarchical organization (sub- and suprasystems), holistic view, cognitive approach, system equifinality, autopoiesis and dynamic balance in system context.

The paper is structured as follows: in Sect. 2 the main characteristics of SSS and their emergence are introduced; the third section deals with the distinction between structural and system traits of SSS. Section 4 copes with the issue of describing static pre-conditions and dynamis determinants of value co-creation in SSS according to the cathegories of Sect. 3; in Sect. 5 the concept of potential and effective value enabled by fitting, engagement and innovation and the Value co-creation as acid test for viability of SSS are presented; the final remarks follow.

Findings confirm a link among traits of SSS and value co-creation processes, especially regarding (i) multi-part specialized contribution, (ii) unopportunistic behaviours and (iii) working together for a common goal (under a win-win logic). All of these evidences foster the smartness of SSS as the basis of their viability and the way to not lose the own identity over the time.

Specifically, SSS typical behaviour calls for engagement to include and involve costumer in collaborative actions able to improve business performances and an increase the perception of value-in-use after any exchange and effective value. Then, in order to meet the customers' expectation, every SSS works to better fit with their needs and wants. Moreover, dynamic interactions between actors operating in the same context (just like a service eco-system as defined in SS) and influencing each other (consciously or not) are exploited to catch a common final goal.

Limits of this work depends on the theoretical nature of reflections introduced up-to-know. This framework will stimulate future lines of research in service by deepening case studies or empirical surveys on stage.

2 All Together Working for a Smarter Planet

In the last decade, through the development of modern interpretations of service systems, we have seen the centrality of customers in the process of service creation and delivery, reinforcing the iterative and cyclical mechanism of contemporary service provision [14]. System studies played an important role, by including studies on resource allocation, advantages of collaboration, alliances and cooperative strategies [15] helpful to manage and conduct operations in service.

In order to stimulate thinking about smart concepts, soon referred to as the Specific, Measurable, Agreed, Realistic and Timely phenomenon [16, 17], SS researchers have investigated every potential evidence of service research 'on stage', referring to something really iterative, interactive, instrumented, interconnected, intelligent [2]. In this direction, a new generation of service systems which are capable of describing and analysing situations occurring and making decisions based on the available data in a predictive or adaptive manner could be defined as Smart [18, 19]. In fact, recent developments in SS Research have led to a new concept of service systems, leading to the definition of SSS.

SSS are then based upon interactions, ties and experiences among actors and facilitators, important parts of SSS [11].

In SSS, service is seen as the final goal, rather than a normal throughput. SSS are a kind of human-centred service system, meaning that knowledge and capabilities are determined by people [20]. Advances in SSS enhanced a shared vocabulary among disciplines, which is one of the main goals for the development of a unified SS [21], crossing different perspectives on SSS. In this sense, the intelligence of SSS derives not from intuition or chance, but from systemic methods of learning, service thinking, rational mode in actions, social responsibility and networked governance [22, 23].

3 The SSS Insights

3.1 Defining 'Structural' Traits of SSS

Organizations are smart if they react to some circumstances and make a rational and efficient use of resources [2]. System studies introduced the notion of equifinality, based on the achievement of different and mutual goals for each entity. All sub-systems (level L−1) of any organization share a common goal, a determined finality pursued by the system (level L) as a whole [24], similar to the way in which SSS are oriented towards enduring performance and satisfaction of all the involved actors. All supra-systems (level L+1) usually are able to influence and shape plans and strategies of organizations [10].

SSS structure concerns with several elements interacting with each other and working together for a common final goal (just like system equifinality calls for). Using the lens of VSA, structural traits of SSS could be clustered into a number of macro-categories focusing on several levels of investigation for different hierarchical organization we want to focus on. VSA FCs [11–13, 25] help us to point out SSS structural traits. VSA scholars stated that every system emerges from a defined structure respecting the strong believes and the capacities of each organization showed in every situation they operate [10]. According to the structure/system dichotomy (FC6), we assume that the set of structural components provides such assistance to user; a business structure, as configuration of elements, counts a number of distinctive resources, information and static relationships. The context analysis highlights consonant relationships among interested actors operating in the same path (FC7) designed for a right complementarity and completeness.

We can intend SSS structure as characterized by a static condition, in which all elements are connected, possibly convergent, following consonant relationships and aimed to catch the same proposition of survival over time, in the long-run.

Structural traits, evident, visible, understandable and measurable, ought to include physical or intangible elements needed for doing well the planned operations; they include all resources and assets available in a stated period of any organization operating like a SSS; but a simple set of static elements is not yet enough to foster real and effective co-creative situations.

In sum, structural traits are fundamental but not sufficient to reach the business final goal and so they seem to be the useful pre-conditions for a business success and, consequently, for any kind of co-creation experience.

Using insights of VSA it is possible to build a sort of thank helpful to cope with, so we can identify the following structural variables frameworking SSS, as showed below.

Distinctive Resources. According to SS, everything that has a name and is useful can be viewed as a *resource*. In SS, systems are essentially dynamic configurations of resources (people, technology, organisations, and shared information) that create and deliver value between the provider and the customer through service provision [21]. All actors are resources, and all service tools are considered useful instruments for business activities [26].

Relevant Information. Any system is an open entity able to catch handy *information* from outside. System thinking assists us to have a key interpretation for the observation of complex phenomena by focusing on the analysis of information among socio-economic entities, in search of news and suggestions [9, 27].

Static Relations. We can consider, as static relationships, internal relationships (among sub-systems), according to a common purpose, shared rules and roles, legitimated government body, all for a potential consonance; and external relationships (among supra-systems), according to the links with all the relevant players active in the surrounding context, as subjectively perceived by any organization.

According to a relational optic [28–31], SS suggests that all actors are considered as dynamic, operant and active, enabling reticular/networked connections [32, 33] oriented to balanced customer centricity [34]. Therefore, activities and entities are not associated to dyadic relations, but always close to many-to-many relationships [35] that seldom can be limited to relationships among main business actors, and have to be considered within a wider set of actors, which include many other involved parts. These relations are consciously determined and finalized to a necessary mutual satisfaction [36] as a function of systems consonance and competitiveness [10].

3.2 Defining 'System' Traits of SSS

Today business scenarios are characterized by globalization, social and political evolutions, technological innovation and other factors that determine a rising hyper-competition. This, in turn, leads to complexity. These conditions, examined by system theories, imply the inability of organizations to act in an unstable context, in which rules are not 'a priori' defined and the risk is high. To tackle with this uncertainty, firms should broaden their boundaries and establish relationships with other entities operating in the same context [37, 38].

The evolving and increasing role of interconnections, enablers, measures, standards and procedures represent the theoretical evidence of evolution of SSS. Intelligent utility networks and metering, intelligent transportation, consumer driven supply chains, intelligent oilfields, manufacturing productivity, instead, are the practical applications [11]. SSS are able to foster connections and interactions between the various actors involved in the process of exchange, following several channels of communication between businesses, consumers, and stakeholders [26]. In this light, SSS combine advances in IT tools with the evolution in thinking about system dynamic interactions, adaptive skills, sustainable development, enhanced learning, reconfiguration capacities and service innovation [39] in complex situations [20].

VSA FCs can help us to point out SSS system traits. VSA scholars stated that physical and not physical components of any structure might relate potentially in many different ways. The 'other' part of each organization is the system emerging by the activation of relations (i.e. interactions), the dynamic resource integration and the diffusion of existing information.

System *viability* (FC8) faces with the ability of any organization to sustain its own activity over time effectively. System *adaptation* (FC9) deals with a steady alignment among critical available resources and mode in actions, able to make the offer fitting with users' expectations and needs over time. This leads to an adaptive impulse of workers in search of positive interactions with each other.

The stronger are mutual interest, shared values and schemes, the higher will be the chances to identify the determinants of the exchange. The decision maker of SSS, according to System studies, is the strategic guide and coordinator of the actions taken by the organization, in order to realize its vision. The decision maker has the mission of continuously analysing the evolving conditions of the specific context (as subjectively perceived and interpreted), trying, as far as possible, to prevent and avoid negative contingencies. By means of change and adaptation, SSS decision-making passes through a continuous learning process of re-organization of the owned knowledge [40].

SSS appear today as resource integrators, socially constructed and knowledge-based. System traits of SSS aligned with the vision all business organization want to rely on. Of course among actors, customers play a key role, since they demand a personalised product/service, high-speed reactions, and high levels of service quality; despite customer relevance, SSS have to deal to every other actor's behaviour, who's expectations, needs and actions directly affect system's development and future configurations [11].

In sum, under the VSA traits of SSS affect the effective participation of actors in the co-operation, and then seem to be the real determinants of value co-creation.

Using insights of VSA it is possible to build a sort of thank helpful to cope with, so we can identify the following system variables frameworking SSS, as showed below:

Resource Integration. In SSS, all actors with the correct attitude, and thanks to their shared purposes, are motivated and most likely willing to develop harmonic interactions offering and integrating resources needed for 'viable' service exchanges. Examples of resource integration are: mode in actions (useful experiences, competitive advantages, re-plans); doing specialization (education, training on the job, learning by doing and failing); products and processes enhancing (innovative process, increased knowledge, aim to change); service approach emergence (adaptive behaviour, scalable re-configuration, fitting customer needs, complexity management).

In our opinion, service research, and specifically SS, offers intriguing insights on the issue because of the underlying theoretical framework based on resource integration.

Information Sharing and Decision Making. VSA stated that, in presence of information variety, categorical values are the strongest and deepest believes responsible for the acceptance or refusal of change that support the application of specific Interpretation Schemes.

In SSS is required to achieve own main objectives, in doing this IT obviously plays the major role. Such IT-based SSS are understood as specifically designed for the prudent management of their assets and goals while being capable of self-reconfiguration to ensure that they continue to have the capacity to satisfy all stakeholders over time [11].

Dynamic Interactions. The interactions and tie among actors represent an important aspect of any SSS. These may be internal interactions (empathy, supportive actions, problem solving, brain storming); and external interactions (networking, mutual interests, co-creative processes).

SSS are socially constructed collections of service events in which participants exchange benefits through a knowledge-based strategy that captures value from a provider-client relationship.

4 Looking for Static Pre-conditions and Dynamic Determinants Featuring Value Co-creation in SSS

4.1 The Static Side of Value Co-creation in SSS

In today economy, any supply-chain could be re-conceptualized as SSS, and for this reason being changeable, adaptable and evolving in relation to changing contextual conditions [41], as context is subjectively felt [9, 10]. The multi-part contributions of knowledge, the application of skills, the ability to configure and re-configure, and the desire to maintain long-terms relationships, all represent the elements of a systemic way of being adaptive in SSS and link to several leverages of value generation, decision making and sustainability.

This concerns RQ1; in particular, the process of value co-creation deserves to be more investigated under a system view because it matches with many assumptions in literature worldwide. Nevertheless, value co-creation [31, 42, 43] is difficult to measure and predict in SSS for its emergent style of going on. As discussed, the VSA structure/system dichotomy implies the identification of both a structural dimension—which is static and considers the parts and the relationships that exist among them—and a systemic dimension, which is dynamic and concerned with the identification of the interactions, while taking into account the structural components. In this system perspective, problem solving issues are usually linked to routines and known paths of resolution, while the decision-making is more linked to emergency and strategic thinking and emphasizes the fundamental role of the governing body/decision maker. This is fundamental for value co-creation processes in SSS and the VSA lens can help in defining the structural traits, as listed before, which represent static pre-conditions for co-creative processes, as detailed in what follows.

Distinctive Resources for Co-creation. Resource setting strongly affects the context in which service is experienced. The socio-economic context may contribute to facilitating or hindering system conditions [44] that lead to the development of the users' potential (e.g., awareness of their rights and expectations, self-determination and sense of responsibility) and, consequently, facilitate or hinder the integration of resources as an opportunity to further promote processes of value co-creation. As highlighted in Service mainstream, the user through participation may broaden and transform his/her resources into skills that can be positively activated while resource integration takes place and ensure that the potential compatibility among the actors who operate in the same context [25].

Relevant Information for Co-creation. Any SSS is considered as an open system that establishes relationships (for exchanging flows of energy, resources, commodities and information), not only with the sub-systems, which contains and manages, but with supra-systems also in which it is included [45, 46]. As the world is becoming smarter [2], SSS must be people-centric, information-driven, and e-oriented to adapt and mutually satisfy any participant involved within the same service eco-system, while the community should encourage and cultivate people to collaborate and innovate [47]. With respect to the attitude to respond to specific needs of the market or to create new ones, the skill to understand the needs (by rightly interpreting critical information) is an intermediate solution that is most likely effective in terms of sustainability of SSS. These are basic elements for a co-creative interface.

Static Relations for Co-creation. SSS are temporally and spatially defined. In an attempt to point out that most of the qualification of a specific actor operating therein, it is necessary to dwell on the relationship that binds to other actors, with which he/she shares the ultimate goal, some resources and information. Only deepening the role and the relevance of these relations, without dwelling too much on how the actor is concerned, you can try to reinforce the logic of win-win cooperation. The sense of responsibility of actors and the intensity of their participation determine the development of positive commitment and a spontaneous participation to the generation of value [12] and definitively allow a co-creative situation. From this point of view, it does not matter to qualify the operators involved: the distinction between supplier, customer or user becomes irrelevant, defining all relations as Actors-to-Actors (A2A) [25].

4.2   The Dynamic Side of Value Co-creation in SSS

In the same way, VSA lens can be the tool-kit by which we are able to deepen also the system aspects of SSS dealing with value co-creation. The system traits, as listed before, could represent dynamic determinants for co-creative experience, as detailed in the following.

Resource Integration for Co-creation. In markets, customers are crucial for enriching the product and are therefore essential. However, customers are not isolated, and the provider-customer relationship is not only bilateral [48] the result is the co-creation of value pursued through resource integration among actors of the exchange. Thus, no single actor or provider can provide a complete co-creative experience. However, consumers do not obtain value directly from the product itself, but from its use, processing or consumption and by comparing it with other entities interested in the building process [49]. The value derives from both the benefit of underlying service and the process of co-production, co-design, and co-marketing, involving multiple contributions, thanks to the sharing of information, resources, skills, needs and risks.

Information Sharing and Decision Making for Co-creation. We should note that decisions are influenced a lot by information, highlighting the role of big data, elaboration of new contents, and intense analysis of feedback; nowadays, managers (in public or private sectors) make many informed decisions, more than in the past. The variety and variability of information about possible connections between SSS promote new forms of cooperation, interpreted as interactions between cognitively aligned actors [40]. Big Data supports both smart decisions in SSS organizations [50], and SSS governance related to the definition and the pursuing of a formal big data strategy, warranting data quality, leveraging information and maintaining its value during the process of its co-creation.

Dynamic Interactions for Co-creation. Value is co-determined by providers and users at the time of purchasing, through a personal 'consumption' process favoured by constant interaction with other parts of the SSS in which users operate [51, 52]. The value co-creation logic is defined in a win-win sense, considering the interaction between different entities and the desire to gain a collective mutual satisfaction. The win-win interactions are developed only through the promotion and maintenance of active relationships with stakeholders or through a common wants to encourage the process of co-creation by fostering not-opportunistic behaviour, long-term relationships and shared values.

5   Discussion

5.1   Potential and Effective Value Through Fitting, Engagement and Innovation

The value of products is perceived by the customer on the base of value-in-use (through the previously defined consumption process). Value, intended as positive effect of the exchange, is then an improvement of the SSS; generally, value creation takes place when a potential resource becomes an effective specific benefit. Specifically, value co-creation follows a dynamic flow while it considers interactions among different SSS possessing critical resources and the desire to reach collective mutual satisfaction, in which the active contribution is multiple, the integration is the highest, and complementarity is fundamental.

In theory, since 'harmony' between actors can be understood as a fusion of listening skills, consideration, dialogue, recognition and respect in intra- and inter-systemic relationships, we can verify how system consonance/resonance qualify competitiveness in business [53] and then how SSS viability is encouraged by value co-creation. This concerns RQ2, as synthesis of all remarks showed above.

Value co-creation implies the active multi-actor contribution, by all the protagonists of the exchange in a particular offer (up to that time only "potential") and the concept of value certainly follows the logic of end-user's "effective" perception and therefore subjective [22].

Direct consequences in practice are related to fitting, engagement and innovation processes, as detailed below.

Fitting. In the generation of value, all actors become real co-creators of value, and consequently, SSS are observed only as integrators (and managers) of resources needed for co-creation exchanges. To follow this, every SSS has to plan, conduct and audit a lot of operations, affecting many elements of their structure; this leads to new governance needs, iteratively improving the ability to react, manage and act to the external contingencies [54, 55]. Fitting, as an adaptive set of actions, can converge on different level of the SSS structure, with different level of depth; it depends on a combination of factors, regarding the strategies of the system (decisional area) and the constraints coming from the outside. Fitting helps to cover the cited distance existing between demand and supply in a stated moment (t0) and consists of the actions made by the Supply side to interpret and manage the needs of the Demand side, modifying something in the initial value proposition. This upgrade produces even new levels (t0 + 1) in production/provision [22, 56].

Engagement. The need to expand the traditional boundaries of the enterprise, to include all the different actors relating with, and use an approach to value subjectively defined and context-specific. In this sense, the 'involvement' follows the service-centred view of exchange [57] and indicates participation in co-production [51]. Instead, the term 'engagement' is very often used by SS scholars to indicate the active, equal and reciprocal participation of users and providers in the co-creation of value [58].

Innovation. At the same time, the opportunity to explore the processes of value co-creation in the context of service eco-system, in which everything is collected, identified, and active, organizations can survive in a particular context only if they improve their ability to evolve and make their operations corresponding to external changes [52, 59]. SSS develop their knowledge and competences needed to compete through innovation processes; innovation indeed may result as an experimental process during which continuous learning obtainable by doing, by using, by failing, by interacting is fostered.

5.2 Value Co-creation as Acid Test for Viability of SSS

In last years, due to the resources crisis in each field, it discovered the necessity of resources optimization, which has slowly led to the idea of a Smarter Planet, now considered as more than a slogan. Indeed, in doing business many problems and mistakes can occur. Organizations employ resources to exchange service with the risk that these resources are dispersed and/or badly used with the consequence that the needs of who has required those services are not fully satisfied, and making the system inefficient. As a consequence, it is necessary to develop organizations as dynamic entities capable of adaptively evolving in order to reduce the mismatch of the service exchange process, thus making it an interactive process, which involves both the provider and the client in the design, creation, application and delivery of the service.

Due to the fact that the SS paradigm is today used to model each economic, managerial, organizational, industrial or computer system, the idea of a Smarter Planet can be easily pursued by focusing on the design of SSS, that evolves by adapting to the changing conditions and thus reducing the mismatch and loss of resources [19]. As discussed, the governance of SSS should direct the system towards a final goal, transforming static structural relationships into dynamic interactions with other entities.

According to the logic of the VSA, SSS are capable of simultaneously optimizing the use of resources and improving the quality of the provided service. Viability appears the final result of all service operations taking place, by merging many contributions coming from all system parts. So this, every actors make an effort to increase their own skills and abilities strongly, in order to allow a viable service exchange.

Viability represents an essential and necessary prerequisite to operate in the contemporary context, and smartness is a trait of proactive behaviour; then viable organizations (or viable systems, VS) are cybernetic, cognitive, autopoietic, aligned. If VS are always smart, not all SSS are viable [60]. Between smartness and viability there is a hierarchical relationship in which smartness is a necessary but not sufficient condition for viability.

In this sense, the smartness strictly connects to the system viability.

Furthermore, SSS have to take into account not only which the wants are today, but also their evolution; supplies follow the changing demands in order to obtain a dynamic equilibrium. The active participation previously analysed highlights different levels of effort/intensity of interaction by the customer in the process of the co-creation of value, but also a different role in the managing of interactions. SSS are value co-creators because they integrate resources and enable connections among actors, strategically involved in the value generation process. Smartness calls for the ability to make offered solutions scalable and functional, and then to facilitate co-creation itself in terms of wide, motivated and focused participation. This lets to improve the quality of results and the value perceived by users finally increasing chances of surviving for all SSS able to do this.

Value co-creation is the main leverage for competitiveness today and for the possibility to survive over time in markets. Value co-creation is the evidence of SSS capacity to maintain their own market-share during the time and then to foster their presence, by adapting to external changes that casually can occur.

Value co-creation is a sort of acid test or acid test of SSS viability as it is frameworked by VSA and system studies.

6 Final Remarks

This work aimed to deepen some aspects of SSS (as intended by SS scholars and practitioners) in terms of system and structural traits in order to answer to research questions (RQ1 and RQ2) and specifically understand what the drivers for value co-creation processes are, as clear demonstration (acid test) of viability as ability of surviving over time, by pointing out pre-conditions and determinants of the co-generation of value. The lens used was a mix of the System Thinking assumptions in general and the FCs of VSA.

Findings bridge toward the co-creation in SSS as leverage for their survive in the long run and then for their viability, regarding especially the Engagement of Actors, the Fitting with customer needs and the Innovation as direct spillovers, all matching with win-win logic, multi-part contribution and effective perceived value. Organizations seen as SSS ought to be competitive over time maintaining a great appeal of their own value proposition by involving users in production/provision process; moreover, the capacity to react to external changes by adapting their behaviour, without losing vision nor identity, makes smart any organization. The habitude to 'read' the context around facilitates the interactions with all the other organizations operating within it; the ability to defend the reached market-share leads to survive in the long run. The mode in action enabling SSS to survive over time fosters the viability of SSS themselves besides.

In sum, there is a great link between the value-co-creation process and viability in SSS. Nevertheless, this is a starting point; in the next future we want to continue this work by making an empirical survey on stage and defining all variables helpful to the scope.

References

1. Spohrer, J., Anderson, L., Pass, N., Ager, T.: Service science and service dominant logic. Otago Forum **2**, 4–18 (2008)
2. Maglio, P.P., Baumol, W.J.: Handbook of Service Science. Springer, Berlin (2010). https://doi.org/10.1007/978-1-4419-1628-0
3. Von Bertalanffy, L.: General System Theory: Foundations, Development Applications. George Braziller, New york (1968)
4. Beer, S.: Brain of the Firm. Penguin Press, London (1972)
5. Von Foerster, H.: Observing Systems. Intersystems Publication, Seaside (1981)
6. Checkland, P.: Systems Thinking, Systems Practice. Wiley, Chinchester (1981)
7. Espejo, R., Harnden, R.J.: The Viable System Model. Wiley, London (1989)
8. Golinelli, G.M.: L'approccio sistemico al governo dell'impresa. L'impresa sistema vitale, vol. I, 3rd edn. Cedam, Padova (2005)
9. Barile, S. (ed.): L'impresa come sistema. Contributi sull'approccio sistemico vitale, 2nd edn. Giappichelli, Torino (2008)
10. Barile, S.: Management sistemico vitale. Giappichelli, Torino (2009)
11. Barile, S., Polese, F.: Smart service systems and viable service systems. Serv. Sci. **2**(1/2), 21–40 (2010)
12. Barile, S., Pels, J., Polese, F., Saviano, M.: An introduction to the viable systems approach and its contribution to marketing. J. Bus. Mark. Manag. **5**(2), 54–78 (2012)
13. Barile, S., Polese, F., Saviano, M., Pels, J., Carrubbo, L.: The contribution of VSA and SDL perspectives to strategic thinking in emerging economies. Manag. Serv. Quality **24**(6), 565–591 (2014)
14. Maglio, P.P., Srinivasan, S., Kreulen, J.T., Spohrer, J.: Service systems, service scientists, SSME, and innovation. Commun. ACM **49**(7), 81–85 (2006)
15. Alter, S.: Service system fundamentals: work system, value chain, and life cycle. IBM Syst. J. **47**(1), 71–85 (2008)
16. Maglio, P.P., Spohrer, J. (eds.): Fundamentals of service science. J. Acad. Mark. Sci. **36**(1), 18–20 (2008)
17. Maglio, P.P., Spohrer, J. (eds.): Special issue on service science, management, and engineering. IBM Syst. J. **47**(1) (2008)
18. Demirkan, H., Spohrer, J., Krishna, V. (eds.): Service Systems Implementation. Springer, New York (2011). https://doi.org/10.1007/978-1-4419-7904-9
19. Demirkan, H., Spohrer, J., Krishna, V. (eds.): The Science of Service Systems. Springer, New York (2011). https://doi.org/10.1007/978-1-4419-8270-4
20. Maglio, P.P., Vargo, S.L., Caswell, N., Spohrer, J.: The service system is the basic abstraction of service science. Inf. Syst. e-Bus. Manag. **7**(4), 395–406 (2009)
21. Spohrer, J., Maglio, P.P., Bailey, J., Gruhl, D.: Steps toward a science of service systems. Computer **40**(1), 71–77 (2007)
22. Barile, S., Carrubbo, L., Iandolo, F., Caputo, F.: From 'ego' to 'eco' in B2B relationships. J. Bus. Mark. Manag. **6**(4), 228–253 (2013)

23. Napoletano, P., Carrubbo, L.: Becoming smarter: towards a new generation of service systems. Impresa Ambiente Manag. **4**(3), 415–438 (2010)
24. Parsons, T.: The System of Modern Societies. Prentice-Hall, Englewood Cliffs (1971)
25. Wieland, H., Polese, F., Vargo, S., Lusch, R.: Toward a service (eco)systems perspective on value creation. Int. J. Serv. Sci. Manag. Eng. Technol. **3**(3), 12–25 (2012)
26. Mele, C., Polese, F.: Key dimensions of service systems: interaction in social & technological networks to foster value co-creation. In: Demirkan, H., Spohrer, J., Krishna, V. (eds.) The Science of Service Systems, pp. 37–59. Springer, New York (2011). https://doi.org/10.1007/978-1-4419-8270-4
27. Ng, I.C.L., Badinelli, R., Polese, F., Lobler, H., Halliday, S., Di Nauta, P.: S-D logic research directions and opportunities: the perspective of systems, complexity and engineering. Mark. Theory **12**(2), 213–217 (2012)
28. Hakansson, H., Snehota, I.: Developing Relationship in Business Network. Routledge, London (1995)
29. Nahapiet, J., Ghoshal, S.: Social capital, intellectual capital and the creation of value in firms. Acad. Manag. Rev. **23**(2), 242–266 (1997)
30. Barabasi, A.L.: Linked: The New Science of Networks. Perseus, Cambridge (2002)
31. Prahalad, C.K., Ramanswamy, V.: The Future of Competition: Co-creating Unique Value with Customers. Harvard University Press, Cambridge (2004)
32. Lovelock, C., Gummesson, E.: Whiter services marketing? In search of a new paradigm and fresh perspectives. J. Serv. Res. **7**, 20–41 (2004)
33. Achrol, R., Kotler, P.: The service-dominant logic for marketing a critique. In: Lusch, R.F., Vargo, S.L. (eds.) The Service-Dominant Logic of Marketing: Dialog, Debate, and Directions, pp. 333–343. ME Sharpe, Armonk (2006)
34. Gummesson, E.: Total Relationship Marketing, 3rd edn. Butterworth-Heinemann/Elsevier, Oxford (2008)
35. Gummesson, E.: Extending the new dominant logic: from customer centricity to balanced centricity. J. Acad. Mark. Sci. **36**(1), 15–17 (2008)
36. Lusch, R.F., Vargo, S.L., O'brien, M.: Competing through service: insights from service-dominant logic. J. Retail. **83**, 5–18 (2007)
37. Maturana, H.R., Varela, F.J.: Autopoietic systems. BLC report 9, University of Illinois (1975)
38. Lazlo, E.: The Systems View of the World: A Holistic Vision for Our Time. Hampton Press, New York (1996)
39. IFM and IBM: Succeeding Through Service Innovation: A Service Perspective for Education, Research, Business and Government. University of Cambridge Institute for Manufacturing, Cambridge (2008)
40. Barile, S., Saviano, M., Polese, F., Di Nauta, P.: Reflections on service systems boundaries: a viable systems perspective: the case of the London borough of Suton. Eur. J. Manag. **30**, 451–465 (2012)
41. Carrubbo, L.: La co-creazione di valore nelle destinazioni turistiche. In: Rirea, C. (ed.) Opera Prima, no. 2 (2013)
42. Ballantyne, D., Varey, R.J.: Creating value-in-use through marketing interaction: the exchange logic of relating, communicating and knowing. Mark. Theory **3**, 335–348 (2006)
43. Grönroos, C.: Adopting a service business logic in relational business-to-business marketing: value creation, interaction and joint value co-creation. Proc. Otago Forum **2**, 269–287 (2008)
44. Chandler, J.D., Vargo, S.L.: Contextualization and value-in-context: how context frames exchange. Mark. Theory **11**(1), 35–49 (2011)
45. Hall, A.D., Fagen, R.E.: Definition of system. In: General Systems (Yearbook of the Society for the Advancement of General Systems Theory), vol. 1, pp. 18–28 (1956)

46. Hannan, M.T., Freeman, J.: The population ecology of organizations. Am. J. Sociol. **82**(5), 929–964 (1977)
47. Qiu, R.G., Fang, Z., Shen, H., Yu, M. (eds.): Towards service science, engineering and practice. Int. J. Serv. Oper. Inform. **2**(2), 103–113 (2007)
48. Gummesson, E., Polese, F.: B2B is not an island! J. Bus. Ind. Mark. **24**(5/6), 337–350 (2009)
49. Katzan, H.: Foundations of service science concepts and facilities. J. Serv. Sci. **1**(1), 1–22 (2008)
50. Wang, H., Xu, Z., Fujita, H., Liu, S.: towards felicitous decision making: an overview on challenges and trends of big data. Inf. Sci. **367–368**, 747–765 (2016)
51. Vargo, S.L., Lusch, R.F.: Service-dominant logic: continuing the evolution. J. Acad. Mark. Sci. **36**(1), 1–10 (2008)
52. Vargo, S.L., Lusch, R.F.: Institutions and axioms: an extension and update of service-dominant logic. J. Acad. Mark. Sci. **44**, 5–23 (2016)
53. Polese, F., Carrubbo, L., Russo, G.: Managing business relationships: between service culture and a viable system approach. Esperienze d'Impresa, no. 2 (2009)
54. Pels, J., Polese, F.: Configurational fit: pathway for successful value co-creation. In: Impresa, Ambiente, Management, Annoiv, no. 3, pp. 355–373 (2010)
55. Mele, C., Pels, J., Polese, F.: A brief review of systems theories and their managerial applications. Serv. Sci. **2**(1/2), 126–135 (2010)
56. Carrubbo, L., Iandolo, F., Pitardi, V., Calabrese, M.: The viable decision maker for CAS survival: how to change and adapt through fitting process. J. Serv. Theory Pract. **27**, 1006–1023 (2016)
57. Vargo, S.L., Lusch, R.F.: Envolving to a new dominant logic for marketing. J. Mark. **68**, 1–17 (2004)
58. Polese, F., Carrubbo, L.: Eco-sistemi di servizio in sanità (2016)
59. Polese, F., Carrubbo, L.: The service dominant logic ed una sua interpretazione al fenomeno turistico. In: Impresa, Ambiente, Management, vol. Ii, no. 1 (2008)
60. Badinelli, R., Barile, S., Ng, I., Polese, F., Saviano, M., Di Nauta, P.: Viable service systems and decision making in service management. J. Serv. Manag. **23**(4), 498–526 (2012)

Towards Enabling Cyber-Physical Systems in Brownfield Environments
Leveraging Environmental Information to Derive Virtual Representations of Unconnected Assets

Sebastian R. Bader[1(✉)], Clemens Wolff[2], Michael Vössing[2],
and Jan-Peter Schmidt[2]

[1] Fraunhofer Institute IAIS, Schloss Birlinghoven, 53757 Sankt Augustin, Germany
`sebastian.bader@iais.fraunhofer.de`
[2] Karlsruhe Institute of Technology, Kaiserstraße 89, 76131 Karlsruhe, Germany
{`clemens.wolff,michael.voessing,jan-peter.schmidt`}`@kit.edu`

Abstract. The digital transformation based on internet technologies comprises huge potentials but also challenges for the production industry. Even though some design characteristics are generally accepted for the digitized integration of machines, applications and surrounding components the inherent complexity and variety of interaction protocols, data formats and interdependencies of existing deployments in so called brownfield environments hampers the data-driven manufacturing of the future.

We propose an iterative approach where existing context data is used to encapsulate the specific complexity of each resource in order to create a flexible integration layer. Nearly all relevant resources are modeled as self-descriptive cyber-physical systems or Virtual Representations according to the setting of the physical production environment, therefore drastically reducing the required access barriers. We present a reference implementation and discuss its business implications by the example of industrial maintenance.

Keywords: Brownfield deployments · Industrial internet
Distributed systems

1 Introduction

Within the industrial sector, manufacturers of industrial machines have realized the potential of offering services [4]. Today, driven by the process of Servitization, many manufacturers offer complementary services to their products, which are usually referred to as *Product-Service-Systems* [3,16]. Even more, some authors like Vargo and Lusch argue that services are the main reason for purchasing products at all [24]. The trend of Servitization is further complemented by the digital transformation in the manufacturing sector. Currently, the Internet of Things (IoT) and cyber-physical systems (CPS) are new paradigms how involved

© Springer Nature Switzerland AG 2018
G. Satzger et al. (Eds.): IESS 2018, LNBIP 331, pp. 165–176, 2018.
https://doi.org/10.1007/978-3-030-00713-3_13

machines, their components, and any other related actor – even related services – communicate with each other in an industrial internet. Indeed, the development of the IoT and CPS are seen as core enablers for new smart services, and thus, the further development of Servitization.

Inspired by the great success of the Web – and the internet as its underlying infrastructure – its basic design principles are now generally accepted. Furthermore, one can note a common agreement that future communication patterns for an industrial internet [15] or a platform for industrial data exchange [6] in the manufacturing industry need to follow these established and mature techniques. In contrast, existing setups of production lines are usually grown environments driven by short-term requirements rather than a long-term design and compliance to internet standards. Therefore, strategies transforming mostly hard-wired communication lines towards a flexible, internet-inspired state requires in-depth analysis.

Basically, two approaches need to be separated. First, a so called *greenfield deployment* starts at the design of a production site without any preliminary conditions by already deployed plants. Any communication pattern can be optimally designed and protocols, interfaces and interaction methods aligned according to the current best practices.

Nevertheless, in the majority of cases, the production site is already in place and running. Neither can potentially inappropriate devices be replaced nor the production process be interrupted for a significant amount of time. The transformation process therefore needs to be aligned to the same extend with the given setup as well as the desired architecture. At the same time, a successful transformation requires an iterative approach as established processes must stay fully operable until the successors are in place their capability is sufficiently tested. Such a scenario is called a *brownfield deployment* where established parts build the basis and previous actions affect the conditions for new developments.

Our contribution to the topic is an iterative deployment technique within a brownfield approach to gain an internet-powered integration architecture. The lack of suitable hardware for observations is tackled by flexible data integration techniques making use of context information and related data sources in order to create a detailed, consistent digital representation of the physical shop floor in the form of *Virtual Representations*. We outline methods how this virtual shop floor enables the deployment of analysis at runtime which have not been foreseen at design time and can be adjusted as necessary.

The remainder of this work is structured as follows: In Sect. 2, we elaborate on fundamentals and related work. In Sect. 3, we introduce a use-case frequently referred to in this work. In Sect. 4, we present a technology stack relying on wide-spread and well supported internet and web technologies in order to enable communication within brownfield approaches. The core of this work, the proposed framework, is presented in Sect. 5. In Sect. 6, business cases enabled by the presented framework are highlighted. Finally, in Sect. 7, this work is concluded.

2   Fundamentals and Related Work

The recent developments in miniaturizing electronics and the enhancement in (wireless) network technologies paved the way for equipping more and more *things* like machines, components, and even sensors with internet connectors. Gubbi et al. [10], similarly to others, define the IoT as the combination of large scale sensing or actuating capability of devices with sharing information based on internet standards. Well-established technologies like URIs for identification and TCP/IP for data exchange lay the foundation for higher level integration. Their main benefit is the loose coupling of clients and servers, allowing high scalability in distributed networks.

Cyber-physical systems (CBS) as e.g. discussed by Lee et al. [13] combine the connected device with a virtual dimension. The enhancement of the physical IoT resource with software-based counterpart enables additional information of the resource and its characteristics and features. Both the IoT and CBS mainly focus on connected objects. That means physical components or devices equipped with e.g. Ethernet or Wi-Fi connectors to send or receive digital messages. In contrast, a regular production line contains of a high number of not connected or even not connectable objects that cannot be equipped with internet-capable devices with reasonable investments. Information on these components is at least as important than the one represented through cyber-physical systems but not yet regarded by the paradigm.

The connected, data-driven manufacturing is often referred to as the industrial internet or industry 4.0 (after steam powered manufacturing, mass production, and digitalization). All relevant players are continuously exchanging information on the state of regarded products, production units and materials over the whole supply chain and product lifecycle [21]. IoT devices and cyber-physical systems form global networks and flexibilize the production. International organizations like the Industrial Internet Consortium or the Platform Industrie 4.0 drive the development of standards to enable the seamless integration of machines, software applications and products. The target is to reach a secure but at the same time flexible integration of any kind of production related unit based on the internet. Main advantages are the reduction of applied protocols, formats and interaction patterns to simplify the digital information exchange and to support a plug-and-play like deployment. This will not only allow faster adjustments to existing production processes but also to apply analysis driven by existing information and not hampered by previously designed interfaces, data silos or interaction patterns. Yet, the current specifications are still high level proposals how a connected production shall be implemented. Commonly agreed technology stacks and transaction formats are still missing, therefore a seamless connection is yet not possible.

Smart Manufacturing [8] comprises efforts to establish a reference architecture with nodes representing physical components in the manufacturing facility to ease the integration and create a generic platform. The promoted modularized approaches model virtual resources similar to their physical counterpart in order to enable rapid deployments and portability. But even though they outline a reference architecture, the targeted integration aspect is still unclear and directly implementable specifications are missing. Hedengren and Eaton [11] further discuss time based mathematical simulation and optimization on highly dynamic measurements. They discuss various types of update frequencies and how to derive predictions. All of the discussed models require decent preprocessed and, most of all, accordingly synchronized input data. Especially in brownfield scenarios, such a state is a major accomplishment and not a prerequisite.

Data Lakes, as e.g. discussed in [20] or [23], are one concept to make data from heterogeneous sources and in different formats accessible. Established technologies like Apache Hadoop provide solutions for NoSQL clusters and enable queries also on dynamic data without a fixed schema. The Data Lake concept is only partly scalable in terms of the underlying cluster but also forms a single point of failure and potentially another data silo with tight coupling which will hamper the data usage in future cases. The not required data format enables the simple data storage but makes an effective data integration without previous knowledge on each data object a challenging task.

The Industrial Data Space [17] provides solutions on how to exchange data between organizations with the focus on data security and sovereignty. While specifying the connectors, gateways and architectures it doesn't provide features to connect the data to the physical world. Information on unconnected resources are only implicitly given in data flows but not explicitly described. Interpretation of the data therefore still requires deep insights in the actual production setting and its dependencies.

3   Use Case

We illustrate the basic concept of our integration approach by presenting how the components of an industrial metal saw can be brought to a Virtual Representation. A Virtual Representation [2] is a digital resource which acts on behalf of an unconnected component or device. It represents its current state by providing descriptions in the semantically defined Resource Description Format (RDF) and serves as a generic container to provide all known information on the otherwise not describable object. In particular, implicit knowledge on e.g. local processes can be made explicit by delivering and executing according algorithms whenever the state of the Virtual Representation is requested.

One goal of an analysis might be a cost calculation of a current cut on a sawing machine or an estimation of its current abrasion state. Unfortunately, the high speed of the blade itself make it impossible to directly observe the abrasion without stopping the whole sawing unit. On the other hand, information from different components of the machine, e.g. cutting parameters and energy consumption of the engines, is available. To grab such data, we are using the Virtual Representation integrates the context information from cyber-physical resources and other internet-accessible sources and computes its state on the fly. We show how Virtual Representations can encapsulate all necessary information at the level of the internet-based integration layer and thereby encapsulate required knowledge at the resource itself, even though the physical object has no direct connection to the internet itself.

Our prototype is divided into two sub projects. The core project contains the Virtual Representations and a resource manager as a cloud server hosting them[1]. Additionally, we provide a web project that serves as a UI[2] for different communication protocols for a simple communication testing with the server of the core project.

4   Iterative Brownfield Deployment

Several architecture approaches are possible in order to digitally connect production machines with an organization's control systems. In the most basic scenario, a document-based information exchange (e.g. relying on proprietary formats, emails or even office documents) transfers jobs either directly to a customized interface. Pulling and polling, varying protocols and data formats, differing data syntax and identifiers hamper a direct data exchange. In our scenario, a script-based application could periodically query the data from some databases and send an email to a certain account. Any change in the setup, the used databases or new information on the abrasion process would then require a manual adjustment of the script, a text-based interpretation of the received email and sufficient information on the requirements of the downstream applications.

Moreover, the connection of machines from different manufacturers requires a deep understanding of each deployed device, its characteristics and limitations and the design, creation and maintenance of highly customized interfaces and control software. In a point-to-point wiring approach like the one mentioned, any necessary change in either its features or at the composition of the production line (like introducing a new unit or replacing an outdated one) leads to mandatory and very complex adjustments at any related device and application. The thereby created organic growing results in inconsistent data models, varying interaction patterns and applied protocols is hardly maintainable. Especially small and medium companies do not have the possibilities to employ the according staff to cope with the thereby created consequences.

Hybrid approaches relying on combinations of field bus architectures (like e.g. the CAN bus) and Ethernet connections combine polling information from a data producer with pulling data to a consuming system. Installing both paradigms is mostly the case when one subset of machines is initially designed for bus communication whereas other systems require a request-response pattern. The resulting environment makes it even harder to understand the existing dependencies and data flows.

On the other hand, industry 4.0 promises consistent, well-structured architectures with plug-and-play like deployment patterns. Even though a majority of

---

[1] https://github.com/sebbader/VirtualRepresentationFramework/tree/master/VirtualRepresentationCore.

[2] https://github.com/sebbader/VirtualRepresentationFramework/tree/master/VirtualRepresentationWeb.

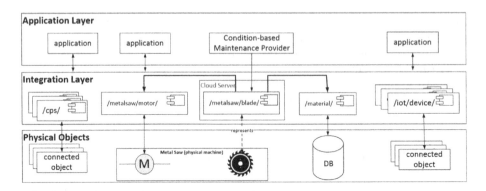

Fig. 1. The integration layer contains both connected objects (as e.g. cyber-physical systems) and unconnected objects (as Virtual Representations).

responsible managers agree on its relevance for future manufacturing [5] a common agreement on its technical characteristics and real world manifestations is still missing. Especially in scenarios where existing production lines need to be upgraded to industry 4.0 standards, well established strategies and procedures are in place yet. So called Brownfield Deployments are common use cases as only in rare cases the gained benefits of a newly created fabric from scratch justifies its investment costs.

We propose a self-descriptive integration approach where each physical object is represented by a digital resource, either a cyber-physical resource if an object can send or receive messages and a Virtual Representation otherwise. Every resource is identified and accessible by an URI, allowing its referencing natively through the internet without additional efforts. Every resource contains information on itself which describes its category, location, functionality and capabilities together with Web links to additional information. The data format for these descriptions are provided in RDF which provides syntactically and semantically defined statements on the resource itself, its characteristics and its current state. Other formats like e.g. XML or JSON lack the semantic part out of the box. The self-descriptive aspect is essential as only the close provisioning of information together with the regarded resource itself guaranties a true modular landscape (see Fig. 1).

Additionally, we restrict the data interaction pattern to four basic methods, namely *create, read, update* and *delete* (CRUD). Limiting calls to these methods drastically reduces the possibilities to model actions. Nevertheless, we argue that only the reduction towards the basic operations has the chance to make as much of the implicit assumptions behind more powerful interfaces explicit. Higher level functionality always requires a deep understanding of the existing dependencies and design decisions which are not possible to explain in a suitable interface description.

As the integration layer forms a distributed network using internet protocols and identification and access mechanisms, it can be connected to the global internet without any adjustments. For security purposes a gateway with an appropriate access control is mandatory but the technical communication will work without any adjustments. Furthermore, with the same mechanisms proven in the Web, new data provider and resources can be added, updated or replaced as the loose coupling of data producers and consumers guarantees a future-proof architecture. The iterative characteristic takes affect that any necessary change can be directly introduced at the concerning resource with only a minimal effect on others. The scalability of the network is directly provided by same features as any number of new resources, servers or cyber-physical systems can be added, similarly to the well-known Web.

5   Framework for Virtual Representations

Restricting the interaction methods to CRUD operations restricts but also simplifies the data management. Another relevant challenge in industrial environments are different communication protocols. Our prototype supports the nowadays most common protocols namely HTTP, WebSockets and OPC UA. HTTP is the most commonly used protocol in the internet and the basis for its most popular domain, the Web. It plays a fundamental role in the success of the internet as the dominant worldwide communication infrastructure and its characteristics are well known. Its low entrance barrier and the broad dissemination make it the protocol of choice for a fast and reliable decentralized communication. Its clear client-server separation is one of its major success factors. As showed in [2] Virtual Representations are solely relying on RESTful interactions on Virtual Representations which adds the clear semantics of CRUD operations to compliant APIs. We propagate this pattern to the other protocols in order to keep interactions consistent and gain a loosely coupling of producers and consumers of data.

WebSockets rely on HTTP but allow bidirectional message exchange. Thus, on event occupation the server can push information to subscribed clients. Another advantage is the higher efficiency due to data compression. Though WebSockets are an extension of HTTP only GET operations are supported. To enable CRUD operations, we use WebSockets sub protocols for operation distinction. For every operation type one connection is established. Sub protocols cannot be changed after connection establishment. Hence we need to keep four connection open for each virtual representation as long as we want to interact with it.

The third protocol that we examine is OPC UA. It is announced as the coming standard communication technique for industry 4.0 applications [19]. OPC UA ships with two different communication protocols named HTTP/SOAP and UA TCP. We already have a HTTP communication implemented so we decided to use UA TCP. Other common IoT protocols like CoaP, MQTT or XMPP basically follow similar patterns and are therefore not yet implemented.

The core project contains various virtual representations that are managed by the virtual representation manager. It is responsible for every incoming operation

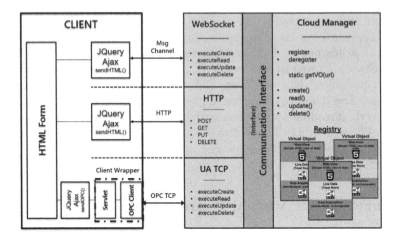

Fig. 2. Cloud Manager for Virtual Representations and communication scheme

request. The platform itself has a backend for each protocol. These backends implement the communication interface that ensures equal effects on the targeted resource. Every implementation has two main jobs: First, convert request to CRUD methods on the virtual representation manager. Second, translate given response in protocol typical response (e.g. exchange status codes or add header information).

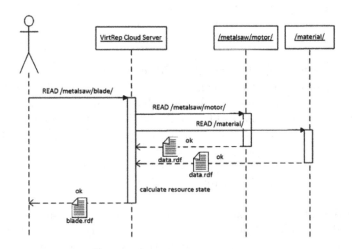

Fig. 3. Virtual Representations are computed at request time

If a client starts a query on a virtual representation data aggregation is started. As stated in [2] Virtual Representations collect their data dynamically at request time. This avoids data inconsistency on different application layers and solves the problem of time synchronization. The acquisition algorithm is configurable for now by a configuration file in N3 syntax but shall be enhanced to script-based solutions (e.g. Python) and Machine Learning models. It is also possible to define direct connections to resources, e.g. a OPC UA Server where the parts of the necessary context data are provided. We provide different serializations due to different possible requirements.

With our approach it is easy to access data that are stored in the virtual representation framework. Nonetheless it is necessary to manually add these representations to the framework, because one has to define where the representation should take the data from. Even if virtual representations could discover data for themselves there is a need to make for example sensors data somehow accessible.

6   Business Case

In the industrial sector, industrial maintenance is a prominent example for an industrial service [9]. Given its importance, concepts behind industrial maintenance have evolved over time. Traditionally, the maintenance business was organized in a transactional fashion [25]. This indicates that a maintenance service was provided as a reaction of an event—usually a machine failure, therefore, allowing equipment to run until failure [18]. Such maintenance actions are commonly referred to as reactive maintenance. However, researchers recognized the potential of preventing a machine failure in the first place and developed proactive maintenance strategies [22]. Within proactive maintenance, researchers differentiate between time-based (TBM) and condition-based maintenance (CBM) [1]. Under TBM, maintenance actions are triggered periodically based on a failure time analysis [14]. Under such scenario, historic data is analyzed in order to estimate a mean time until failure which is used to schedule periodic maintenance actions. Therefore, the core assumption behind TBM is that failure behavior of equipment follows a behavior that is expressed in historic data. Within this approach, researchers further differentiate between simple time or actual machine usage time based maintenance [1]. In CBM, maintenance actions are undertaken as soon as a certain condition threshold has been met [12]. Whilst not only aiming at preventing a failure in the first place, CBM furthermore tries at minimizing costs as maintenance is only performed when necessary and not after a certain time has passed.

Whilst pro-active maintenance strategies bring additional performance to manufacturers [22], they also require data. Especially CBM strategies rely heavily on real-time machine data [7]. Unfortunately, the dependency on data harms the broad application of CBM, as not all machines have the required connectivity or sensors installed and, as a matter of fact, only cyber-physical resources or otherwise observable components can be regarded. As, for example, the previously introduced example of the metal saw, data on the motors allows a CBM approach. However, the saw does not have any sensors equipped and thus is not available for CBM. Using the proposed solution above, we are able to use the data of surrounding machines encapsulated in order to create a virtual representation of the metal saw. Using the provided virtual representation, the CBM provider is able to estimate the current condition of the metal saw and provide an according maintenance service without needing to know the details of the metal saw, its job history or the details of the production line. Therefore, the above proposed approach is an enabler for CBM in brownfield environments. Once the virtual representation is created, it may also be used for other services. For example, the metal saw producer may analyze certain behavior and provide consulting services for the industrial customer.

It is important to note that using CPS, we are able to create the virtual representation of the metal saw on premise, thus indicating that data is only used locally and not sent to a remote site. This is a very important condition within the industrial sector, as data is usually seen as being proprietary and doesn't want to be shared among others.

7   Conclusion

This work presents a concept for an iterative deployment layer to transform production facilities to industry 4.0 setups. We explain methods to reduce existing interaction patterns and how available information can lead to a use case driven enhancement of the virtual model of the shop floor. Additionally, we show how the reached information gain can create economic benefits. We explained how the new possibilities of a holistic integration approach can change the way the industry is organized and how new business models can be implemented. Therefore, the presented work contributes to ongoing research on the introduction of IoT within brownfield environments.

The main challenge in our approach is the consequent transformation towards a state-based model and the restriction to the CRUD operations. Existing interfaces commonly do not implement these requirements especially when based on proprietary protocols. Even though major trends towards RESTful interactions simplify the communication, existing APIs require wrapper modules in order to translate and transform messages.

We argue that a sustainable transformation strategy for digital integration of manufacturing systems needs to be aligned with well-established practices of the internet. Only the distributed, loose coupling of systems and the encapsulation of relevant information at the resource itself can create a sustainable IT landscape. The outlined framework for various IoT protocols provides one step for a plug-and-play behavior in future data-driven manufacturing.

Future works will include a detailed examination of notifications and event-based interactions in general. Whereas a high-frequent polling approach can comply the requirements to some extent, an efficient integration technique needs to natively comply with the characteristics of current event and push-based systems. In addition, access control policies and reliability of information have not been yet discussed at all. We will further investigate how current state-of-the-art data security and provenance solutions can be adapted to our distributed integration setting and how a transparent and trustworthy mechanisms can comply to the specific requirements in brownfield environments.

Acknowledgement. The research and development project that forms the basis for this report is funded under project No. 01MD16015 (STEP) within the scope of the Smart Services World technology program.

References

1. Ahmad, R., Kamaruddin, S.: An overview of time-based and condition-based maintenance in industrial application. Comput. Ind. Eng. **63**(1), 135–149 (2012)
2. Bader, S.R., Maleshkova, M.: Virtual representations for an iterative IoT deployment. In: Companion of The Web Conference 2018, pp. 1887–1892. International World Wide Web Conferences Steering Committee (2018)
3. Baines, T.S., et al.: State-of-the-art in product-service systems. Proc. Inst. Mech. Eng. Part B: J. Eng. Manuf. **221**(10), 1543–1552 (2007)
4. Baines, T., Lightfoot, H., Benedettini, O., Kay, J.: The servitization of manufacturing. J. Manuf. Technol. Manag. **20**(5), 547–567 (2009)
5. Bauer, H., et al.: Industry 4.0 after the initial hype-where manufacturers are finding value and how they can best capture it. McKinsey Digital (2016)
6. Bedenbender, H., et al.: Industrie 4.0 Plug-and-Produce for Adaptable Factories: Example Use Case Definition, Models, and Implementation (2017). http://www.plattform-i40.de/I40/Redaktion/EN/Downloads/Publikation/Industrie-40-%20Plug-and-Produce
7. Campos, J.: Development in the application of ICT in condition monitoring and maintenance. Comput. Ind. **60**(1), 1–20 (2009)
8. Davis, J., Edgar, T., Porter, J., Bernaden, J., Sarli, M.: Smart manufacturing, manufacturing intelligence and demand-dynamic performance. Comput. Chem. Eng. **47**, 145–156 (2012)
9. Gitzel, R., Schmitz, B., Fromm, H., Isaksson, A., Setzer, T.: Industrial services as a research discipline. Enterp. Model. Inf. Syst. Archit. **11**(4), 1–22 (2016)
10. Gubbi, J., Buyya, R., Marusic, S., Palaniswami, M.: Internet of Things (IoT): a vision, architectural elements, and future directions. Future Gener. Comput. Syst. **29**(7), 1645–1660 (2013)
11. Hedengren, J.D., Eaton, A.N.: Overview of estimation methods for industrial dynamic systems. Optim. Eng. **18**(1), 155–178 (2017)
12. Jardine, A.K., Lin, D., Banjevic, D.: A review on machinery diagnostics and prognostics implementing condition-based maintenance. Mech. Syst. Sig. Process. **20**(7), 1483–1510 (2006)
13. Lee, J., Bagheri, B., Kao, H.A.: A cyber-physical systems architecture for industry 4.0-based manufacturing systems. Manuf. Lett. **3**, 18–23 (2015)
14. Lee, J., Ni, J., Djurdjanovic, D., Qiu, H., Liao, H.: Intelligent prognostics tools and e-maintenance. Comput. Ind. **57**(6), 476–489 (2006)
15. Lin, S.W., et al.: The Industrial Internet of Things Volume G1: Reference Architecture (2017). http://www.iiconsortium.org/IIRA.htm
16. Meier, H., Roy, R., Seliger, G.: Industrial product-service systems-IPS2. CIRP Ann. - Manuf. Technol. **59**(2), 607–627 (2010)

17. Otto, B., Lohmann, S.: Reference architecture model for the industrial data space. Technical report, Fraunhofer-Gesellschaft zur Förderung der angewandten Forschung e.V (2017)
18. Paz, N.M., Leigh, W.: Maintenance scheduling: issues, results and research needs. Int. J. Oper. Prod. Manag. **14**(8), 47–69 (1994)
19. Rauen, H., Mosch, C., Niggemann, O., Jasperneite, J.: Industrie 4.0 kommunikation mit OPC UA. Leitfaden zur Einführung in den Mittelstand. Hg. v. VDMA und Fraunhofer-Anwendungszentrum Industrial Automation. Frankfurt am Main (2017). ISBN 978-3-8163-0709-9
20. Stein, B., Morrison, A.: The enterprise data lake: better integration and deeper analytics. PwC Technol. Forecast: Rethink. Integr. **1**, 1–9 (2014)
21. Stock, T., Seliger, G.: Opportunities of sustainable manufacturing in industry 4.0. Procedia CIRP **40**, 536–541 (2016)
22. Swanson, L.: Linking maintenance strategies to performance. Int. J. Prod. Econ. **70**(3), 237–244 (2001)
23. Tanuska, P., Spendla, L., Kebisek, M.: Data integration for incidents analysis in manufacturing infrastructure. In: Computing Conference, 2017, pp. 340–345. IEEE (2017)
24. Vargo, S.L., Lusch, R.F.: Service-dominant logic: continuing the evolution. J. Acad. Market. Sci. **36**(1), 1–10 (2008)
25. Wolff, C., Vössing, M., Schmitz, B., Fromm, H.: Towards a technician marketplace using capacity-based pricing. In: Proceedings of the 51th Hawaii International Conference on System Sciences, pp. 1553–1562. Waikoloa (2018)

Market Launch Process of Data-Driven Services for Manufacturers: A Qualitative Guideline

Achim Kampker, Marco Husmann[(⊠)], Philipp Jussen,
and Laura Schwerdt

Institute of Industrial Management (FIR) at RWTH Aachen University,
Campus-Boulevard 55, 52074 Aachen, Germany
{achim.kampker,marco.husmann,
philipp.jussen}@fir.rwth-aachen.de,
laura.schwerdt@rwth-aachen.de

Abstract. Traditional manufacturing companies increasingly launch data-driven services (DDS) to enhance their digital service portfolio. Nonetheless, data-driven services fail more often than traditional industrial services or products within the first year on the market. In terms of market launch, their digital characteristics differ from traditional industrial services and thus need specific structures and actions, which companies currently lack. Therefore, a process guideline for a six-month market launch phase of DDS is developed. The guideline relies on analogies from product, service and software launches based on the latest literature from service marketing and successful practices from various industries. Finally, the guideline is evaluated within five industrial case studies. Thus, the guideline provides scientific research insights regarding the market launch process of DDS and adds to the research of service marketing. It provides practical guidance for manufacturing companies by serving as a reference process for the market launch and offering a collection of successful practices within this area.

Keywords: Data-driven services · Service marketing · Market launch
Launch tactics · Manufacturing · New service development

1 Introduction

As a result of the progressive digitization, existing industrial service business models of manufacturers are changing and might disappear, while new opportunities for DDS occur at the same time [1]. DDS use digital components to generate, analyze and link data in order to archive new benefits [2]. Although in research there is no common definition for the term DDS yet, most authors agree that data-driven services are services that rely on data streams and are either linked to physical products as data suppliers and complement them in a meaningful way, or services that are detached from products and based on data [3, 4].

In particular, manufacturing companies in the mechanical engineering industry often develop data-driven business models in order not to become an exchangeable

© Springer Nature Switzerland AG 2018
G. Satzger et al. (Eds.): IESS 2018, LNBIP 331, pp. 177–189, 2018.
https://doi.org/10.1007/978-3-030-00713-3_14

supplier [5]. In practice, DDS fail more often than traditional services or products within the first year after the market launch [6]. On the one hand, this is caused by an increased perceived intangibility of DDS among the customers, which leads to a greater uncertainty and risk perception than in case of familiar services [7]. On the other hand, customers now play an active role in the implementation process of DDS, which they are often not capable of. Furthermore, a transformation in the business respective revenue models of DDS occurs by shifting from a classical purchase of a service towards a license or usage rate model [8]. As a further cause, successful companies in the field of DDS identified the market launch phase as one of the greatest challenges. The companies lack structures and measures for a successful market launch of DDS [9].

For this paper, DDS are defined as follows: Data-driven services are services which are characterized by a digital component and build on data from intelligent and connectable products. Data-driven services create benefits for companies and/or customers through generation, collection, analysis and/or combination of internal and external data. With regard to the market launch, in particular the novelty and increased awareness of DDS's immateriality among customers lead to increased uncertainty and higher adoption barriers [7]. Therefore, the market launch process of DDS differs from traditional services or products and calls for individual structures and activities. Within this paper, the market launch period is considered to begin with the preparation of the market launch during the development process. It continues beyond the time of market entry and ends when a first growth in the market, which is called the take-off [10], is reached. For a basic understanding, a market launch in this work is defined as follows: A market launch characterizes the period around the first market entry of a new product or service. The market launch period begins with the preparation of market entry and ends with the take-off of the product or service in the target market. It includes all relevant market, product or service and company-related measures.

The academic literature focuses on product and traditional service market launch in general. Thus, a specific market launch process of DDS or digital services has not been developed in previous models before and there are only few concrete recommendations for action and specific measures in the current literature. The present paper therefore develops a guideline for the market launch of DDS which includes actions and measures for a successful market launch with regard to the specific characteristics of DDS. The present work thus contributes to the successful market launch in practice. The research question of this paper therefore is 'How should the market launch process of data-driven services in the sector of mechanical and plant manufacturing be designed in the short and medium term to achieve a successful service market launch?'. In order to support the research question, the following question will also be analyzed: 'What specific recommendations for action are to be suggested for the market launch of data-driven services in mechanical and plant manufacturing?'

The paper is structured as follows: First, relevant literature and existing theoretical approaches are discussed. Secondly, the research methodology and the study design are described. Thirdly, the development of the research model with regard to two identified dimensions of the market launch (time and content) is explained. Fourthly, key findings are presented. Thereafter, key contributions and implications for practice are given. Lastly, a conclusion and an outlook for further research are presented.

2 Related Work

2.1 Delimitation of Data-Driven Services

Although digital services are frequently addressed in academic literature, there is no generally accepted definition of digital services or DDS. In order to ensure a consistent understanding of the terms, DDS are differentiated from other terms used in the literature in the following.

The foundation of data-driven services is built by information and communication technology (ICT), which comprises hardware and software as well as all products and services of ICT, such as landline telephones or computers [11]. DDS, on the other hand, comprises only services that contain a digital component. Hence, DDS are based on ICT [4] and rely on product-centric data as the key resource, which can be transmitted analogously by cable or remotely. In the literature, the terms 'e-services' (Electronic services) and 'e-innovations' (Internet-enabled service innovations) are also used, but they focus mostly on e-commerce services or are nowadays used as a synonym for digital services [12, 13]. Digital services instead focus on product-centric data which are transmitted remotely via the Internet [8]. They are internet-based services and transactions that are either linked to physical products as data suppliers and complement them in a meaningful way, or services that are detached from products and based on data [3, 4]. In manufacturing, only using data does not make a digital service a data-driven service. There is a difference between using data (consuming and/or producing it) and having the data at the very center of a service business model, which is the focus for this paper. Smart Services represent the most advanced development of DDS. They link suitable receivers and products in real time and derive appropriate consequences based on automated data evaluation, i.e. by means of machine learning algorithms [14]. They describe interactive data-based services provided via the internet and are therefore also defined as a category of DDS. As these services are based on data, they also add to the category of DDS. Within this work, DDS refer to the definition given in Sect. 1 and therefore include Smart Services as well as e-innovations and digital services. Fig 1 illustrates the differentiation of DDS based on the benefit of the individual service.

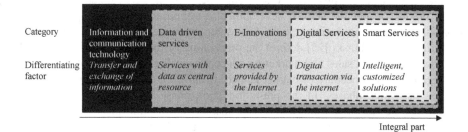

Fig. 1. Differentiation of DDS [15]

2.2   Literature Review

To gain a comprehensive overview of the scientific literature in the area of market launch, a structured and systematic literature review is conducted. The systematic literature review process is based on a structured content analysis in which existing literature content is identified based on pre-defined criteria and sorted into categories according to a defined procedure [16]. Criteria are e.g. the dimension of time (last phase of the innovation process). Therefore, current literature addressing the market launch of DDS as well as product and service market launch is reviewed and analyzed. Selected approaches are presented below.

Market Launch of Digital Services. Kuester et al. [7] develop a framework by defining success factors for ICT service innovations which includes success factors and supporting operational actions within the market launch period. They highlight the importance of trustworthiness and usability signals in order to increase the trust and decrease the insecurity of customers caused by internet-enabled services. The authors thereby focus on digital B2C innovations. B2B ICT technologies are addressed by Schindlbeck [17], who defines five obstacles for the use of these technologies. The analyzed literature does not focus on the market launch phase for DDS or develop guidelines and structures for this time period. DDS, defined as above, are not addressed in the literature presented above. Furthermore, the literature does not develop actions to include necessary requirements for a successful market launch of DDS within the launch phase.

Product and Service Market Launch. In the area of product and service launch, the current literature identifies success factors for market launch. Talke and Hultink [18], for instance, identify factors that support the success of product launch by the reduction of diffusion barriers. Those include internal and external actions such as the information of employees or the cooperation with customers [18]. Storey et al. [19] confirm the importance of these factors, adding the 'organizational design' and the 'involvement of front line staff' as success factors. Additionally, conducted evaluations conducted by different authors show that an intensive marketing, the inclusion of customers and employees and an accurate launch plan support the success of a market launch [20–22]. The importance of the launch plan is also identified by other authors [20, 23]. Cooper [24] defines a launch plan as based on market research and studies and as an integral part of the product development process. It should begin at an early stage of the product respective service development. Furthermore, sufficient personal and financial resources and the integration of sales personnel are crucial for launch success. Conformingly, Jaschinski [25] defines the preparation of the market launch to include the development of a sales and marketing concept as well as a pilot test of the service.

Regarding the procedure of market launch, Bruhn [26] divides the development of services into three phases. In the pre-introduction phase, the service and necessary frameworks are developed and tested. The market introduction phase comprises the first implementation of the service in the real market, and the review phase includes an analysis of the results and necessary adjustments to ensure a long-term market success. Going into more detail, Scheuing and Johnson [27] divide the whole service development process into 15 steps, summing up the last four to be the service 'introduction'.

Most launch process models do not consider the full market launch phase as they end with the market entry of the new product or service [23, 25, 27]. Exceptions to this are for example Lin and Hsieh [28]. They develop a five step process of service development, including 'service commercialization' as a last step after 'service implementation' [28].

Identified Need for Research. Since DDS differ from traditional services (see Sect. 1) a separate consideration of the market launch process is required. In a further instance, they differ through the possibility of fast scalability in the market of DDS, which is caused by the fact that for DDS a separation of the production and consumption of the service is possible and the Uno-actu principle is not valid for DDS anymore [29]. In addition, DDS differ from traditional services in the possibility of software releases after a first market launch. Due to the higher degree of software and therefore a bidirectional communication flow towards customers, DDS can be released remotely, i.e. with upgrades of the DDS.

The conducted analysis shows that the current literature focuses on product and service launch. The launch of DDS in general and the changed and new requirements for companies within the launch phase in particular are not addressed in detail. Overall, the market launch phase is mostly investigated as the final stage of the development process and not as a focused time period. In the area of digital services, the literature focuses on success factors and obstacles of market launch. Thereby, the process of DDS-specific market launch activities is not examined. In order to investigate in this field, this article develops a process guideline for DDS market launch, focusing on the manufacturer's industry.

3 Research Methodology and Study Design

This article follows a case-based research approach [30]. Therefore, the structure of the guideline is developed by an analogy analysis based on the relevant literature in the areas of product and service market launch. In the following, recommendations for action are developed through an analogy analysis based on the analyzed literature. The actions and measures are supplemented by successful practices that are identified within an executed pre-test workshop and three case studies with service experts. The workshop with 26 service experts and the first three case studies were used as an exploration source only. Afterwards, the defined actions and measures are arranged as a process guideline and evaluated in five further qualitative case studies with service experts. Consequently, the guideline integrates theoretical knowledge from literature and successful practices from the industry.

3.1 Analogy Analysis

Analogy considerations aim at developing solutions for a targeted problem by transferring and adapting a solution from a source problem [31]. The prerequisite for this procedure is the conclusion that the similarity of some parts of the situations implies the similarity of other parts [32]. To draw analogy conclusions between product and service launches and launches of DDS, the process of market launch is defined as a valid principle for the source and target situation [33]. Therefore, an analogy conclusion can be drawn based on the mechanism of an abstract principle. Further, this article defines the launch of DDS as a modification of product or service launch, wherefore analogy conclusions can also be drawn by the mechanism of transformation.

3.2 Data Collection and Sample Selection

As a second source, this article draws recommendations for action from qualitative case studies with service experts from the industry [30]. For this purpose, a pre-test workshop was conducted with 26 After Sales experts with a minimum of three years of professional experience from various manufacturing companies originating in Europe. Further, three semi-structured interviews (see Table 1, companies 1–3) are carried out in order to identify successful practices for the market launch of DDS and to gain further insights into the process of market launch [34]. The semi-structured interviews are conducted on the basis of a differentiated guideline. The guideline aims at gaining new insights into the actions and measures of the market launch of DDS and therefore consists of open questions and narrative requests [35]. Conclusively, the developed model is evaluated within five case studies (see Table 1, companies 4–8) with service experts from corporate practice with a minimum of three years of professional experience in service management. The experts evaluate the actions with regard to frequency of use in their company, relevance for the success of market introduction and implementation difficulty. Furthermore, the temporal positioning and duration of the actions as well as the completeness of the guideline are evaluated. An overview of the three semi-structured interviews to identify successful practices and the five case studies for validation purposes are illustrated in Table 1. All participants of the pre-test workshop are currently dealing with the topic of market launch of digital products and services within their companies. The participants of the case studies are chosen according to their knowledge and experiences in the field of market launch of DDS and the success of their companies in this area. All companies are equipment manufacturers which operate in different industry sectors and countries. The sample only considers equipment manufacturers because this segment of companies struggles most when developing and launching DDS. The transformation from the established product business towards a business with digital services and products seems to be an undiscovered field of research.

Table 1. Overview of the conducted interviews

Companies	Manufacturing sector	Number of employees	Total turnover p.a.
Company 1	Equipment/packaging	>5.000	>1.5 bn EUR
Company 2	Equipment/agriculture	>11.000	>3.5 bn EUR
Company 3	Equipment/sensors	>8.000	>1.5 bn EUR
Company 4	Equipment/steel	>150.000	>43 bn EUR

(continued)

Table 1. (*continued*)

Companies	Manufacturing sector	Number of employees	Total turnover p.a.
Company 5	Equipment/laser	>12.000	>3 bn EUR
Company 6	Equipment/cleaning	>12.000	>2.5 bn EUR
Company 7	Equipment/printing	>11.000	>2.5 bn EUR
Company 8	Equipment/bearings	>90.000	>14 bn EUR

4   Model Development

4.1   Time Structure and Limitation of the Developed Guideline

The development of the guideline as a time structure model requires a temporal limitation of the market launch phase. In scientific literature, time periods between one month and several years are defined [36]. Overall, digital products have a shortened life cycle by factor 2 or 3 compared with traditional products [37]. Services in general have shorter life cycles than products [38]. Transferring these finding on DDS in comparison to traditional services, the life cycles of DDS are shorter. This article therefore assumes a duration of six months for the defined period of DDS market launch. The exact duration varies depending on the service and industry [26]. The evaluation shows that most experts assume a longer duration for this period. On the other hand, the case studies show that some DDS are launched within two launch phases of four months each. The guideline structure is developed by analogy conclusions based on the previously executed literature review. Analogous to Kuhn [22], the guideline consists of a time and a content-related perspective. Thus, each action and measure is a two-dimensional activity and contains a defined time and a detailed description of the activity. Within the time perspective, the market launch period is divided into five stages based on analogy conclusion from the previous literature analysis. The five stages are subordinated to the three phases of Bruhn [26]. In line with Cooper [39], the individual phases are not defined as a sequential process, but occur partially in parallel. Accordingly, the timing of the activities is not a linear process [25]. The guideline's timeline is iterative, which allows jumping forward and backward between individual actions. The five stages of the launch period are shown in Fig. 2.

Fig. 2. Defined market launch steps based on [23, 24, 26]

The first step includes the development or finalization of a market launch plan that contains all required activities and resources as well as the related time constraints [27].

Similar to other models, a detailed market launch plan comprises sales, communication and branding concepts based on market and target group analyses [24, 26]. Due to the novelty of the DDS, testing the DDS as well as the required internal infrastructure is of particular importance [7]. For this reason, the second phase is defined as the implementation of a pilot to test and validate the DDS and the prerequisites for the company [28]. In analogy to previous models, the training of employees takes place after the execution of the pilot [22]. The increasing uncertainty of DDS among customers has a negative effect on the adoption of DDS, which is why the third phase serves as an intensive promotion of DDS to reduce the uncertainty [7]. The fourth stage comprises the initial implementation and provision of the DDS as well as the use of all installed and/or modified technical and organizational structures [26]. After market entry, the 'After launch stage' includes continuous diffusion monitoring and feedback analysis, which are especially necessary for a new service [23]. In this stage, communication serves to build up a reputation and to promote the diffusion process. These stages have been confirmed within the carried-out case studies with service experts.

4.2   Content Structure of the Developed Guideline

In order to successfully shape the market launch, the content-related perspective of the developed guideline includes a strategic as well as an operational level as defined by Bleicher et al. [40]. Bleicher et al. further define a normative management. The present work focuses on the short- and medium-term management of market launch and thus only has limited influence on the constitutive framework, which is why normative management is not considered. Strategic activities aim for the development, maintenance and exploitation of success potentials whereas operational activities serve the operational execution of strategic goals.

Relevant factors for product and service launch as well as DDS launch have been identified in the literature analysis. Based on these factors and supplemented by the findings from case studies, seven success factors for the launch of DDS are determined, all of which companies can influence in the launch period. The selection process for the seven success factors as well as for individual actions and measures was based on an iterative cross-case approach. All identified success factors/actions/measures have been reflected in detail concerning their practical relevance for the interviewed companies and the participants of the workshop, respectively. Only the success factors/actions/ measures which have practically proven to be successful in terms of the market launch in the five companies interviewed for validation (see Table 1, company 4–8) or are going to be implemented in the future have been integrated into the model. The success factors are implemented in the guideline through three levels. The success factors themselves are the defined goals which positively influence the market launch of DDS. They are realized by actions on the second level. Actions are put into practice by measures on the third level. Within the guideline, the developed actions are classified as strategic or operational actions. Since measures are always an execution action, they are generally defined as operative. The whole structure is presented in Fig. 3.

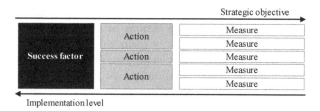

Fig. 3. Content levels of the guideline

5   Findings

Figure 4 provides an overview of the guideline and some of its defined actions. Since the research process is not completed yet, only parts of the guideline are presented in. Overall, the guideline includes seven success factors, 15 actions and 30 measures.

Employees	Top-Down Information	Apps	Playbook	...
	Integration of employees	Employee feedback system	Sales Force Group	...
	Motivation of employees	Incentive system for employees	Incentive system for sales force	...
	Trainings	Webinar	...	...
Organisation	Coordination	Coordinator	Reference process	...
	Organizational structures	Structural adjustment	Interdisciplinary Teams	...
Feedback	Customer integration	Feedback incentive system	Customer feedback system	...
	Internal feedback system	Employee feedback system	...	...
External information	Explanation of data-driven services	Video-Tutorials	Information dur. machine handover	...
	Explanation of data protection	Transparency regarding data usage	Certificates/standards	...
	Communication of customer benefits	Success Stories	...	...
Trust	Customer recommendations	Customer references	...	...
	Presentation of performance	Prove of capability	Certificates/standards	...
Visualisation	Material representation	Success Stories	Apps	...
	Experiential design	Free trail version	Segmentation	...
Implementation speed	Motivation of employees	Incentive system for employees	Incentive system for sales force	...
	Coordination	Coordinator	Coordination meeting	...
	Customer integration	Customer feedback system	Feedback incentive system	...
	Customer recommendations	Customer references	Incentive system for customers	...

Fig. 4. Overview of the developed guidelines for market launch of DDS

The evaluation with experts from the industry based on a written questionnaire confirmed the temporal structure of the guideline and the relevance of the developed activities for a successful launch of DDS. The actions 'coordination', 'customer integration' and 'trainings' have the highest relevance for the success of DDS launch. Together with the 'communication of the customers' benefits", they are also the most used actions. The interviewed companies use 26 out of 30 defined measures, which emphasizes the practical benefit of the defined measures and functions as a qualitative validation. Transferring the valuations on the success factors, this highlights the relevance of the factors 'organization', 'feedback' and 'external information' for the success of the market launch. The actions belonging to the success factors 'external information' and 'organization' are used most frequently, whereas the difficulty of implementation was rated highest for the actions of these two success factors and lowest for the actions of the success factor 'visualization'. The evaluation also clearly showed the individuality of the market launch process. Only for 3 out of 15 actions, all experts agreed on the time of implementation and the duration of the actions. In the other cases, the experts contradict each other partly regarding the implementation duration or time. Solely regarding the action 'organizational structures', all experts agree on a longer duration than the six-month period of the guideline. Overall, the temporal evaluation showed that the market launch phase of DDS is an individual process and cannot be standardized across companies and DDS. Thus, the developed guideline serves as a reference process that contains recommendations for practical action in terms of time and content.

As the research process has not yet been completed, one exemplary action will be presented here. Within the evaluation, the measure of 'customer references" generated the greatest difference concerning its relevance for the success of the market launch. Whereas three of the interviewed experts rank the relevance of this action as quite low, the two other experts identify this action as one of the most relevant ones. The executed pre-workshop showed that the novelty and immateriality of DDS lead to greater perceived uncertainty and higher adoption barriers among customers compared to traditional services. By creating trust between companies and customers, customer acceptance barriers are overcome and perceived risks are reduced [12]. As a result, stronger measures for building confidence and trust are necessary. Customer confidence can be strengthened by the use of customer recommendations, as customers believe other users more readily than the company's conventional advertising [41]. Recommendations partly compensate customers' uncertainties and thus have a strong influence on the purchase decision, especially in case of intangible assets such as DDS [42]. The acquisition of new customers through existing customers thus weakens diffusion barriers and increases the probability of adoption of the new service [18]. Three of the five case studies use customer references within the market launch. The customer references are used as soon as possible within the launch process not only to win and convince future customers, but also to convince and motivate the sales and execution team. For example, first positive feedback quotes from customers are sent to the involved stakeholders per email only four weeks after the first pilot projects started. The expert states that word of mouth is often more effective than a marketing campaign to convince stakeholders of the DDS.

6   Key Contributions and Implications

The central research question (see Sect. 1) was answered in this paper with the developing of a guideline for the market introduction of DDS. To do so, seven success factors were identified which can be influenced by companies during the market launch phase. These success factors were implemented through 15 strategic and operational actions and 30 operational measures. Their relevance was confirmed by the evaluation of the five case studies and by the use of the individual actions and measures in practice. In addition, the service experts confirmed the accuracy of the five time phases of market launch defined in this paper. The evaluation showed that the market launch phase is an individual process and that the standardization of time structure is not possible across companies and DDS. The developed action guideline therefore serves as a reference process which contains recommendations for practical application in terms of time and content. A first validation has been conducted for middle sized/big manufacturing companies (>10.000 employees and >2.5 bn EUR p.a.). For this group, some patterns have been discovered for the launch process of DDS, e.g. that the product management service was responsible for the development and market launch of DDS from an organizational perspective; in these companies, most of the effort and experience made in launch tactics was put into established product manufacturing formally. With the development of the guideline, this work addresses the research gap identified in the field of market launch processes of DDS and provides fundamental contributions to the research in this area. The guideline offers practical benefits for companies in the manufacturing sector through identified actions and measures and the developed structures for the market launch. In addition, this work offers added value through identified successful practices and in-depth insights into the market launch process of successful companies. The findings are limited by the number of case studies, the focused industry and the time frame. Consequently, the guideline does not claim to be complete. For instance, there has not been a differentiation of traditional and digital-native customers using DDS. The new generation of consumers may have a different perception of the technology and different consumption habitudes concerning digital services, which sensibly affect the analysis that has been made. For instance, the degree of trustworthiness or the level of insecurity might not be the same in the future for non-millennial customers.

7 Conclusion and Further Research

The purpose of this work was to develop a guideline for the market launch process of DDS. To reach this goal, recommendations on content and timing were developed and structured as reference process. Therefore, this work defines recommended actions and measures by analogy analysis of current literature regarding product and service market launch. These procedures and the results stemming from the research were further evaluated with the help of successful practices who put them to the test as part of a pre-workshop and three qualitative interviews with service experts. The evaluation with five service experts confirms the relevance and practical benefit of the developed actions for a successful market launch.

As for future research, further investigations are required focusing on an empirical validation of the actions regarding their relevance for a successful market launch and the costs of implementation. Furthermore, insights into different industries would enable a more detailed guideline. Therefore, a large-scale empirical analysis must be conducted within different industries to gain further knowledge regarding the market launch of DDS. In the eight conducted case studies, a further need for research regarding possible revenue models for DDS was detected.

References

1. Uhlmann, E., Hohweiler, E., Geisert, C.: Intelligent production systems in the era of Industrie 4.0 - changing mindsets and business models. J. Mach. Eng. **17**(2), 5–24 (2017)
2. Dotzel, T., Shankar, V., Berry, L.L.: Service innovativeness and firm value. J. Mark. Res. **50**(2), 259–276 (2013)
3. Acatech: Smart Service Welt: Umsetzungsempfehlungen für das Zukunftsprojekt Internet-basierte Dienste für die Wirtschaft. Abschlussbericht (2015)
4. Huang, M.-H., Rust, R.T.: IT-related service: a multidisciplinary perspective. J. Serv. Res. **16**(3), 251–258 (2013)
5. Bitkom: Business models in Industrie 4.0 utilising and actively shaping opportunities and potentials: fact paper (2017)
6. Demirkan, H., Bess, C., Spohrer, J., et al.: Innovations with smart service systems: analytics, big data, cognitive assistance, and the internet of everything. Commun. Assoc. Inf. Syst. **37**(35), 733–752 (2015)
7. Kuester, S., Konya-Baumbach, E., Schuhmacher, M.C.: Get the show on the road: go-to-market strategies for e-innovations of start-ups. J. Bus. Res. **83**, 65–81 (2018)
8. Williams, K., Chatterjee, S., Rossi, M.: Design of emerging digital services: a taxonomy. Eur. J. Inf. Syst. **17**(5), 505–517 (2008)
9. Baumbach E.: The launch of E-innovations: an analysis of go-to-market strategies and the consumer adoption decision-making process. Inauguraldissertation, Universität Mannheim (2016)
10. Tellis, G.J., Stremersch, S., Yin, E.: The international takeoff of new products: the role of economics, culture, and country innovativeness. Mark. Sci. **22**(2), 188–208 (2003)
11. Oye, N.D., Shallsuku, Z.K., et al.: The role of ICT in education: focus on university undergraduates taking mathematics as a course. Int. J. Adv. Comput. Sci. Appl. **3**(2), 136–143 (2012)
12. Schuhmacher, M.C., Kuester, S., Hanker, A.-L.: Investigating antecedents and stage-specific effects of customer integration intensity on new product success. Int. J. Innov. Manag. **22**(4), 1850032 (2018)
13. Hull, R., Benedikt, M., Christophides, V., et al.: E-services: a look behind the curtain systems. In: Neven, F., Beeri, C., Milo, T. (eds.) Proceedings of the Twenty-Second ACM SIGMOD-SIGACT-SIGART Symposium on Principles of Database Systems, pp. 1–14. ACM, New York (2003)
14. Dobre, C., Xhafa, F.: Intelligent services for Big Data science. Future Gener. Comput. Syst. **37**, 267–281 (2014)
15. Acatech: Smart service welt: recommendations for the strategic initiative web-based services for businesses. Final report (2015)
16. Fink, A.: Conducting Research Literature Reviews: From the Internet to Paper. Sage Publications, Thousand Oaks (2013)
17. Schindlbeck, B.: Verbreitung und Durchdringung von Business-to-Business Technologien: Interaktive Formulare als alternative Technologie zur Unterstützung des Informationsaustauschs zwischen Unternehmen. Logos Verlag, Berlin (2015). (in German)
18. Talke, K., Hultink, E.J.: Managing diffusion barriers when launching new products. J. Prod. Innov. Manag. **27**(4), 537–553 (2010)
19. Storey, C., Cankurtaran, P., Papastathopoulou, P., et al.: Success factors for service innovation: a meta-analysis. J. Prod. Innov. Manag. **33**(5), 527–548 (2016)
20. Barczak, G., Kahn, K.B.: Identifying new product development best practice. Bus. Horiz. **55**(3), 293–305 (2012)

21. Wang, K.-J., Lestari, Y.D.: Firm competencies on market entry success: evidence from a high-tech industry in an emerging market. J. Bus. Res. **66**(12), 2444–2450 (2013)
22. Kuhn, J.: Markteinführung neuer Produkte. Deutscher Universitätsverlag, Mannheim (2007). (in German)
23. Garcia, R.: Creating and Marketing New Products and Services. Taylor & Francis Group, Abingdon (2014)
24. Cooper, R.G.: New products - what separates the winners from the losers and what drives success. In: Kahn, K.B. (ed.) The PDMA Handbook of New Product Development, 3rd edn, pp. 3–35. Wiley, Hoboken (2013)
25. Jaschinski, C.M.: Qualitätsorientiertes Redesign von Dienstleistungen. Dissertation. Shaker Verlag, Aachen (1998)
26. Bruhn, M.: Markteinführung von Dienstleistungen: Vom Prototyp zum marktfähigen Produkt. In: Bullinger, H.-J., Scheer, A.-W. (eds.) Service Engineering, pp. 227–249. Springer, Berlin (2006). https://doi.org/10.1007/3-540-29473-2_9. (in German)
27. Scheuing, E.E., Johnson, E.M.: A proposed model for new service development. J. Serv. Mark. **3**(2), 25–34 (1989)
28. Lin, F.-R., Hsieh, P.-S.: A SAT view on new service development. Serv. Sci. **3**(2), 141–157 (2011)
29. Riedl, C., Leimeister, J.M., Krcmar, H.: New service development for electronic services - a literature review. In: Association for Information Systems (ed.) 15th Americas Conference on Information Systems 2009, pp. 5243–5251. Curran, Red Hook (2009)
30. Eisenhardt, K.M.: Building theories from case study research. Acad. Manag. Rev. **14**(4), 532–550 (1989)
31. Gentner, D.: Are scientific analogies metaphors? In: Miall, D.S. (ed.) Methaphor. Problems and Perspectives, pp. 106–132. Harvester Press, Sussex (1982)
32. Kroes, P.: Structural analogies between physical systems. Br. J. Philos. Sci. **40**(2), 145–154 (1989)
33. Clement, J.: Observed methods for generating analogies in scientific problem solving. Cognit. Sci. **12**(4), 563–586 (1988)
34. Qu, S.Q., Dumay, J.: The qualitative research interview. Qual. Res. Acc. Man. **8**(3), 238–264 (2011)
35. Bryman, A.: Integrating quantitative and qualitative research: how is it done? Qual. Res. **6**(1), 97–113 (2016)
36. Schmalen, C.: Erfolgsfaktoren der Markteinführung von Produktionnovtionen klein- und mittelständischer Unternehmen der Ernährungsindustrie. Herbert Utz Verlag (2005). (in German)
37. MEDEA: The Medea + Design Automation Roadmap: Design automation Solutions for Europe, 3rd edn. (2002)
38. Hsu, L.-F.: E-commerce model based on the internet of things. Adv. Sci. Lett. **22**(10), 3089–3091 (2016)
39. Cooper, R.G.: Invited article: what's next?: after stage-gate. Res. Technol. Manag. **57**(1), 20–31 (2014)
40. Bleicher, K., Abegglen, C.: Das Konzept integriertes Management: Visionen - Missionen - Programme, 9th edn. Campus Verlag, Frankfurt am Main (2017). (in German)
41. Hoyer, W.D., Chandy, R., Dorotic, M., et al.: Consumer cocreation in new product development. J. Serv. Res. **13**(3), 283–296 (2010)
42. Solnet, D.: Antecedents and dimensions of service orientation: a conceptual framework. In: O'Mahony, B., Whitelaw, P. (eds.) CAUTHE 2006 Conference: To the City and Beyond, p. 320. Victoria University, Melbourne (2006)

Service Business Models

Success Factors of SaaS Providers' Business Models – An Exploratory Multiple-Case Study

Sebastian Floerecke[(⊠)]

Chair of Information Systems II, University of Passau, Passau, Germany
sebastian.floerecke@uni-passau.de

Abstract. Market studies have revealed major differences in the level of performance among providers of Software as a Service (SaaS). The literature's understanding of the underlying success factors and thus, the reasons for this performance discrepancy is, however, still limited. The goal of this research paper is therefore to investigate the success factors of SaaS providers' business models by conducting an exploratory multiple-case study. 21 expert interviews with representatives from 17 cloud providers serve as central method of data collection. The study's result is a catalogue of 27 success factors. In particular, a SaaS service should be developed as a system comprising modular microservices in order to meet the desired requirements in terms of cost advantages, performance and scalability. Overall, established SaaS providers obtain a reference framework to compare, rethink and innovate their present business models. Companies that are planning to offer SaaS in future gain valuable insights which should directly feed into their business model design process.

Keywords: Cloud computing · Software as a Service (SaaS) · Business model
Success factors · Exploratory multiple-case study · Expert interviews

1 Introduction

Organizations have significantly increased the use of cloud computing services, which offer virtualized IT resources in terms of infrastructure, data and applications, over the last years [1]. This development is set to continue: according to a current study of Gartner [2], the worldwide public cloud service market is projected to grow 16.6% in 2018 to total USD 287.82 billion, up from USD 246.84 billion in 2017. The largest segment will remain Software as a Service (SaaS) with USD 55.14 billion.

However, cloud providers face considerable challenges, as they generally require profound expertise with regard to both technical infrastructure concepts and the design and management of service-oriented business models [3]. In practice, it can be particularly observed that several cloud providers encounter difficulties to effectively design suitable business models. This is why many are still experimenting with a variety of business models aiming to put themselves in a sustainable and profitable position within the cloud computing ecosystem [4, 5]. Indeed, market studies have revealed major differences in the level of performance between cloud providers [6, 7]. The literature's understanding of success factors of cloud providers' business models and thus, the reasons for this performance discrepancy is, however, still limited [8, 9].

© Springer Nature Switzerland AG 2018
G. Satzger et al. (Eds.): IESS 2018, LNBIP 331, pp. 193–207, 2018.
https://doi.org/10.1007/978-3-030-00713-3_15

Especially the largest segment, SaaS, has only been covered in a few isolated studies with regard to success factors [10–12]. These contributions mainly focus on the value proposition, while other business model components are disregarded, or are solely literature-based, trying to summarize the fragmented literature on SaaS or transferring success factors of related fields. Against this background, the goal of this research paper is to investigate the business models' success factors of SaaS providers by conducting an exploratory multiple-case study. 21 expert interviews with representatives from 17 cloud providers offering SaaS services serve as main data collection method.

2   Related Work

Cloud computing is defined as *"[...] a model for enabling ubiquitous, convenient, on-demand network access to a shared pool of configurable computing resources (e.g., networks, servers, storage, applications, and services) that can be rapidly provisioned and released with minimal management effort or service provider interaction"* [13]. The literature distinguishes between three main service models – Infrastructure as a Service (IaaS), Platform as a Service (PaaS) and Software as a Service (SaaS). These three service models form layers which are interrelated, each building upon the former. A further differentiation is made between four deployment models: public, hybrid, private and community [1]. The characteristics of cloud services, including on-demand self-service, broad network access, resource pooling, rapid elasticity and measured service, distinguish it from its traditional counterpart the on premise IT [4].

Whereas research has primarily emphasized the technical aspects of cloud computing, significantly less consideration has been given to the substantial changes within the business perspective [1, 6]. A literature research of Herzfeldt, Floerecke, Ertl and Krcmar [6] revealed that three classes of publications from the cloud provider's business perspective have been advanced so far: papers that are (1) proposing models or algorithms focusing on the optimization of costs and other resources, (2) dealing with business models and ecosystem models, or (3) discussing fundamental business benefits and challenges of cloud computing from different perspectives and for various industries. Concerning (2), the introduction of cloud computing has radically changed the way IT resources are produced, provided and consumed [1]. Hence, it is considered as a co-evolution of computing technology and business models [14].

A business model, in general, is regarded as a tool for depicting, innovating and evaluating the business logic of a firm [15]. Even if no commonly accepted definition of the term *"business model"* has been established yet, the component-based view dominates the research [16]. Accordingly, a business model is a system comprising a set of components and the relationships between them [15]. There is, however, no consensus on the specific set of relevant components [17]. Nevertheless, a large number of cross-industry and industry-specific business model frameworks provide possible design options for selected components [16]. One comprehensive and widespread cross-industry framework is the Business Model Canvas [18], which includes nine components: key activities, key resources, partner network, value propositions, customer segments, channels, customer relationships, cost structure and revenue streams.

Specific research on cloud computing business models is nascent [4, 19]. Giessmann and Stanoevska-Slabeva [20] proposed a classification model for PaaS providers' business models and hypotheses regarding future directions. Giessmann and Legner [21] published a set of possible design principles that guide software providers to define PaaS business models. So far, the only holistic cloud-specific business model framework was developed by Labes, Erek and Zarnekow [22]: their morphological box entails various categories representing the basic components of a business model, each broken down into design features that show design options [22]. Based on this, Labes, Hanner and Zarnekow [8] analyzed the business models of selected cloud providers, matched them with the framework and identified four common patterns of cloud business models. Another research strand is the analysis of the impacts of the shift from an on premise to a cloud computing business model (e.g. [4, 5, 23–26]). Other scholars have dealt with the process itself of transforming an on premise to a cloud business model [27, 28]. For supporting this business model development process, Ebel, Bretschneider and Leimeister [29] developed and evaluated a software tool. A literature study of Labes, Erek and Zarnekow [19] shows that several further contributions have dealt with one or a small quantity of business model components, such as the revenue [30] or the resource model [6], while a holistic approach remains an exception. Such a strict separation, however, contradicts the logic of business models as the single components are understood as interrelated [15].

For a long time, business models have played a central role in explaining a firm's performance and deriving success factors [16, 17]. According to Rockart [31], success factors are defined as *"[...] the limited number of areas in which results, if they are satisfactory, will ensure successful competitive performance for the organization"*. Success factors are applicable to all companies in an industry with similar objectives and strategies [31, 32]. A distinction can be made between generic success factors, which are valid for all kind of companies, and domain-specific success factors, in this case cloud-specific success factors [8]. Therefore, it is difficult to transfer the success factors from adjacent research areas to the cloud computing ecosystem without prior examination. The literature's understanding of success factors of cloud providers' business models is limited [8, 9]. Beside a recent study focusing on success factors that relate to the providers' relationship with the consumer in the end consumer market [9], research has only provided one extensive analysis of the characteristics of cloud computing business model components and their link to business success: Labes, Hanner and Zarnekow [8] derived abstract success factors by relating publicly available characteristics of the business model components to a firm's web visibility and profit. However, both studies disregard that the cloud computing ecosystem contains providers fulfilling various roles and thus, is characterized by a high degree of heterogeneity [33]. Focusing on the ecosystem role of SaaS providers, Ernst and Rothlauf [12] proposed potential success factors by transferring success factors from related research fields. Walther, Plank, Eymann, Singh and Phadke [11] summarized single success drivers based on a review of general SaaS literature. Wieneke, Walther, Eichin and Eymann [10] conducted expert interviews with employees of a SaaS provider with particular focus on the value proposition, while neglecting the other business model components.

3 Research Design

3.1 Research Methodology

To investigate the success factors of SaaS providers' business models, a multiple-case study approach was selected [34, 35]. Yin [35] defines a case study as "*[...] an empirical inquiry that investigates a contemporary phenomenon (the "case") in depth within its real-life context, especially when the boundaries between phenomenon and context may not be clearly evident*". Case study research is suitable (1) to undertake research in a field in which few previous studies have been carried out, (2) to answer "*how*", "*why*" and "*what*" questions in order to understand the nature and complexity of the processes taking place and (3) to learn about the state of the art and generate theories from practice [34]. All of these three factors apply to the study at hand.

This case study follows the positivistic tradition [36]. More precisely, an inductive research approach was applied with the aim to reach predominantly exploratory conclusions [35]. In line with that, this research did not start with a specific hypothesis being tested. However, Eisenhardt [37] argues that a priori specification of constructs can be a helpful tool in shaping the initial design of a study. Therefore, this research used existing concepts from the literature on cloud computing business models and success factors for initial framing. But "*[...] no construct is guaranteed a place in the resultant theory*" [37], because of the study's exploratory nature.

3.2 Site Selection

Case studies can be categorized as single and multiple-case studies [36]. For this research, a multiple-case study approach was chosen as the intent was hypothesis and theory building. In addition, a multiple-case design allows cross-case analysis which yields more general research results and ensures internal validity [34, 35].

Multiple-case designs depend on careful case selection to maximize insights [36]. The units of analysis are individual SaaS providers. These were chosen for enabling literal (conditions of the case lead to predicting the same results) and theoretical replication logics (conditions of the case lead to predicting contrasting results) [35]. As a prerequisite, the respective provider had to be listed in a ranking as a successful provider by a leading research and advisory company like Gartner or Forrester. In this way, it was ensured that the basic settings of the providers were fundamentally the same (literal replication). For theoretical replication, the aim was to include providers with different experience, size, geographic coverage, number of occupied ecosystem roles, target markets, served industries, and assessment of the importance of cloud compared to on premise. Due to the heterogeneity of the cloud providers, it was possible to draw cross-case results, as it replicated findings across all cases and helped to detect similar and contrasting results, which lead to generalizable conclusions [35]. The selection procedure resulted in seven SaaS providers. This is in line with Eisenhardt [37] who recommends four to ten cases as a reasonable number to reach external validity. Besides, ten successful IaaS and PaaS providers that additionally offer SaaS services were chosen using the same case selection procedure.

3.3   Data Collection and Analysis

As data triangulation is highly recommended in case study research [34, 36], data collection relied on more than one source. The study started with a screening of the websites of the respective providers to gain background information. This initial data collection was directly incorporated in the following expert interviews with one or two representatives from each organization. The interview partners were selected based on the following criteria: the person should hold a managerial position and have responsibility for overseeing the organization's business models and strategies. Seven interviews across seven SaaS providers were conducted. Besides, fourteen interviews with IaaS and PaaS experts from ten cloud providers which also offer SaaS were performed. The 21 experts stemmed from twelve large and five medium-sized cloud providers, had between three and ten years' experience in the cloud field and held leading positions within their companies (board members, portfolio, product, sales, marketing and IT managers, and senior consultants). This research paper presents the integrated results as the discussions were not limited to the chosen cloud layer. Interviewees often drew on their experience gathered in SaaS. This may be due to the fact that the three cloud layers build up on one another. That proved to be a strong advantage as it increased the data pool and thus, the generalizability of the findings.

The interviews were based on a pre-tested interview guide, encompassing open-ended questions. When designing the interview guide, it had to be considered, whether to ask with a framework such as the Business Model Canvas [18] in mind or without a given set of business model components concerning success factors. In line with the study's exploratory character, it was decided not to push the experts into a given framework. Instead, they should be given the chance to think and answer freely, independent of predefined components. This approach was also deemed more appropriate by the two pre-test sessions. The resulting interview guide (available upon request from the author) focused on identifying the success factors of SaaS business models from various perspectives (Fig. 1). These perspectives were derived from the literature on collecting success factors [32] and on the characteristics of business models.

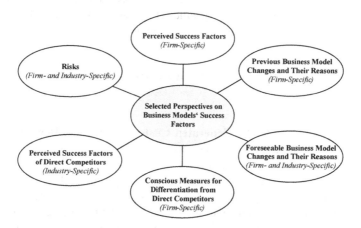

Fig. 1. Selected perspectives on success factors of cloud computing business models

The interview guide was not sent to the experts in advance as a spontaneous response was desired. The 21 interview sessions took place from June to November 2017 in a face-to-face manner or via telephone in German. The duration of the interviews ranged from 30 to 100 min. Whenever appropriate, the laddering technique was applied, which follows a process of digging deeper by asking further questions [38]. Thus, the interviews were conducted as guided conversations rather than structured interviews. In order to facilitate the data analysis, all interviews were recorded with the permission of the participants. Each interview was subsequently transcribed and proof read. As the experts were ensured anonymity, the data acquired was sanitized so that no individual person or organization can be identified. The data collection was undertaken until theoretical saturation was reached [39]. This was the case, when the answers of the experts were repeated many times and no new aspects were added, so that further data would have not provided additional insights. This was achieved relatively quickly as the fourteen experts for IaaS and PaaS often also referred to SaaS.

The data analysis process was conducted in two phases based on the recommendations of Corbin and Strauss [39] using MAXQDA. The first phase consisted of open coding – a line-by-line analysis of the transcribed data. The components of the Business Model Canvas [18] served as the basis categories of the coding scheme. The resulting codes were discussed iteratively with colleagues of the chair until consensus was obtained. Within this step, the number of codes was reduced. In the second phase, the codes were further consolidated via axial coding technique. The whole data analysis was an iterative process of coding data, splitting and combining categories, and generating new or dropping existing categories. A factor was classified as *"success"*, if it was expressed as such by various experts or if, in case of a single notion, convincing arguments were provided, or if it emerged as an important aspect during analysis of an individual case or if it was identified as such when conducting cross-case comparisons. In line with Corbin and Strauss [39], no attempt was made to statistically evaluate the importance of the results. Theoretical relevance of the concepts was established by their repeated presence or notable absence. As a SaaS provider is dependent on the IaaS and PaaS layers, only success factors that can be influenced actively by a SaaS provider were incorporated. Every success factor was further classified by the author as generic *(g)*, similar to other domains, or SaaS-specific *(s)*. In the following, the classification is noted in brackets after the designation of each factor.

4   Success Factors of SaaS Providers' Business Models

4.1   Revenue Streams

Setting Prices According to the Generated Customer Business Benefit *(s)*

For SaaS providers, a monthly subscription fee is the common pricing practice. A traditional basis for price calculation is the provider's incurring costs. A much more promising approach is pricing according to the generated customer benefit: the provider tries to tip into the additional value his service yields for the customer's business – direct cost savings or additional revenues. This has the benefit of being very transparent and comprehensive for clients which increases their willingness to pay. A prerequisite is that the provider maintains a close and long-term relationship with the client and understands his business processes. This approach has proved most suitable for applications related to production processes, but less for standard office software.

Offering Flexibility Within the Pricing Model *(s)*
Besides offering predefined pricing models, an open attitude towards individual requests from customers concerning the design of the price model can have a strong effect on customer satisfaction. According to the experts, a SaaS provider should not insist on a standard pricing model. Instead, a price model should take the characteristics of the business model and the situation of different customers into account. Therefore, a SaaS provider should, in certain cases, be capable to adapt the pricing model according to specific customer preferences.

Supporting of "Bring Your Own License" *(s)*
In order to attract new customers or to move long-term on premise customers into the cloud, it is crucial to allow bringing previously purchased licenses of on premise software for a corresponding cloud service. This is challenging in case the license stems from a third-party provider. An on premise license has a specific calculation basis, most commonly the customer's CPU cores. But due to virtualization of hardware, it is difficult to ensure that the service solely runs on the allowed number of CPUs in the cloud. Nevertheless, it is vital to comply with the license conditions.

4.2   Key Resources

Building Up Domain Knowledge and Industry Expertise *(g)*
SaaS providers must aim to accumulate a broad domain knowledge and industry expertise in the fields they operate in. Despite the importance of technical aspects of SaaS service implementation, it is a success factor to have a deep understanding of the customers in their situation within their industry. If providers are not able to build this up in-house, they must rely on partners that possess such knowledge. This is a prerequisite to ensure that the developed SaaS service delivers added value to customers.

Possessing the Leading Certificates *(s)*
A SaaS provider has to obtain the relevant certificates. These are commonly demanded within tendering processes. The importance of certificates varies in relation to client size and type of SaaS service: whereas small firms may partly consider certificates as dispensable, they are vital in an enterprise environment. SaaS services working with insensitive data commonly do not demand certifications. As the procedure of obtaining a certificate is time-consuming and expensive, smaller providers are unable to compete in this aspect. Providers offering SaaS exclusively are strongly advised to choose a certified IaaS provider.

Having a Multitude of Highly Qualified Employees *(g)*
The experts stressed the importance of having a multitude of highly qualified employees. This particularly includes software developers who implement the cloud service portfolio. Besides that, employees with technical know-how are needed to support the customers in case of uncertainties and problems over the whole cloud service lifecycle. A high volume of qualified employees generates speed and innovation

which is required in the rapidly growing and changing cloud computing ecosystem. However, the interviewees reported that it is becoming increasingly difficult to attract and retain skilled developers as the market for developers is running dry.

Owning a Large (Pre-cloud) Customer Base *(g)*
A large customer base that a provider established before the emergence of cloud computing is considered a high-valued resource. A successful, established business relationship provides easy and fast access to potential cloud customers: First of all, long-standing and satisfied customers have confidence in the qualities of the provider. Second, the customers know the provider's support processes and do not want to contact another provider when problems occur. Third, customers are used to the provider's applications, so they are commonly reluctant to change the provider.

4.3   Value Propositions

Offering Cloud Native Applications (Microservices) *(s)*
Traditional on premise software is not cloud ready, per se. This means, it is unable to meet the high expectations in terms of cost advantages, performance and scalability. Hence, it is mandatory to transform the architecture or to rebuild the whole application for cloud purposes. Specifically, SaaS services should be built as a system comprising modular microservices. These microservices are characterized by their ability to manage autonomously which and how much IaaS resources they require for their current operation. Deployed into an open source container, they can be directly run at different IaaS providers. In addition, these single modules can be combined according to the individual requirements of a specific customer.

Parallel Offering of Cloud and on Premise Applications *(s)*
Many customers have their core IT systems, e.g., manufacturing systems, internal and want to keep them as is over a longer period. This is mainly due to security, but also performance concerns. In total, the demand for on premise applications remains strong. Therefore, providers that have a high amount of clients within the on premise segment should in no case stop offering these systems and their support. Providing both SaaS and on premise applications is especially important for global players as they can thereby address the varying cloud adoption rates between countries.

Providing Adaptability of the Application on the Customer's Side *(s)*
SaaS customers are by definition expected to give up their desire for individual adaptions. Nevertheless, clients want to be able to parameterize and configure the SaaS service to their requirements and flavor. Otherwise, the probability that the SaaS service is used is low. A SaaS service should consequently cover a broad spectrum of best practices for the clients to choose from. This must be possible without any programming skills as SaaS is usually utilized by business users. For customers, it is not tolerable that each small adaption requests a change project.

Achieving a High User Experience of Cloud Services *(g)*
The experts agreed that the topic of user experience is highly significant in the SaaS field. The handling of SaaS services must be self-explanatory and simple, to enable immediate use. Traditional concepts which have been utilized within large on premise

software systems are less suitable. Customers desire an uncomplicated user experience, which they are accustomed to in their private environment. This has the additional advantage to minimize training course expenditures for customers' employees.

Offering Private Cloud Deployment Models *(s)*
Currently, the customer demand is significantly higher for private than for public cloud deployment models. Many clients consider private clouds as an interim solution on their way to public cloud. Private clouds are preferred due to data protection, security, regulation and compliance reasons. In addition, a private cloud allows a higher degree of customization which customers often demand. Ultimately, it is mandatory for SaaS providers to include private cloud offerings in their portfolio.

Guaranteeing a High Availability *(g)*
It is absolutely imperative to ensure a high availability of the offered SaaS services. However, negative incidents in the past prove how challenging this undertaking is. A long-lasting service failure comes along with a significant loss of customer confidence, leading to adverse effects on the provider's economic situation. Therefore, the specification and compliance of the service level agreements is very relevant.

Ensuring the SaaS Service to Run on All Leading Platforms *(s)*
A SaaS service has to be developed to be compatible with all leading IaaS/PaaS platforms. For this purpose, a SaaS provider has to be careful in utilizing proprietary PaaS services as these vary between providers. Otherwise, the SaaS service is tied to that specific PaaS platform making it hardly portable. Hence, the experts recommended to rely on standard open source services and to use platform-specific services only when absolutely necessary. This way, SaaS providers can ensure that their services are, if necessary with slight adaptions, compatible with most IaaS/PaaS platforms.

Offering Customer-Specific Service Individualization *(g)*
SaaS services are characterized by a high degree of standardization. Nonetheless, it is important to preserve a certain degree of flexibility. Some customers have more individual requirements that cannot be entirely met with the standard offering. Of course, each customization leads to additional costs. But customers are willing to pay for the considerable added value. Offering customer-specific adaptions is a way to differentiate from competitors. Smaller providers have an advantage as their organization structure is more flexible allowing them to address individual demands more easily.

Offering a Broad SaaS Service Portfolio *(g)*
The experts found offering a multitude of SaaS services important to succeed. The reasons for this are the following: First, a broad service portfolio achieves increased attractiveness of a provider as customers feel more prepared for the future. Second, customers always look for an answer to their individual requirements. By means of a broad portfolio a provider can better respond to that demand. Third, taking a broad approach offers a way to distinguish oneself from competitors.

Providing Extensive Customer Support *(g)*
Providing extensive support over the whole SaaS service lifecycle is essential. This includes support services related to selection and composition, as well as usage and operation of SaaS services. SaaS providers should commit to work very closely with

customers. A lot of customers appreciate a personal contact partner and are willing to pay extra for qualitative support. They want to call if a problem should occur. The customer support process has to be integrated, meaning that customers should not have to talk to more than one employee to identify and solve a problem.

4.4   Customer Relationships and Channels

Offering Personal Sale Beside Self-service *(g)*
Some SaaS providers have the illusion that each type of cloud service can be sold by self-service without considering the size of the client company. However, the acceptance of self-services decreases from IaaS toward SaaS. A reason for this is that SaaS services often have to be integrated in the customers' business processes and IT landscapes. Whereas small firms intensively use the self-service option, medium-sized and large enterprises commonly insist on personal contact to the provider combined with individual contract negotiations. Therefore, depending on the target group, it is mandatory to offer the additional possibility of engaging in negotiations.

Conducting Marketing Activities *(g)*
To stand out from the large number of SaaS providers, marketing is of great significance. In addition, due to the short contractual periods in the SaaS field, the customer royalty is decoupled. Further, the decision on the client's side in favor of a specific provider is often not based on performance features. On the contrary, the provider's image is a decisive factor. The importance of marketing is expected to grow further as the decision-making power moves increasingly towards business units. Examples for promising marketing channels are presentations at conferences, publication of articles, cooperation with universities, use of social media and buildup of communities. Moreover, the value of reference customers for newly developed services was emphasized.

Initial Explaining of the Cloud Computing Concept *(s)*
Many potential cloud customers have considerable doubts whether they should move into the cloud or not. This is especially the case for customers from German-speaking countries. The central concern is data security. Oftentimes the source of these doubts is a lack of knowledge. For SaaS providers, it is crucial to address these fears and unknowingness at the beginning of the customer relationship by explaining the general conditions of cloud computing in detail. This is a valuable contribution to establish trust.

Establishing a SaaS-Specific Incentive System for the Sales Division *(s)*
The traditional incentive system for the sales division which has worked well for on premise solutions is not directly transferable to SaaS. A long time it has been possible for a salesman to make deals at regular intervals within one customer relationship whereby he received his commissions regularly. In the case of SaaS, clients usually pay a subscription fee for a longer period of time. But from then on, no salesman is needed any more. By selling SaaS the salesman takes away his future up-selling options as customers get an all-inclusive deal. This makes it mandatory to develop a SaaS-specific incentive system to promote motivation within the sales division.

4.5   Key Activities

Utilizing Agile Software Development Models *(s)*
Traditional, sequential software development models such as the Waterfall Model [40] are not appropriate for the development of SaaS services. Instead, agile methods like Scrum [40] are substantial to realize short innovation cycles. Development speed is an important factor: new services or additional features have to be delivered continuously and fast on the platform in order to improve the portfolio. In the context of agile development, DevOps plays an important role, meaning there has to be a close connection between development and operation of any cloud service.

Conducting Research and Development *(g)*
The interviewed experts pointed out that research and development at a high level is a deciding factor in the rapidly changing cloud computing ecosystem. The risk of lagging behind in technological development is high, which is why many cloud providers are not only carrying out in-house research and development, but also acquire smaller cloud providers to increase their service portfolio and knowledge.

Involving Customers in the Development of New SaaS Services *(g)*
It is considered a serious mistake to develop a new SaaS service internally for an anonymous market. Instead, one should work closely together with customers. A promising source for new SaaS services is direct customer feedback. When customers request a feature or suggest a new SaaS service, it should be analyzed whether other customers might be interested, too. Another source for new SaaS services is to scale from customer-specific projects to other customers. Further, the establishment of customer workgroups for constant exchange with regard to customers' problems and expectations and to forecast future trends is seen as a valuable asset.

4.6   Partner Network

Building Up a Partner Ecosystem *(s)*
A thriving partner ecosystem firstly serves as a sales and marketing channel: cloud services are highly scalable, but it is impractical to interact personally with each client as the sheer volume of sales staff for this cannot be supported. Hence, partners can act as resellers of a provider's SaaS services. Thereby, a closer proximity to clients, further geographic coverage and outreach customer segments that do not exclusively fall into the provider's scope can all be achieved. Secondly, a SaaS provider cannot perform all additional services outside his core business himself. These services include training and support which are fundamental to enable optimum use of a SaaS service.

4.7   Customer Segments

Focusing on Ambitious Medium-Sized and Large Companies *(s)*
The target customer segment is mainly determined by the type of application: small firms commonly search for small and isolated SaaS services, larger firms require applications to be well integrated in their processes. The experts recommended focusing on a specific segment. The more sophisticated medium-sized and large companies are

considered as the most valuable one: (1) Providers can further sell on premise solutions. By focusing on small start-ups, one would miss this opportunity as they tend to obtain their whole IT from the cloud. (2) Many small firms lag behind in technological advances and business trends. (3) Larger firms have more financial resources.

Offering the Opportunity for Firms of All Sizes to Become a Customer *(g)*
Although it is primarily important to focus on a special target group, SaaS providers should also cater other possible clients. Small firms, in particular, should be given the opportunity to order a SaaS service and pay with credit card. Another option is to distribute special SaaS services for smaller companies through partner firms. To reach larger companies, a SaaS provider must mandatorily offer personal support as well. The overall target should be to reach as many customers as possible.

5   Discussion

This study revealed that several SaaS providers simply transfer their traditional on premise applications to a virtual server without modification and offer it as SaaS. However, this way, the requirements in terms of cost advantages, performance and scalability cannot be met. Instead, a SaaS application should be developed as a system comprising modular microservices that are characterized by the ability to autonomously manage their need of IaaS resources for operating a specific task. As a consequence, the main concern of IaaS providers will no longer be to solely offer a virtual server because it is becoming less common for customers to order a pre-defined amount of compute, storage and network resources. The current IaaS business model is therefore expected to change drastically in the near future.

Moreover, it turned out that demand for on premise solutions will remain strong over the next years. However, many providers have emerged on the basis of the cloud concept and thus do not have any existing on premise offering. On the other hand, a lot of established IT firms have just begun developing a cloud version of their applications. Thus, providers who are already able to offer applications for both worlds have a great advantage. A promising strategy is to develop each new application as a cloud native version because it must only be slightly adapted to be sold as on premise.

In addition, private clouds are currently very popular. They are mostly seen as an interim solution on the way to public clouds. The experts hence predict a significant rise in demand for public clouds in the coming years. Nonetheless, clients from certain sectors such as the public or the banking sector are expected to remain skeptical. Also, only a few firms are likely to transfer sensitive data into public clouds.

Furthermore, this study showed the major advantage of openness towards client-specific requests, e.g., regarding the pricing model or the design of the SaaS service. Related to that, ordering a SaaS service via self-service is hardly accepted particularly among medium-sized and large companies. Personal contact is preferred. However, addressing client-specific requests contradicts the SaaS providers' goal of achieving economies of scale.

Concerning the business model development and innovation process as such, it became obvious that a large majority of the considered cloud providers does, contrary to the recommendation of scholars [15, 16], not undergo a systematic, phase-oriented process. It is rather conducted within occasional workshops in which new ideas regarding the business models are collected and then released for implementation.

6 Summary, Limitations and Outlook for Future Work

In order to investigate the success factors of SaaS providers' business models, an exploratory multiple-case study was conducted. This resulted in a catalogue of 27 success factors. Most success factors relate to the value proposition, whereas the cost structure was not addressed by the experts. Besides being a promising starting point for further research, the results are particularly useful for practitioners. Established SaaS providers get a reference framework to compare, rethink and innovate their present business models. Companies that are planning to offer SaaS in future gain valuable insights that should directly feed into their business model design process.

The limitations of this study include the relatively small number of SaaS providers. However, through the integration of the insights of fourteen interviews with IaaS and PaaS experts who also gathered experience in the SaaS field, the data pool and hence, the generalizability of the findings was significantly increased. A second limitation may be the geographic scope being centered on Germany. But as the majority of the selected cases consisted of internationally operating providers, this influence is regarded as low. Despite the valuable results achieved, there remains a considerable need for further research: First, as not all derived success factors are of equal importance, their relevance has to be ranked. Second, it should be investigated to which extent the success factors are currently covered in practice. Third, the focus of this study was laid upon the isolated effects of individual success-driving business model characteristics. As a business model is a system comprising a set of components and the relationships between them, the combined effect of specific characteristics has to be examined. Fourth, as success factors may change over time, they have to be reassessed in future regarding their ongoing relevance. Thereby, researchers might use a given business model framework to overcome the unequal distribution of success factors. Finally, due to the applied research design, the direct comparability of the study's results with the existing, isolated contributions on success factors of SaaS business models is only possible to a limited extent. Against this background, there is the necessity to conduct a systematic, comparative and integrative meta-study.

Acknowledgements. The author sincerely thanks Prof. Dr. Franz Lehner, holder of the Chair of Information Systems II (Information and IT Service Management) at University of Passau, for making this study possible. A further thank you goes to the interview partners, whose names cannot be mentioned without violating their guaranteed anonymity, for their effort, time and openness. The author also wants to thank Svea Buttgereit, student assistant at the chair, for her valuable support and input during the interview analysis.

References

1. Marston, S., Li, Z., Bandyopadhyay, S., Zhang, J., Ghalsasi, A.: Cloud computing – the business perspective. Decis. Support Syst. **51**(1), 176–189 (2011)
2. Pettey, C.: Gartner Says Worldwide Public Cloud Services Market to Grow 18 Percent in 2017. Gartner, Stamford (2017)
3. Iyer, B., Henderson, J.C.: Business value from clouds: learning from users. MIS Q. Exec. **11**(1), 51–61 (2012)
4. Clohessy, T., Acton, T., Morgan, L., Conboy, K.: The times they are a-chaning for ICT service provision: a cloud computing business model perspective. In: European Conference on Information Systems, Istanbul, Turkey (2016)
5. DaSilva, C.M., Trkman, P., Desouza, K., Lindič, J.: Disruptive technologies: a business model perspective on cloud computing. Technol. Anal. Strateg. Manag. **25**(10), 1161–1173 (2013)
6. Herzfeldt, A., Floerecke, S., Ertl, C., Krcmar, H.: The role of value facilitation regarding cloud service provider profitability in the cloud ecosystem. In: Khosrow-Pour, M. (ed.) Multidisciplinary Approaches to Service-Oriented Engineering, pp. 121–142. IGI Global, Hershey (2018)
7. Henkes, A., Heuer, F., Vogt, A., Heinhaus, W., Giering, O., Landrock, H.: Cloud vendor benchmark 2016: Cloud Computing Anbieter im Vergleich. Experton Group (2016)
8. Labes, S., Hanner, N., Zarnekow, R.: Successfull business model types of cloud providers. Bus. Inf. Syst. Eng. **59**(4), 223–233 (2017)
9. Trenz, M., Huntgeburth, J., Veit, D.: How to succeed with cloud services? Bus. Inf. Syst. Eng. 1–14 (2017). https://doi.org/10.1007/s12599-017-0494-0
10. Wieneke, A., Walther, S., Eichin, R., Eymann, T.: Erfolgsfaktoren von On-Demand-Enterprise-Systemen aus der Sicht des Anbieters – Eine explorative Studie. In: International Conference on Wirtschaftsinformatik, Leipzig, Germany (2013)
11. Walther, S., Plank, A., Eymann, T., Singh, N., Phadke, G.: Success factors and value propositions of software as a service providers – a literature review and classification. In: American Conference on Information Systems, Seattle, USA (2012)
12. Ernst, C.P.H., Rothlauf, F.: Potenzielle Erfolgsfaktoren von SaaS-Unternehmen. In: Multikonferenz Wirtschaftsinformatik, Braunschweig, Germany (2012)
13. Mell, P., Grance, T.: The NIST Definition of Cloud Computing. NIST, Gaithersburg (2011)
14. Iyer, B., Henderson, J.C.: Preparing for the future: understanding the seven capabilities of cloud computing. MIS Q. Exec. **9**(2), 117–131 (2010)
15. Wirtz, B.W., Pistoia, A., Ullrich, S., Göttel, V.: Business models: origin, development and future research perspectives. Long Range Plan. **49**(1), 36–54 (2016)
16. Zott, C., Amit, R., Massa, L.: The business model: recent developments and future research. J. Manag. **37**(4), 1019–1042 (2011)
17. Lambert, S.C., Davidson, R.A.: Applications of the business model in studies of enterprise success, innovation and classification: an analysis of empirical research from 1996 to 2010. Eur. Manag. J. **31**(6), 668–681 (2013)
18. Osterwalder, A., Pigneur, Y.: Business Model Generation: A Handbook for Visionaries, Game Changers, and Challengers. Wiley, New Jersey (2010)
19. Labes, S., Erek, K., Zarnekow, R.: Literaturübersicht von Geschäftsmodellen in der Cloud. In: Internationale Tagung Wirtschaftsinformatik, Leipzig, Germany (2013)
20. Giessmann, A., Stanoevska-Slabeva, K.: Business models of platform as a service (PaaS) providers: current state and future directions. J. Inf. Technol. Theory Appl. **13**(4), 31–55 (2012)

21. Giessmann, A., Legner, C.: Designing business models for cloud platforms. Inf. Syst. J. **26** (5), 551–579 (2016)
22. Labes, S., Erek, K., Zarnekow, R.: Common patterns of cloud business models. In: Americas Conference on Information Systems, Chicago, IL (2013)
23. Hedman, J., Xiao, X.: Transition to the cloud: a vendor perspective. In: Hawaii International Conference on System Sciences, Manoa (2016)
24. Boillat, T., Legner, C.: From on-premise software to cloud services: the impact of cloud computing on enterprise software vendors' business models. J. Theor. Appl. Electron. Commerce Res. **8**(3), 39–58 (2013)
25. Korpela, K., Mikkonen, K., Hallikas, J., Pynnönen, M.: Digital business ecosystem transformation – towards cloud integration. In: Hawaii International Conference on System Sciences, Manoa (2016)
26. Morgan, L., Conboy, K.: Value creation in the cloud: understanding business model factors affecting value of cloud computing. In: Americas Conference on Information Systems, Chicago (2013)
27. Khanagha, S., Volberda, H., Oshri, I.: Business model renewal and ambidexterity: structural alteration and strategy formation process during transition to a cloud business model. R&D Manag. **44**(3), 322–340 (2014)
28. Kranz, J.J., Hanelt, A., Kolbe, L.M.: Understanding the influence of absorptive capacity and ambidexterity on the process of business model change – the case of on-premise and cloud-computing software. Inf. Syst. J. **26**(5), 477–517 (2016)
29. Ebel, P., Bretschneider, U., Leimeister, J.M.: Leveraging virtual business model innovation: a framework for designing business model development tools. Inf. Syst. J. **26**(5), 519–550 (2016)
30. Al-Roomi, M., Al-Ebrahim, S., Buqrais, S., Ahmad, I.: Cloud computing pricing models: a survey. Int. J. Grid Distrib. Comput. **6**(5), 93–106 (2013)
31. Rockart, J.F.: Chief executives define their own data needs. Harv. Bus. Rev. **57**(2), 81–93 (1979)
32. Leidecker, J.K., Bruno, A.V.: Identifying and using critical success factors. Long Range Plan. **17**(1), 23–32 (1984)
33. Floerecke, S., Lehner, F.: Cloud computing ecosystem model: refinement and evaluation. In: European Conference on Information Systems, Istanbul, Turkey (2016)
34. Benbasat, I., Goldstein, D.K., Mead, M.: The case research strategy in studies of information systems. Manag. Inf. Syst. Q. **11**(3), 369–386 (1987)
35. Yin, R.K.: Case Study Research: Design and Methods. Sage Publications, Thousand Oaks (2014)
36. Dubé, L., Paré, G.: Rigor in information systems positivist case research: current practices, trends, and recommendations. Manag. Inf. Syst. Q. **27**(4), 597–636 (2003)
37. Eisenhardt, K.M.: Building theories from case study research. Acad. Manag. Rev. **14**(4), 532–550 (1989)
38. Corbridge, C., Rugg, G., Major, N.P., Shadbolt, N.R., Burton, A.M.: Laddering: technique and tool use in knowledge acquisition. Knowl. Acquis. **6**(3), 315–341 (1994)
39. Corbin, J., Strauss, A.: Basics of Qualitative Research: Techniques and Procedures for Developing Grounded Theory. SAGE Publications, Thousand Oaks (2008)
40. Ruparelia, N.B.: Software development lifecycle models. ACM SIGSOFT Softw. Eng. Notes **35**(3), 8–13 (2010)

End-to-End Methodological Approach for the Data-Driven Design of Customer-Centered Digital Services

Jürg Meierhofer[1,3]([⊠]) and Anne Herrmann[2,3]

[1] School of Engineering, ZHAW Zurich University of Applied Sciences,
Winterthur, Switzerland
juerg.meierhofer@zhaw.ch
[2] School of Applied Psychology,
FHNW University of Applied Sciences and Arts Northwestern Switzerland,
Olten, Switzerland
anne.herrmann@fhnw.ch
[3] Expert Group Smart Services, Swiss Alliance for Data-Intensive Services,
Thun, Switzerland

Abstract. The collection, analysis, and interpretation of digital data has become an important factor for the provision of services. However, there is a lack of methodologies for using data analytics systematically in an end-to-end process for designing services. Therefore, in this paper, we develop a conceptual approach covering the innovation funnel from idea generation to market deployment. In particular, we describe how qualitative approaches alternate with quantitative approaches along the innovation process. We pay special attention to the design of data-driven value propositions including the analysis and modeling of the customer needs, a phase in which the concept of hidden needs and pains is applied. To conclude, we propose the development of a tool to support and industrialize the approach discussed in this paper.

Keywords: Data-driven value creation · Data-driven innovation process
Service innovation

1 Data-Driven Services

This paper describes the concept of a new approach and a tool for data-driven service innovation. The goal is to develop a methodological approach and to conceptualize a tool for the systematic procedure along the innovation process starting from finding the appropriate innovation field to testing and deploying the services in the field and in the market.

Data and analytics have become a relevant differentiator for service-based business models [1]. The application of data science in services is considered the next frontier of innovation and customer-oriented value creation [2]. Most literature sources dealing with data-driven services and products consider the data-driven value creation by the actual product or service. That means they are not taking into account that data and analytics can also create value in the course of the entire service innovation process.

© Springer Nature Switzerland AG 2018
G. Satzger et al. (Eds.): IESS 2018, LNBIP 331, pp. 208–218, 2018.
https://doi.org/10.1007/978-3-030-00713-3_16

Data-driven value propositions make use of data-based insights during the usage of the service. In [3] and [4], for example, a wide range of data-driven applications is discussed. These applications are classified into domains like services for personal relationships, family and life, for marketing and advertisement, financial and risk management, healthcare, law enforcement, fault detection, safety, logistics, manufacturing and process management, governmental and public services, human resources management etc. According to [5], the value chain for data-driven services can be depicted as shown in Fig. 1. Using analytics and data science, insights can be created from data. However, insights by themselves do not yet make up a service because it is not clear whether they provide value to a user. Many practical cases have shown that insights generated by analytics do not automatically generate business value. Instead, these insights need to be transformed into value for users in order to contribute to a service. Providing a relevant value to users increases their willingness to pay for that value, and hence the service. This payment can also be provided in non-financial forms (e.g., by providing data or other forms of personal investment). In Sect. 2 we will discuss in more details how to determine the appropriate service values for users in a given context.

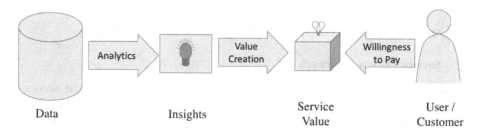

Fig. 1. The data-driven service value chain

For a given service design challenge, it is often not clear how to apply data science for creating service value for specific customers. [6] and [7] postulate that data-driven services must create actionable outcomes that help customers to reach their goals. As shown in the previous sections, there are sources in the literature describing the application of data for value generation in products or services. However, there is still a lack of methodology about data-driven innovation and design procedures for the development of such new value propositions. The support of the service innovation process with data science-based approaches is discussed in the literature. Still, the focus is primarily on the application of data science for analyzing customer behavior or create a customer segmentation (e.g., [8–10]).

Further research for data-driven service design is relevant and required [11, 12]. The goal is to design and engineer services that are consistently customer-centered, i.e., derived from the customer's requirements, and that leverage the potential of data science. In [5] and [13], the application of data science tools along the service design process is described, whereby a categorization of data science in nine different business-oriented outcomes is applied according to [14]. In [15], the data-based elicitation of customer needs based on the analysis of social media data is described for the detection of unsatisfied needs from individuals. However, this does not yet encompass a comprehensive end-to-end approach for data-driven service innovation, which would also cover the early innovation stages like defining the innovation field and ideation, continuing with customer analysis and the design of the actual data-driven value proposition, and finally concluding with the customer testing of the service and its market launch.

The approach in this paper differentiates itself from the existing literature in the following ways: it is based on [5] and [13], but considers the end-to-end service innovation process taking into account all phases from defining the innovation field to testing and launching the service. A special focus is put on the phase for designing data-driven value propositions including the systematic assessment of hidden customer pains.

In Sect. 2, the customer-centered service design and its challenges are discussed. This creates the basis for the further development of a data-driven innovation process which is then described in Sect. 3. We conceptualize a data-driven process approach for the end-to-end design of services and suggest the development of a tool to support this process.

2   Customer-Centered Service Design and Its Challenges

2.1   Service Design Approach

The approach discussed in this paper is aligned with the concept of the service-dominant (S-D) logic, which defines service as the application of competences (knowledge and skills) for the benefit of another entity or the entity itself [16–18]. According to the S-D logic, value is always co-created by the customer in this context. Value creation is deployed over a period of time which exceeds the discrete moment of sales and distribution [18]. E.g., the service value creation may start with the customer evaluating the service, continue with learning to use it, paying, renewing it etc. This can be described by the concept of the customer journey, which is the interaction journey of the user [19, 20]. With respect to the customer benefit, each use of a service creates a different experience and results in an assessment by the customer, depending on his individual situation and context. When designing a new service, it is therefore essential to first define the target customer and to explore the needs for service in the specific context. This is often done using the concept of the persona [20].

The next step is then to design a value proposition that provides relevant service benefits for the chosen persona. Approaches for the integrated design of service experiences that are methodologically based on the theory of service science are described in the literature [18, 19, 21, 22]. Here, the co-creation of value and service experiences plays a central role. The research-based approaches for service design have been transferred into a set of methodologies for practitioners. They have been applied and tested in practice for designing relevant value for customers (e.g., [20, 23, 24]). Service is designed in an iterative process consisting of several phases in which there are specific design challenges to be solved. With the data available today and the tools to analyze it, better solutions for these challenges can be found. Still, there is no systematic procedure available to do this.

The customer receives a service along the steps of the customer journey. Hence, a service is an experience over time covering all phases of the customer journey [19, 21]. Thus, service design takes care of how and in which context potential customers become aware of the service, how they buy, install and start using the service, how they use and benefit from the service and what they experience at the end of the customer journey when they leave or renew the service. The evaluation of the customer needs represents a central starting point. Value propositions are designed, tested and improved together with the customer in order to provide the right benefit in the right context.

To start, we now put the focus on the problem of understanding the customer's needs and designing value propositions to meet those needs. The design of service value propositions needs to take into account not only functional, but also emotional and social needs of the individuals. The literature [23] provides templates and procedures for the systematic design of value propositions according to which the customer needs are assessed in terms of customer jobs, pains, and gains (see Fig. 2).

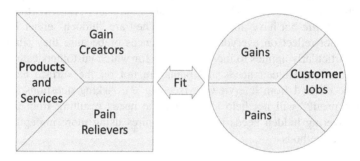

Fig. 2. Value proposition canvas (adapted from [23])

Services are designed in iterative procedures in close co-creation with the customer: rapid service prototypes are implemented in very short cycles and tested with customers [19]. This yields feedback about the assumptions made about customer jobs, pains, and gains and about the presumed relevant value proposition. Based on the feedback, new assumptions are made and implemented again using rapid service prototyping. After a sufficient number of such iterations, the design process is expected to converge to a solution that is good enough to be brought to market [21, 23].

2.2 Challenges of the Service Design Approach

The service design approach as described in Sect. 2.1 has proven to be very effective in practice for designing new service value propositions that are relevant for users or customers [25]. However, there are also a couple of unresolved problems and challenges.

First, although the iterative design process is effective in the sense of avoiding costly implementations of irrelevant solutions thanks to the "fail early, fail cheap" principle, it lacks quantitative evaluation techniques and is mainly based on qualitative methods for testing solutions [26]. In other words, the process from assessing the user needs by interviews to translating them into solutions by value proposition design and testing them in the field with test users is highly qualitative and thus takes long to converge towards a solution.

Second, due to the qualitative approaches in assessing and testing the user needs, the conclusions drawn from the experiments depend on the people that are involved. I. e., the characteristics of the people and teams involved play an important role [27]. In practice, too, the authors have observed some kind of arbitrariness in the conclusions of service design workshops. Giving the same design challenge to different design teams may result in completely different outcomes. The literature reports that service design-based approaches may not look scientific and the results often have the character of visions or wish scenarios [28].

Third, although the design-based approach aims at assessing the underlying motivational needs of the users or customers, it is often not obvious whether the real needs have been detected. When referring to needs, a common concept is this of "hidden needs and pains" [29]. Hidden pains are issues and problems the customers face in everyday life but have not yet realized. They are "hidden" either because the customers do not reflect on everyday inconveniences or because they simply accept those. This particularly applies to inconveniences for which customers cannot imagine a practical solution. Hidden needs and pains are not on the surface hence cannot necessarily be elicited from interviewing (see Fig. 3). Asking questions such as "so what do you need?" will not help to identify the needs resulting from these pains. Instead, discovering hidden needs and pains requires using more in-depth observant and explorative methods.

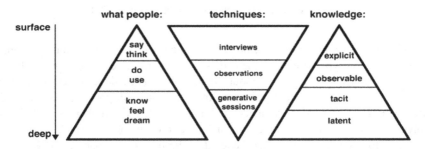

Fig. 3. Levels of knowledge about experience and their assessment [35]

These are used to deduce how products or services have to be designed to be perceived as useful and convenient by customers, thus increasing the likelihood of a successful launch of the innovation.

3 Data-Driven Process Approach for the End-to-End Design of Services

3.1 The Innovation Process Funnel

The innovation funnel is a typical form to represent the structured procedure of the innovation process. There are several forms reported in the literature (e.g., [30–33]). They all have in common that they start with a widely open exploration and converge over time towards a commercial solution that is brought to market launch (see Fig. 4). The exploration starts with scanning market and technology opportunities and transitions to idea generation, which is strongly based on creativity techniques. In the subsequent phase, customer-centered service design approaches as discussed in Sect. 2.1 are applied. After successful service design, the solution is tested in the market and, if again successful, finally launched commercially. During this process, the search field for service innovation is narrowed down by continuous iterations of "design-test-improve" cycles, as represented by the converging funnel in Fig. 4. By testing and discarding ideas and rapid service prototypes, the number of candidate services (dots inside the funnel in Fig. 4) is continuously reduced until there is ideally one tested and approved service left over for launch.

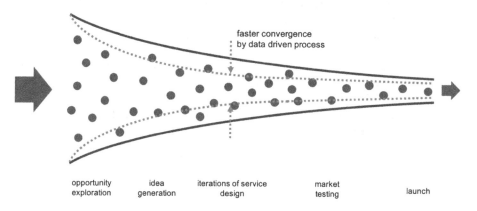

Fig. 4. Innovation process funnel

The procedure along the innovation process funnel shown in Fig. 4 is effective in the sense that it systematically provides relevant results. However, as discussed in Sect. 2.2, the process of exploring opportunities, generating ideas, designing and testing service prototypes and market testing is largely based on qualitative methodologies and therefore highly variable and lacks scientific and data-based approaches [26–28]. Co-creating with people, leveraging their mind and intuition is indispensable for generating new ideas and validating them. However, with the broad availability of data in all kinds of domains today, it is possible to shorten the cycles and speed up the end-to-end innovation process over all. Leveraging data and technology is a research priority that needs to be interlinked with service design and service innovation and represents an emerging area [34]. An approach to leverage data along the innovation funnel is described in [33]. In the innovation funnel of Fig. 4, this means that the funnel converges faster (dotted line), meaning also that the stages of the process move to the left on the time axis, which allows for an earlier market launch.

3.2   Data-Driven Process Along the Innovation Funnel

As stated in the previous section, data and analytics have the potential to advance the steps along the innovation funnel much faster by complementing the qualitative and analogous approaches thanks to data-based decision support. In the exploration of opportunities and generation of ideas, for instance, uncovered needs can be identified by mining data of social media and other publicly available sources (as described in [15]).

For the iterative service design process, the leverage of data has been described in [5] and [13]. A simplified version of the concept discussed in these references is shown in Fig. 5.

– The horizontal axis in Fig. 5 depicts the stages of a typical service design process for a given innovation idea. From left to right, this process typically has the steps "customer insight research", "customer modelling" (right hand side of Fig. 2), the design of the actual value proposition (left hand side of Fig. 2), and the testing of the value proposition with the customer in several iterations.
– The vertical axis in Fig. 5 outlines potential outcomes contributed by the data available in the given business situation. This may be, for example, finding associations with other entities or users, or predicting a numerical value for, e.g., the intensity of the service usage, or assigning an entity or a user to a certain class, e.g., a segment.
– The dots in the coupling matrix in Fig. 5 indicate at which stage of the service design process which data analytics outcome can be applied. These dots in the matrix depend on the specific case, its context, and the available data.

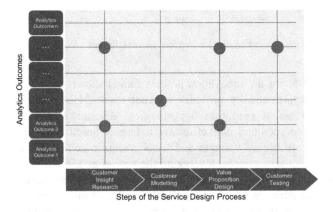

Fig. 5. Coupling matrix mapping analytics outcomes to the stages of the service design process

Examples:

- One block in the vertical axis may be based on data from a large pool of customers and their jobs, pains, and gains from another market. We may then apply this data to model the behavior of these customers in our own market with a different context and how they would interact with our new value proposition.
- We may have data available of the usage patterns of customers of a service that has sufficient similarity with our new service idea. This may then allow for modelling the take-rate and intensity of usage of our new product.
- In the customer testing phase, data may be used for optimizing the test sample and to draw conclusions to the larger customer base, for example.

As said, in the service design phase, a special focus is put on the identification of the hidden needs and pains. Discovering these hidden needs and pains requires more in-depth observational and explorative methods. This is part of the service engineering processes. Several methods can be utilized to identify hidden needs and pains

- **Repertory grid technique (RGT):** Originally, the RGT was developed to understand individuals' cognitive maps and their tacit knowledge [36]. This makes it a particularly useful technique to identify knowledge, convictions and expectations that are "below the surface" (see Fig. 3). Specifically, using indirect questions and asking users to compare and contrast their experiences of existing products and services allows uncovering customers' hidden needs and pains. Consequently, RGT has been used more frequently in the recent past to develop new product or service ideas [29, 37].
- **Ethnographic market research (EMR):** Conventional interviewing has its limitations when it comes to uncovering hidden needs and pains. Instead, EMR uses contextual inquiry, i.e. combining classical interview techniques with (1) contextual interviewing while the participant is acting in the relevant context and (2) with observations [38]. For the latter, video ethnography is applied frequently to allow for a more detailed, post-hoc analysis of the behavior in the relevant context. Video ethnography has the advantage of being less intrusive, hence making the results of the research less researcher-biased [39]. Through a combination of interviewing and observing the blind spots can be reduced by "expressing what cannot be observed" and "observing what cannot be expressed". Thus, it provides a deeper understanding of the behavior in the relevant context, making it more likely to uncover hidden needs and pains in the process.
- **Lead user technique [40]:** Lead users are defined as individuals who have a highly differentiated usage behavior. This makes it more likely that they have specific needs that are already on the surface (see Fig. 3) while these same needs are still "below the surface" amongst typical users. Furthermore, due to the lead users' extensive usage and more specific needs, they might modify products or services or use them in unforeseen ways to meet their needs [41]. Understanding the underlying motivations for these modifications provides insights into currently hidden needs of the typical users. Consequently, lead users are considered valuable sources to identify currently hidden needs amongst the typical users [42], functioning as a "need-forecasting laboratory" [40, p. 791].

Identifying the hidden needs and pains in service engineering has several benefits: First, it provides insights into current barriers preventing people from doing certain things and using certain services. Second, it provides guidance on how services have to be designed or may have to be changed to be more useful or more pleasant to use. Third, when viewed according to the Kano Model, providing solutions for hidden needs allows going beyond delivering basic or performance features and offering "excitement features", hence leading to higher customer satisfaction.

Identifying the hidden needs and pains as part of the Customer Insight Research is the first step in the service design process (see Fig. 5). These can be combined with analytics outcomes to generate insights that can be used in the subsequent steps in the process.

Towards the end of the innovation funnel (referring back to the innovation process funnel shown in Fig. 4) the service design phase is followed by the market testing and launch phase. Here, market simulation tools can be applied in order to model and anticipate the growth rate under different scenarios. For the launch phase, data and analytics can be used to identify the appropriate segments for targeted marketing.

4   Conclusions and Outlook

In this paper, the design of customer-centered services was investigated from the perspective of the end-to-end service innovation process. The systematic approach for the design of service value propositions that are derived from the customers' needs is shown. It was discussed that the challenges of this approach lie mainly in the fact that the process is slowly converging, that the results are subject to a certain arbitrariness and that the hidden user needs are often not yet uncovered. Hence, a new approach addressing these limitations is needed.

As an alternative, a new approach for the innovation and the design of services is suggested that is based on leveraging data along the entire process. This enables a more systematic development process and shorter development cycles, which results in a faster convergence of the innovation funnel. In addition, the systematic and data-driven approach allows for a better identification of hidden user needs, which in return provides more relevant service value propositions and, in the end, increases the economic value of the service.

The new approach outlined in this paper should be made available to be used in the service innovation process by service practitioners. Further research will therefore focus on the industrialization of the new approach by implementing it in the form of a tool. This tool will include a mixture of data-driven, partially automated steps that interchange with manual design steps. It will support the service designer by means of a prescribed workflow. Parts of this workflow will link to analytics components for supporting design decisions, whereas other parts will instruct the designer in the application of non-digital design steps.

References

1. Schüritz, R., Satzger, G.: Patterns of data-infused business model innovation. In: 2016 IEEE 18th Conference on Business Informatics (2016)
2. Schüritz, R., Seebacher, S., Satzger, G., Schwarz, L.: Datatization as the next frontier of servitization – Understanding the challenges for transforming organizations. In: Thirty Eighth International Conference on Information Systems, South Korea (2017)
3. Siegel, E.: Predictive Analytics. Wiley, Hoboken (2016)
4. Marr, B.: Big Data: Using Smart Big Data, Analytics and Metrics to Make Better Decisions and Improve Performance. Wiley, Chichester (2015)
5. Meierhofer, J., Meier, K.: From data science to value creation. In: Za, S., Drăgoicea, M., Cavallari, M. (eds.) IESS 2017. LNBIP, vol. 279, pp. 173–181. Springer, Cham (2017). https://doi.org/10.1007/978-3-319-56925-3_14
6. Loukides, M.: The Evolution of Data Products. O'Reilly, Sebastopol (2011)
7. Howard, J., Zwemer, M., Loukides, M.: Designing Great Data Products. O'Reilly, Sebastopol (2012)
8. Scherer, J.O., Kloeckner, A.P., Duarte Ribeiro, J.L., Pezzotta, G., Pirola, F.: Product-service system (PSS) design: using design thinking and business analytics to improve PSS design. Procedia CIRP **47**, 341–346 (2016)
9. Wang, B., Miao, Y., Zhao, H., Jin, J., Chen, Y.: A biclustering-based method for market segmentation using customer pain points. Eng. Appl. Artif. Intell. **47**, 101–109 (2016)
10. Kwong, C.K., Huimin, J., Luo, X.G.: AI-based methodology of integrating affective design, engineering, and marketing for defining design specifications of new products. Eng. Appl. Artif. Intell. **47**, 49–60 (2016)
11. Peters, C., Maglio, P., Badinelli, R., Harmon, R.R., Maull, R., Spohrer, J.C., et al.: Emerging digital frontiers for service innovation. Commun. Assoc. Inf. Syst. **39**(1), 136–139 (2016)
12. Spohrer, J., Demirkan, H., Lyons, K.: Social value: a service science perspective. In: Kijima, K. (ed.) Service Systems Science. TSS, vol. 2, pp. 3–35. Springer, Tokyo (2015). https://doi.org/10.1007/978-4-431-54267-4_1
13. Meierhofer, J.: Service value creation using data science. In: Gumesson, E., Mele, C., Polese, F. (eds.) Service Dominant Logic, Network and Systems Theory and Service Science: Integrating Three Perspectives for a New Service Agenda. Youcanprint Self-Publishing, Rome (2017)
14. Provost, F.P., Fawcett, T.: Data Science for Business. O'Reilly, Sebastopol (2013)
15. Kuehl, N., Scheurenbrand, J., Satzger, G.: Needmining: identifying micro blog data containing customer needs. Research papers, p. 185 (2016)
16. Vargo, S., Lusch, R.: Evolving to a new dominant logic for marketing. J. Mark. **68**(1), 1–17 (2004)
17. Vargo, S., Maglio, P., Akaka, M.A.: On value and value co-creation: a service systems and service logic perspective. Eur. Manag. J. **26**(3), 145–152 (2008)
18. Jaakkola, E., Helkkula, A., Aarikka-Stenroos, L.: Service experience co-creation: conceptualization, implications, and future research directions. J. Serv. Manag. **26**(2), 182–205 (2015)
19. Patrício, L., Fisk, R.P., Falcão e Cunha, J., Constantine, L.: Multilevel service design: from customer value constellation to service experience blueprinting. J. Serv. Res. **14**(2), 180–200 (2011)
20. Polaine, A., Løvlie, L., Reason, B.: Service Design: From Insight to Implementation. Rosenfeld Media, Brooklyn (2013)
21. Teixeira, J.G., Patrício, L., Huang, K.-H., Fisk, R.P., Nóbrega, L., Constantine, L.: The MINDS method: integrating management and interaction design perspectives for service design. J. Serv. Res. **20**(3), 240–258 (2016)

22. Andreassen, T.W., et al.: Linking service design to value creation and service research. J. Serv. Manag. **27**(1), 21–29 (2016)
23. Osterwalder, A., Pigneur, Y., Bernarda, G., Smith, A.: Value proposition design: how to create products and services customers want. Wiley, Hoboken (2014)
24. Brenner, W., Uebernickel, F. (eds.): Design Thinking for Innovation. Springer, Cham (2016). https://doi.org/10.1007/978-3-319-26100-3
25. Kimbell, L.: Rethinking design thinking: part I. Design and culture. J. Des. Stud. Forum **3** (3), 285–306 (2011)
26. Mueller, R., Thoring, K.: Design thinking vs. lean startup: a comparison of two user-driven innovation strategies. In: International Design Management Research Conference (2012)
27. Badke-Schaub, P., Cardoso, C.: Design thinking: a paradigm on its way from dilution to meaninglessness? In: Proceedings of the 8th Design Thinking Research Symposium (DTRS8) (2010)
28. Hentschel, C., Czinki, A.: Design thinking as a door-opener for TRIZ - paving the way for systematic innovation. In: Aoussat, A., Cavallucci, D., Tréla, M., Duflou, J. (eds.) Proceedings of the European TRIZ Association (ETRIA)'s TRIZ Future Conference TFC 2013: Towards Systematic Inventive Processes in Industry (2013)
29. Goffin, K., Lemke, F., Koners, U.: Repertory grid technique. In: Goffin, K., Lemke, F., Koners, U. (eds.) Identifying Hidden Needs, pp. 125–152. Palgrave Macmillan, London (2010)
30. Cooper, R.G.: Stage-gate systems: a new tool for managing new products. Bus. Horiz. **33**(3), 44–54 (1990)
31. Lazzarotti, V., Manzini, R.: Different modes of open innovation: a theoretical framework and an empirical study. Int. J. Innov. Manag. **13**(4), 615–636 (2009)
32. Chesbrough, H.W.: Open Services Innovation: Rethinking Your Business to Grow and Compete in a New Era. Jossey-Bass, San Francisco (2011)
33. Kayser, V., Nehrke, B., Zubovic, D.: Data science as an innovation challenge: from big data to value proposition. Technol. Innov. Manag. Rev. **8**(3), 16–25 (2018)
34. Ostrom, A.L., Parasuraman, A., Bowen, D.E., Patrício, L., Voss, C.A.: Service research priorities in a rapidly changing context. J. Serv. Res. **18**(2), 127–159 (2015)
35. Visser, F.S., Stappers, P.J., Van der Lugt, R., Sanders, E.B.: Contextmapping: experiences from practice. CoDesign **1**(2), 119–149 (2005)
36. Kelly, G.A.: The Psychology of Personal Constructs, Volume 1: A Theory of Personality. W. W. Norton, New York (1955)
37. Hassenzahl, M., Wessler, R.: Capturing design space from a user perspective: the repertory grid technique revisited. Int. J. Hum. Comput. Interact. **12**(3–4), 441–459 (2000)
38. Holtzblatt, K., Jones, S.: Contextual inquiry: a participatory technique for system design. In: Participatory Design: Principles and Practices, pp. 177–210 (1993)
39. Kumar, V., Whitney, P.: Daily life, not markets: customer-centered design. J. Bus. Strategy **28**(4), 46–58 (2007)
40. Von Hippel, E.: Lead users: a source of novel product concepts. Manag. Sci. **32**(7), 791–805 (1986)
41. Eisenberg, I.: Lead-user research for breakthrough innovation. Res. Technol. Manag. **54**(1), 50–58 (2011)
42. Lilien, G.L., Morrison, P.D., Searls, K., Sonnack, M., Hippel, E.V.: Performance assessment of the lead user idea-generation process for new product development. Manag. Sci. **48**(8), 1042–1059 (2002)

Utilizing Data and Analytics to Advance Service

Towards Enabling Organizations to Successfully Ride the Next Wave of Servitization

Fabian Hunke[(✉)] and Christian Engel

Karlsruhe Service Research Institute (KSRI),
Karlsruhe Institute of Technology (KIT),
Kaiserstr. 89, 76133 Karlsruhe, Germany
fabian.hunke@kit.edu,
christian.engel2@student.kit.edu

Abstract. For decades, servitization served as a strategy to gain a competitive advantage over competitors. However, due to its ubiquitous adoption, it is no longer a viable source for differentiation. In this context, data and analytics bear the potential to create new value and, thus, is believed to drive the next frontier of servitization. Yet, the majority of organizations fail to create new innovative services utilizing data and analytics, while research on this topic is also still very limited. Based on a structured literature review, we derive the following contributions to this research field: First, we provide a general overview over the topic, linking single discussions to a larger discourse. Second, we contribute to the fundamental understanding of the research field by pointing out the gaps in the existing literature. Third, we lay the foundation for future research by opening a research agenda to address the highlighted gaps.

Keywords: Servitization · Service advancement · Big data · Data analytics
Literature review · Research agenda · Data- and analytics-based service

1 Introduction

Today, the concept of servitization is widely acknowledged among researchers and practitioners and is broadly discussed in academic literature. The shift from selling products to selling integrated products and services that deliver value in use has been considered a key strategy for manufactures to capture additional value on top of their existing product portfolios [1, 2]. Particularly in markets that are moving towards commoditization, the implementation of a servitization strategy has been perceived a source for differentiation, ultimately leading to competitive advantage [3].

However, servitization seems to have reached its saturation point in terms of differentiation potential. Recent research emphasizes the omnipresence of the concept in practice [4], as it has found its way into almost every industry on every continent [5]. Instead of providing the ground to build competitive advantage, it has rather turned into a strategic necessity for an organization's sustained success [3].

© Springer Nature Switzerland AG 2018
G. Satzger et al. (Eds.): IESS 2018, LNBIP 331, pp. 219–231, 2018.
https://doi.org/10.1007/978-3-030-00713-3_17

In this context, recent research identifies data and analytics as a way to advance servitization by developing sophisticated and novel service offerings [3, 5]. Massive amounts of data are collected by organizations across all industries [7]. Using analytics to provide customers with meaningful insights from this amount of data is believed to open up a promising path to increase the level of competitive advantage again [8].

The opportunities to create customer-facing value using data and analytics seem manifold. Daimler, with its subsidiary Fleetboard, manages to enhance existing logistics and time management services for fleet owners of trucks and vans. In cooperation with an insurance company, data and analytics enables them to additionally offer their customers highly individualized pricing models for the insurance based on the driving behavior as well as the intensity of vehicle usage [9]. Marshall et al. [10] highlight how Monsanto, an agricultural biotechnology corporation, gains competitive advantage in their market by increasing the farmer's yield. It combines data and analytics to suggest farmers the most suitable seed for their individual fields.

Gathering, storing, accessing, and analyzing data is not new to organizations and has been widely discussed in academic literature [11]; yet the focus of these practices remained intra-organizational, helping business users making better decisions [12]. Rather new is the approach of utilizing data and analytics to generate additional customer value to gain competitive advantage over competitors. The vast maturity of organizations (78%) state to have started investing in data- and analytics-based innovations [13]. Still, in practice it seems that only few organizations have attained the expected benefit from these investments so far and find it difficult to offer their customers new value (e.g. Monsanto) or amplify the value of an existing core offering (e.g. Daimler). In terms of our previously used terminology, few organizations actually "ride" the next wave of servitization, yet.

Research on customer-facing services utilizing data and analytics remains scarce [14], resulting in a crucial mismatch between the importance of the topic for practice and the guidance offered by academic literature [15]. In this research, we address this gap in academic literature by extending the existing body of knowledge on utilizing data and analytics to advance service offerings. First, we seek to consolidate the extant work to understand the prevailing debates on the topic. Second, we strive to investigate what debates are still missing in literature to create actionable insights for practice. Thereby, our article gives structure to the rather young field of 'utilizing data and analytics to advance service' to help novice researchers position their future work more precisely. Third, we present a research agenda to tackle the identified white spots.

In the remainder of this paper, we first describe the method we used to review the existing body of literature. This is followed by the findings of this review, which we structured along the four central debates we revealed within the literature. Drawing upon these findings, we discuss missing links in the research field and outline possible avenues for future research. Finally, we conclude with a summary and an outlook on the emerging field of 'utilizing data and analytics to advance service'.

2   Research Method

We conducted a systematic literature review to grasp the breadth of the research field and to make sense of the extant knowledge on the research topic [16, 17]. The review aimed to identify the central debates in academic literature discussing 'service advancement through data and analytics' as a unit of analysis and helped us to eventually generate insights concerning the meanings of these debates. Therefore, instead of just providing an exhaustive and descriptive overview of existing work, we are able to present an in-depth overview of the discourses in the field.

The systematic literature review consists of the search and selection process for relevant literature as well as the subsequent analysis and synthesis process. To make our approach as transparent as possible, ensuring a high level of credibility and confirmability [17], the steps are profoundly described in the following.

2.1   Search and Selection Process

First, we intended to identify a representative set of academic literature focusing on the understanding, conceptualization, or implementation of services that are advanced through the use of data and analytics. For this purpose, five established scientific databases (AISel, Emerald, ScienceDirect, EBSCOhost, SCOPUS) were included in the search process. This allowed us to achieve a comprehensive coverage of the work published in leading IS journals, but also enabled us to consider debates from related disciplines such as manufacturing or computer science. Additionally, we were able to take into account scientific papers from leading IS conferences, thus, including more recent studies in our review that had not reached a journal outlet so far.

It was expected to approach a rather young research field. Therefore, we did not limit our search process to a specific time span. In order to increase the number of potentially relevant papers for the review, different search strings consisting of the words "data", "analytics", and "service" were compiled by the two researchers to form initial keywords that used to search the databases. Subsequently, the search was conducted on title, abstract, full-text. A full-text search was considered necessary, since it was deemed that not all authors had specified the type of service in the title or abstract directly but during their elaboration. While this increased the initial number of potentially relevant candidates, the irrelevant ones could still be excluded during the selection process.

For the selection process, the identified papers were initially judged based on their title, abstract, and keywords if they could be placed in the research field. Otherwise, they were excluded from our dataset. In a second round, the full text of the remaining papers was investigated. In case there were uncertainties if an article had to be considered for our further investigations, the two researchers discussed whether it was to be placed in the research field under investigation. Accordingly, out of 1341 initially considered research articles 84 were identified as relevant for the analysis phase. See Table 1 for a detailed overview of the results from the search and selection process and the number of papers that were found on each database.

Table 1. Search and selection process

Search string	Database					Sum	Relevant
	AISel	Emerald	Science direct	EBSCO	Scopus		
Data-driven service	23	11	27	44	50	155	33
Data-enabled service	2	0	4	0	0	6	1
Data-enriched service	2	0	4	0	1	7	1
Data-enriched product	1	0	0	0	0	1	1
Data-based service	9	3	174	12	35	233	5
Service analytics	34	6	337	54	38	469	13
Analytics-as-a-service	12	0	64	81	74	231	21
Analytics-based AND service	58	21	33	88	39	239	6
			Sum (w/o duplicates)				71
			Added through forward/backward search				13
			Total sum				84

2.2 Analysis and Synthesis of Literature

The analysis and synthesis phase aimed at finding relevant ideas, findings and contributions within the previously selected papers. Afterwards, this research landscape was synthesized into a compact classification of different research streams to present the reader a intelligible layout of the research field [18]. This layout builds the basis to assess the literature and allows to reveal under-researched problems in the academic discourse.

A concept-centric approach was chosen to analyze the literature with regard to the depicted goals [16]. For this purpose, the analysis of the articles was guided by a framework that considered (1) the central concept the article elaborated; (2) the unit of analysis that was researched; (3) the method that was used to conduct the research attempt. Afterwards, we analytically synthesized the different concepts in the framework iteratively to reveal the central debates held in the literature.

The critical reading of the articles also led to new relevant publications which were included in the review's body of literature as well. Thus, the review was conducted iteratively between the search and selection process and the phase described in this section. Eventually, this contributed to the purpose of the review by concentrating on discussions on concepts, ideas, and findings held in the literature rather than on the discussion of individual articles [16].

3   Four Debates Related to Utilizing Data and Analytics to Advance Service

The review unveiled two central topics in the literature related to the application of data and analytics to advance service. That is, *understanding the potential* of using data and analytics to create new value and *leveraging the potential* of data- and analytics-based services in organizations. First, by understanding the potential, we refer to investigations on new ways of value creation in the era of digitalization and a rapidly transforming ecosystem. Second, by leveraging the potential, we refer to human resources, technical capabilities and new business models that enable organizations to successfully establish data- and analytics-based services.

Again, we did not intend to provide the reader with an exhaustive and descriptive overview of existing work conducted in the field. Instead, we aimed to provide a layout of the research field showing the central debates in the academic discourse. A summary of the four central debates identified in our review and a representative grounding in the literature is depicted in Fig. 1. In the following, each debate is outlined very briefly.

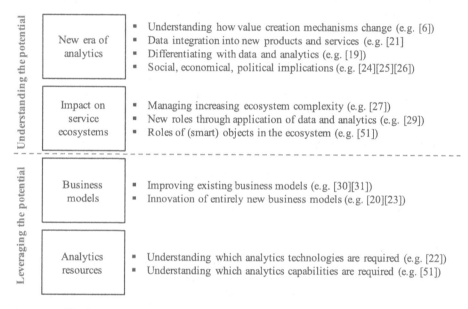

Fig. 1. Summary of the results from the analysis and synthesis process.

3.1 Understanding the Potential

New Era of Analytics. Arguably, the corporate world has reached a new age which is driven by analytics rooted in enormous amounts of data [19, 20]. While preceding research concerning data and analytics focused on the analytical support of internal decision making and mastering to collect and process big data [12], the third wave of analytics concentrates on analytics supporting customer-facing products, services, and features, thus, marking a new era of analytics ("Analytics 3.0" in [19]).

In this context, literature strives to understand how the mechanisms of organizations and whole industries creating value change. Huberty [21] emphasizes the necessity to build data- and analytics- based services in order to live up to the expectations of the (big) data revolution, as so far very few companies managed to transform data into new products or services. Similarly, Alvertis et al. [22] identify the successful development of services building on data and analytics as a key factor of differentiation in the future. Thus, literature agrees, that it is essential to understand how the mechanisms of creating value function in this new era of change [23].

Scholars also investigate social, economic and political implications that come with the exploitation of data and analytics. For instance, Heiskala et al. [24] and Lopez et al. [25] point out how smart city applications based on citizens' location and motion data can contribute to safer, cleaner and less stressful living conditions. Similarly, a game-changing potential is detected in the global food supply chain through a utilization of data and analytics in the agricultural sector [26].

Impact on Service Ecosystem. Another debate focuses on the impact of a data and analytics application on service ecosystems. Specially, new and ambivalent roles customers, organizations but also machines and smart objects can now adopt, increase the ecosystem's complexity [27].

For instance, early work establishes two novel roles in a service ecosystem when using data and analytics, namely Data-as-a-Service (DaaS) and Analytics-as-a-Service (AaaS) [28]. The former is referred to as a provider of raw and aggregated content, whereas the latter service offers a rich set of common analytics components and infrastructure. Additionally, more recent research reports the emergence of multi-sided roles in an ecosystem merging both roles in a single one [29]. In this case, organizations aggregate data from one group of customers and create additional value for another group of customers by gaining insight from the collected data through analytics.

3.2   Leveraging the Potential in Organizations

Business Models. There exists some considerable acknowledgement in the literature of data and analytics holding the potential to leverage an organization's business model. Business models are commonly referred to as representations describing how an organization creates, delivers and captures value and thereby reflect the organization's business strategy [30]. Research related to this debate concentrates on services utilizing data and analytics by investigating how the business model representing the organization's attempt to create value changes.

Two fundamental discussions arise in this context [9]. On the one hand, scholars point out a possible business model improvement of the existing one [31, 32]. This is achieved by assessing existing data sources and analytics techniques with the goal of improving existing processes. The organization still pursues the same business strategy but more effectively. On the other hand, business model innovation is seen as appropriate to create new value from data and analytics [21, 24]. This is achieved when data and analytics lead to new value propositions, ultimately altering the organization's business strategy.

Analytics Resources. At the organizational level, scholars stress the importance of new resources and capabilities necessary to create value in services using data and analytics. Based on our review, we identified this debate to consist of two streams, analytics technology assets and analytics capabilities [33].

With regard to analytics technology assets, literature refers to the infrastructure, software and tools an organization requires to actually offer a data- and analytics-based service [23]. As different types of services require different assets, scholars aim to develop distinction-frameworks in order to classify the assets used in particular services. While software and tools are commonly described as supporting either descriptive, predictive or prescriptive analytics [34, 35], a common framework to distinguish the assets required for data and analytics applications has not yet been acknowledged [35].

Analytics capabilities refer to the employee's competencies organizations need to build in order to perform data- and analytics-based services [36–38]. Thus, literature strives to identify strategic and structural design options, common processes, best-practices of how to cultivate analytics capabilities.

4 Discussion

In this section, we want to critically discuss the literature on 'using data and analytics to advance service', presented in the previous section. Eventually, we aim to shed light on missing links between the existing knowledge in the literature that still hinder organizations to create new services utilizing data and analytics in a systematic manner. By identifying white spots in the research field, we seek to propose a research agenda addressing open topics, thereby laying the foundations for research enabling organizations to establish data- and analytics-based services.

4.1 A Critical Analysis of the Findings

The review of the literature shows that the research field on utilizing data and analytics to advance service is still in a nascent stage. Much of the work we reviewed did not emerge before 2015. Consequently, the literature still seeks to gain a common understanding of how organizations realize value from services that build upon data and analytics.

A number of papers focus on understanding the inherent opportunities organizations can seize. This "path to value" is investigated both from a conceptual [e.g. 3] and an empirical perspective [e.g. 9]. Moreover, research supports the effort to increase the knowledge on the application and, eventually, monetization of data and analytics in organizations by discussing the benefits of data and analytics in service in empirical studies [e.g. 32]. Several positive relations have been identified based on real-world cases between newly build analytical capabilities in organizations and the ability to gain competitive advantage through data and analytics in services (e.g. [23, 39, 40]). Herterich et al. [23], for instance, point out key capabilities for harnessing data streams in an industrial context and link them to specific business benefits they were able to observe.

Yet, from a practitioners' perspective little attention has been paid to bridge the gap between fundamentally realizing the substantial value of data and analytics as a new source of competitive advantage and the implications derived from several studies showcasing the successful application. In particular, we argue that more attention needs to be paid to the integration of data and analytics into the design process to advance service (cf. Fig. 2). Although service design studies have contributed to developing systematic approaches to new value creation in services in general [41], little is known about the design principles when service advancement is fostered through the use of data and analytics [42, 43]. In fact, our literature review did not reveal any methodology or framework that would guide the process of creating a new data- and analytics-based service. We argue, that future research needs to focus on filling this gap in order to provide organizations with the necessary guidance that enables them to systematically create such services for their customers.

Fig. 2. Missing link of existing literature from a practitioners' point of view.

In this sense, we can confirm findings that there is a mismatch in service science between the knowledge to advance service and its importance for practice [15]. In addition, we take a step further towards the root cause for this mismatch and point out the necessity to increase the foundations of service design and innovation in the context of service advancement through utilizing data and analytics. Drawing on this perception, a research agenda tackling this gap in literature is proposed in the following section.

4.2   Avenues for Future Research

Based on the outlined motivation to foster service advancement through data and analytics (cf. Sect. 1) and the identified gap in academic literature concerning its application (cf. Sect. 4.1), a research agenda addressing this issue is proposed. The primary objective of this agenda is to foster a theorizing process regarding the understanding and design of customer-facing services that draw upon data and analytics to generate value. Such a foundation should bridge the existing gap and provide practitioners with the necessary guidance to systematically establish data- and analytics-based services. Below we propose three research streams addressing this objective.

(1) Value creation in data- and analytics-based services. We claim that the design of services utilizing data and analytics differ from previous design genres as the advent of data and analytics results in dramatic changes ranging from the organization's ecosystem to its own required capabilities (cf. Sect. 3). Thus, the objective of this research stream would be to investigate data- and analytics-based services as a unit of analysis to deepen the understanding of the concept's distinctiveness.

Future research should investigate how value creation mechanisms look like in customer-facing services drawing upon data and analytics. A fundamental understanding of the value creation mechanism is essential before investigating the design of such services [44]. The underlying mechanisms are well understood with regard to internal services using data and analytics to support better decision making [12]. However, as depicted in Sect. 3.2 literature points out additional capabilities and practices required to perform such customer-facing services utilizing data and analytics. Thus, it is necessary to understand how services utilizing data and analytics create customer value along key enabling organizational capabilities.

Additionally, this research stream should investigate key characteristics of data- and analytics-based service. We argue that a theoretical framework explaining the general key components of such a service would contribute to enabling a systematic creation of services that leverage data and analytics in practice. Particularly, a taxonomy describing services utilizing data and analytics may help both researchers and practitioners to fill the void unveiling relevant design principles. Taxonomy building is well known in IS research and a legitimate method to obtain a classification scheme for services that are already observable in the market [45]. The contribution of such a taxonomy is twofold. First, it entails central design principles for the service design and therefore constitutes a design space for possible solutions of services utilizing data and analytics. Second, based on the classification scheme, further research can be pursued, e.g. cluster analysis identifying patterns of distinct service types (e.g. [46, 47]).

(2) Design of data- and analytics-based services. As stated above, we believe the design of services utilizing data and analytics differs from previous attempts. Thus, this research stream should focus on the design of such services in organizations. Studies need to investigate which key activities are required to systematically develop customer-facing services drawing upon data and analytics. The service design literature provides a useful basis of generic knowledge. Researchers have developed process models that define key activities required to develop new services [48, 49]. Future research should investigate the unique nature of service design in the specific context of data- and analytics-based services to extend the existing body of knowledge. This would eventually provide practitioners with the necessary guidance needed to systematically develop such services systematically.

Building on an understanding of key activities for a service design process, it might be also interesting to tap into new tools supporting these activities. Tool support guiding a development process has already been applied in related areas such as business model innovation (e.g. [32]). In practice, the development of data- and analytics-based services is a highly interdisciplinary process as project teams often consist of members with a data science, engineering, or business background [36]. Building a common understanding of the team's aspired service solution often is a challenge during such projects. Hence, the development of tools supporting practitioners during distinct design steps is an aspired contribution.

(3) Strategies for data- and analytics-based services. Let us consider Fleetboard and Monsanto, the two introductory examples (cf. Sect. 1), once again. Fleetboard is able to provide an insurance pricing model based on driving behavior and vehicle usage by analyzing vehicle data from Daimler Trucks and Vans. No third party is involved during the value creation process and Fleetboard strictly keeps its data internally. Only the analytically derived insights are forwarded to the insurance company Fleetboard works together with. On the other hand, Monsanto actively drives an open platform approach, where farmers are able to access data and analytics services from third party providers. Even though Fleetboard and Monsanto provide a similar service in terms of aggregating, analyzing and providing additional value to their customers through insight generation, both pursue different strategies within their organizational network with regard to the degree of openness towards third party data and analytics providers.

We could not find any literature during our review addressing strategic considerations with regard to services utilizing data and analytics that could serve as a starting point for this research stream. Data- and analytics-based services allow to create value in organizational networks in novel ways. Especially multi-sided platforms have gained increased popularity in the literature providing valuable insights on their key mechanisms [50]. Research on strategies for services utilizing data and analytics should extend this knowledge by providing insights on how organizations can benefit from applying different strategies in an inter-organizational context.

5 Conclusion

This paper addresses the emerging field of service advancement through data and analytics integration. Using data and analytics to advance service is seen as a key strategy to gain competitive advantage in the future, forming the next frontier of servitization [3, 15, 27]. However, recent research points out that, so far, few organizations actually manage to establish services that use data and analytics to provide new value to their customers and, thus, actually "ride" the next wave of servitization [9].

A structured literature review is conducted to gain a better understanding of the research field, addressing the topic of utilizing data and analytics to advance service. Four central debates dominate the literature in this field, namely required analytics resources, changes in business models, implications on service ecosystems, and general implications from a new era of analytics. Building on these finding, we show that a systematic design of services building on data and analytics constitutes a crucial gap in the literature. Consequently, we propose a research agenda to fill this void, aiming to enable organizations to establish services utilizing data and analytics.

Hence, our research provides three vital contributions to the research field of service advancement through data and analytics. First, having identified the central debates in the literature, we are able to link and map related research to the larger discussion of advancing service through data and analytics. Second, based on our findings, we are able to point out the underdeveloped topic of service design within the research field which needs to be addressed in order to enable practitioners to actually advance service through data and analytics. Third, this paper lays the foundation for future research and opens a research agenda in order to seize the white-space in the research field. Particularly, we point out possible avenues for future research fostering to understand the value creation mechanisms, the design, and strategic implications of customer-facing services using data and analytics.

References

1. Baines, T.S., Lightfoot, H.W., Kay, J.M.: Servitized manufacture: practical challenges of delivering integrated products and services. J. Eng. Manuf. **223**(9), 1207–1215 (2009)
2. Vandermerwe, S., Rada, J.: Servitization of business: adding value by adding services. Eur. Manag. J. **6**(4), 314–324 (1988)
3. Opresnik, D., Taisch, M.: The value of big data in servitization. Int. J. Prod. Econ. **165**, 174–184 (2015)
4. Neely, A.: The servitization of manufacturing: an analysis of global trends. In: Proceedings of the 14th European Operations Management Association Conference, pp. 1–10 (2007)
5. Neely, A.: Exploring the financial consequences of the servitization of manufacturing. Oper. Manag. Res. **1**(2), 103–118 (2009)
6. McAfee, A., Brynjolfsson, E.: Big data: the management revolution. Harv. Bus. Rev. **90**(10), 61–67 (2012)
7. Atzori, L., Iera, A., Morabito, G.: The internet of things: a survey. Comput. Netw. **54**(15), 2787–2805 (2010)
8. Lavalle, S., Lesser, E., Shockley, R., Hopkins, M.S., Kruschwitz, N.: Big data, analytics and the path from insights to value. MIT Sloan Manag. Rev. **52**(2), 21–32 (2011)
9. Schüritz, R., Satzger, G.: Patterns of data-infused business model innovation. In: Proceedings of the 18th IEEE Conference on Business Informatics (CBI), pp. 133–142 (2016)
10. Marshall, A., Mueck, S., Shockley, R.: How leading organizations use big data and analytics to innovate. Strateg. Leadersh. **43**(5), 32–39 (2015)
11. Chen, H., Chiang, R.H.L., Storey, V.C.: Business intelligence and analytics: from big data to big impact. MIS Q. **36**(4), 1165–1188 (2012)
12. Watson, H.J.: Tutorial: business intelligence – past, present, and future. Commun. Assoc. Inf. Syst. **25**(39), 487–510 (2009)
13. Fueling growth through data monetization. https://www.mckinsey.com/business-functions/mckinsey-analytics/our-insights/fueling-growth-through-data-monetization?cid = other-eml-alt-mip-mck-oth-1712. Accessed 30 June 2018
14. Maglio, P., Lim, C.-H.: Innovation and big data in smart service systems. J. Innov. Manag. **4**(1), 11–21 (2016)
15. Ostrom, A.L., Parasuraman, A., Bowen, D.E., Patrício, L., Voss, C.A.: Service research priorities in a rapidly changing context. J. Serv. Res. **18**(2), 127–159 (2015)
16. Webster, J., Watson, R.T.: Analyzing the past to prepare for the future: writing a literature review. MIS Q. **26**(2), 13–23 (2002)
17. vom Brocke, J., Simons, A., Riemer, K., Niehaves, B., Plattfaut, R., Cleven, A.: Standing on the shoulders of giants: challenges and recommendations of literature search in information systems research. Commun. Assoc. Inf. Syst. **37**(1), 205–224 (2015)
18. Boell, S.K., Cecez-Kecmanovic, D.: A hermeneutic approach for conducting literature reviews and literature searches. Commun. Assoc. Inf. Syst. **34**, 257–286 (2015)
19. Davenport, T.H.: Analytics 3.0. Harv. Bus. Rev. **91**(12), 64 (2013)
20. Legner, C., Eymann, T., Hess, T., Matt, C., Böhmann, T., Drews, P., et al.: Digitalization: opportunity and challenge for the business and information systems engineering community. Bus. Inf. Syst. Eng. **59**(4), 301–308 (2017)
21. Huberty, M.: Awaiting the second big data revolution: from digital noise to value creation. J. Ind. Compet. Trade **15**, 35–47 (2015)

22. Alvertis, I., et al.: Challenges laying ahead for future digital enterprises: a research perspective. In: Persson, A., Stirna, J. (eds.) CAiSE 2015. LNBIP, vol. 215, pp. 195–206. Springer, Cham (2015). https://doi.org/10.1007/978-3-319-19243-7_20

23. Herterich, M.M., Uebernickel, F., Brenner, W.: Stepwise evolution of capabilities for harnessing digital data streams in data-driven industrial services. MIS Q. Exec. **15**(4), 299–320 (2016)

24. Heiskala, M., Jokinen, J.P., Tinnilä, M.: Crowdsensing-based transportation services - an analysis from business model and sustainability viewpoints. Res. Transp. Bus. Manag. **18**, 38–48 (2016)

25. Lopez, P.G., Tinedo, R.G., Montresor, A.: Towards Data-driven software-defined infrastructures. Procedia Comput. Sci. **97**, 144–147 (2016)

26. Wolfert, S., Ge, L., Verdouw, C., Bogaardt, M.J.: Big data in smart farming – a review. Agric. Syst. **153**, 69–80 (2017)

27. Schüritz, R., Seebacher, S., Satzger, G., Schwarz, L.: Datatization as the next frontier of servitization – understanding the challenges for transforming organizations. In: Proceedings of the 38th International Conference on Information Systems (ICIS), pp. 1–21 (2017)

28. Chen, Y., Kreulen, J., Campbell, M., Abrams, C.: Analytics ecosystem transformation: a force for business model innovation. In: Proceedings of the Annual SRII Global Conference, pp. 11–20 (2011)

29. Schüritz, R., Seebacher, S., Dorner, R.: Capturing value from data: revenue models for data-driven services. In: Proceedings of the 50th Hawaii International Conference on System Sciences (HICSS), pp. 5348–5357 (2017)

30. Teece, D.: Business models, business strategy and innovation. Long Range Plann. **43**(2), 172–194 (2010)

31. Herterich, M.M., Uebernickel, F., Brenner, W.: The impact of cyber-physical systems on industrial services in manufacturing. Procedia CIRP **30**, 323–328 (2015)

32. Zolnowski, A., Christiansen, T., Gudat, J.: Business model transformation patterns of data-driven innovations. In: Proceedings of the 24th European Conference on Information Systems (ECIS), pp. 1–16 (2016)

33. Krishnamoorthi, S., Mathew, S.K.: Business analytics and business value: a case study. In: Proceedings of the 36th International Conference on Information Systems (ICIS), pp. 1–17 (2015)

34. Delen, D., Demirkan, H.: Data, information and analytics as services. Decis. Support Syst. **55**(1), 359–363 (2013)

35. Sun, Z., Zou, H., Strang, K.: Big data analytics as a service for business intelligence. In: Janssen, M., et al. (eds.) I3E 2015. LNCS, vol. 9373, pp. 200–211. Springer, Cham (2015). https://doi.org/10.1007/978-3-319-25013-7_16

36. Schüritz, R., Brand, E., Satzger, G., Bischhoffshausen, J.: How to cultivate analytics capabilities within an organization ? – design and types of analytics competency centers. In: Proceedings of the 25th European Conference on Information Systems (ECIS), pp. 389–404 (2017)

37. Lim, C.H., Kim, M.J., Heo, J.Y., Kim, K.J.: Design of informatics-based services in manufacturing industries: case studies using large vehicle-related databases. J. Intell. Manuf. **29**(3), 497–508 (2015)

38. Marjanovic, O.: From analytics-as-a-service to analytics-as-a-consumer-service: exploring a new direction in business intelligence and analytics research. In: Proceedings of the 48th Hawaii International Conference on System Sciences (HICSS), pp. 4742–4751 (2015)

39. Dremel, C., Wulf, J., Herterich, M.M., Waizmann, J.-C., Brenner, W.: How AUDI AG established big data analytics in its digital transformation. MIS Q. Exec. **16**(2), 81–100 (2017)

40. Ross, J., Sebastian, I., Beath, C., Mocker, M., Moloney, K., Fonstad, N.: Designing and executing digital strategies. In: Proceedings of the 37th International Conference on Information Systems (ICIS), pp. 1–17 (2016)
41. Patrício, L., Gustafsson, A., Fisk, R.: Upframing service design and innovation for research impact. J. Serv. Res. **21**(1), 3–16 (2018)
42. Lim, C., Kim, M.-J., Kim, K.-H., Kim, K.-J., Maglio, P.P.: Using data to advance service: managerial issues and theoretical implications from action research. J. Serv. Theory Pract. **28**(1), 99–128 (2018)
43. Hunke, F., Seebacher, S., Schuritz, R., Illi, A.: Towards a process model for data-driven business model innovation. In: Proceedings of the 19th IEEE Conference on Business Informatics (CBI), pp. 150–157 (2017)
44. Patrício, L., Fisk, R., Constantine, L.: Multilevel service design: from customer value constellation to service experience blueprinting. J. Serv. Res. **14**(2), 180–200 (2011)
45. Nickerson, R.C., Varshney, U., Muntermann, J.: A method for taxonomy development and its application in information systems. Eur. J. Inf. Syst. **22**(3), 336–359 (2013)
46. Hartmann, P., Zaki, M., Feldmann, N., Neely, A.: Capturing value from big data – a taxonomy of data-driven business models used by start-up firms. Int. J. Oper. Prod. Manag. **36**(10), 1382–1406 (2016)
47. Hunke, F., Schüritz, R., Kuehl, N.: Towards a unified approach to identify business model patterns: a case of e-mobility services. In: Za, S., Drăgoicea, M., Cavallari, M. (eds.) IESS 2017. LNBIP, vol. 279, pp. 182–196. Springer, Cham (2017). https://doi.org/10.1007/978-3-319-56925-3_15
48. Edvardsson, B., Olsson, J.: Key concepts for new service development. Serv. Ind. J. **16**(2), 140–164 (1996)
49. Fitzsimmons, J.A., Fitzsimmons, M.J. (eds.): New Service Development: Creating Memorable Experiences. Sage Publication Inc., Thousand Oaks (2000)
50. Stummer, C., Kundisch, D., Decker, R.: Platform launch strategies. Bus. Inf. Syst. Eng. **60**(2), 167–173 (2018)
51. Porter, M.E., Heppelmann, J.: How smart, connected products are transforming companies. Har. Bus. Rev. **93**, 10 (2015)

Big Data in Services

Forecast Correction Using Organizational Debiasing in Corporate Cash Flow Revisioning

Florian Knöll[✉] and Katerina Shapoval

FZI Research Center for Information Technology, Karlsruhe, Germany
{knoell,shapova}@fzi.de

Abstract. Corporations that employ information systems, such as decision support systems for judgmental forecasts, have business objectives that require accurate forecasts. But the accuracy of these forecasts is most likely biased by organizational and individual structures within the corporation. These biases, such as revenue targets or personal objectives, may alter the forecasters' prediction due to financial incentives in a predefined way. This paper argues that model-driven correction of forecasts – which typically utilizes only statistical methods – should incorporate organizational debiasing methods. In a case of an international corporation, local experts forecast cash flows for corporate risk management. The forecasts are later aggregated on a corporate level with subsequent debiasing techniques for decision support. Empirical results show that considering organizational objectives for debiasing techniques can strongly improve forecast accuracy. The total correctable expert error is reduced by up to 60 % for all forecasts of a month, providing better decision support for managers.

Keywords: Decision support system · Corporate cash flows
Judgmental forecasting · Revision process · Organizational bias
Forecast correction · Service analytics

1 Introduction

Corporations with activities in enterprise's risk management, planning, and decision making usually depend strongly on forecast quality. Cash flow forecasting is therefore of crucial importance for corporate finance of multinational firms [1,7,9,15,17,23], in particular future foreign exchange risks. As forecasts provide the basis for hedging options to cover currency exposures, forecast inaccuracies can result in increased hedging costs or uncovered currency risks for corporate operations.

Therefore, corporations that operate worldwide typically have forecasting processes on a regular basis. For this purpose, the local subsidiaries send thousands of forecasts and revisions in a decentralized fashion to corporate headquarters. These forecasts are aggregated and provide the basis for corporate-wide forecasting and *key performance indicators* (KPIs). Some important KPIs

© Springer Nature Switzerland AG 2018
G. Satzger et al. (Eds.): IESS 2018, LNBIP 331, pp. 235–246, 2018.
https://doi.org/10.1007/978-3-030-00713-3_18

and resulting business key figures (KPIs with financial context in the accounting year) can be found in [16]. An example of such key figure is *Earnings Before Interest, Taxes, Depreciation, and Amortization* (EBITDA) margin, which can be used as one of the primary proxies for a company's operating profitability.

However, most of today's forecasting processes are the result of human judgment [21]. The latter is often prone to individual biases and these latent human influences are often underrated but they entail corporations' forecasting and planning in many ways [8]. [13] provides a comprehensive literature research on cognitive and behavioral biases, while further studies suggest that these biases often result in increased errors of forecasts [14]. *Organizational biases* result from structures and dependencies in the organization. These organizational biases influence the experts' revision behavior, and, therefore the forecast adjustments are entangled by business key figures. In particular, it has been found that meeting important organizational earning targets and incentives alter management activities [5,22]. Further it has been discussed that insufficient information flows can result in organizational biases [6,10].

Organizational biases may affect the corporation's cash flows. Much effort is spent on identification and handling of cash flow biases [4,11]. Corporations aim to reduce forecast error and to enhance corporate forecast support system with model driven support services [2]. Typically, the analyses employ statistical approaches to identify patterns in the forecasting processes, but do not include important business dependencies like organizational targets.

Hereof, the method recently shown in [12] is applied for the proxies of return margins to identify high-level biases in business forecasts. The results therein provide evidence that forecasting heavily depends on organizational structures, specifically the proxies for targets of organization's return margins. In the current study the predictive value of these insights is analyzed. To reduce cash flow biases and thereby to improve business performance this study utilizes different correction models to analyze the impact of organizational biases on forecast correction.

The study provides the following contribution. Arguing that forecasting methods should incorporate organizational debiasing methods, this paper compares the predictive results of a model for forecast correction that utilizes statistical information only to the results of an enhanced model using information of a KPI proxy. Using cash flow forecasts of an international corporation, the analyses of the explanatory power of different models strongly support the thesis that organizational biases must be concerned in forecasting. The improvements of monthly model-driven forecasts show the meaning of business insights for information systems and forecast services.

The remainder of this paper is structured as follows. Section 2 describes the empirical dataset. Section 3 introduces the notation and is followed by the research design for the hypotheses in Sect. 4. The empirical results and interpretation are then presented in Sect. 5. Section 6 discusses the implications and limitations of this work for improvements in the area of forecasting systems.

2   Cash Flow Data: Acquisition and Descriptives

The data used in the analysis stems from a record of cash flow forecasts and realizations provided by a multinational sample corporation. With about 110,000 employees, the company generates annual revenues in the billion Euro range. The corporation is headquartered in Germany, but has worldwide more than 300 separate legal entities. The subsidiaries are grouped into four distinct divisions, based on their business portfolios (D1–D4). Each subsidiary operates officially independently of the corporation, while there are some organizational dependencies. First, based on the set of local plans, the corporation re-adjusts the planning to an overall view, and sets the target requirements for local operations for being rated as "successful" subsidiaries. Second, as the subsidiaries operate independently, they have their own financial information system, a heterogeneous payment structure (e.g., incentivization bonuses), and ensure liquidity for their operations (e.g., with earnings management processes). Third, each subsidiary that is participating in the forecasting process enters their expectations on future cash flow in a digital, corporate-based forecasting system.

Financial risk management is centralized, with the local subsidiaries reporting cash flows to the corporation's central finance department, where these serve as the basis for further actions and services in corporate finance. The forecasts are transferred into the corporate information system, which provides validations on the syntax and the semantic for each single item. The current set of validations also checks if a specific forecast is in line with the predictions of correction algorithms. The algorithms cover time-series and behavioral analysis, such as identification of anchoring pattern [3]. The predictions of the algorithms are fed into a decision support system, which recommends the inspection of specific forecasts (*issue*) to the corporate managers which then contact the subsidiaries' responsible to review the situation. The workload for the managers increases with the number of issues that should be inspected. Also, managerial understanding of each specific issue is required, which increases workload further. Overall, for thousands of forecasts, with potentially one or more issues each, the effort in the corporation is enormous.

The corporate finance department receives cash flow forecasts (*forecasts*) generated by the subsidiaries , denominated in foreign currencies. After the realization date, the corporation receives in every month the cash flow figures for realizations (*actuals*). The forecasts and actual data available for this paper cover item-types of invoices issued and invoices received from the corporate IT system. Delivered by the subsidiaries on a quarterly basis, the forecasts cover intervals with horizons of up to 15 months. The dataset for actual invoices ranges from January 2008 to December 2013 with the corresponding forecasts covering the actuals' period. In total, actuals and forecasts are available for the 99 largest subsidiaries resulting in 44 different currencies for the dataset. Actuals grouped by division, subsidiary, currency and item-type result in 484 actual time series. Overall, the raw dataset consists of 20,472 monthly invoice actuals, with five associated forecasts each. Table 1 gives a brief summary of the dataset.

Table 1. The summary of available cash flow data.

Divisions	Subsidiaries	Currencies	Actual time series	Forecasts
D1	12	16	70	17,010
D2	19	26	146	29,070
D3	13	8	52	13,460
D4	53	37	216	42,820
All	99	44	484	102,360

3   Notation and Forecasting Process

Denoting the *actual* of cash flow item i as $A(i)$, the *lead time* t of a *forecast* $_tF(i)$ for $A(i)$ refers to a quarter of the year until the actual date $(t = 0)$. The initial forecast $_5F(i)$ is delivered with a lead time of five periods and is revised four times until the last one–period–ahead forecast $_1F(i)$ is generated. Figure 1 visualizes the temporal structure of the forecasting process in five steps for an actual $A(i)$. Subscripts m, y, and e denote the realization month, realization year, and the ID of the corresponding subsidiary of the actual. Superscript g denotes the type of the actual ($g \in \{$invoice issued (II), invoice received (IR)$\}$). Therefore, the maximum indexing for an actual is $A_{e,y,m}^{g}(i)$. If an index is irrelevant or obvious in the context, the respective index is omitted for reasons of brevity. The use of F instead of A in the notation refers to individual forecasts.

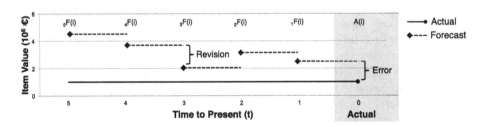

Fig. 1. Temporal structure of cash flow forecasts $_tF(i)$ for the related actual A(i).

Because corporate planning is determined on an aggregate level, the cash flow forecasts are cumulated up to this level. As a proxy for percentage return margin within a fiscal year for a specific subsidiary ($e = E$), the computation of the entity's *ratio* R uses aggregated revenues (II) and expenses (IR). The ratio for in the M-th month of a year Y and the K months ($K < M$) before M is shown in Eq. (1).

$$R^K(A_{Y,M}) = \frac{\sum_{1 \le j \le K} A_{y=Y,m=M-j}^{g=II}}{\sum_{1 \le j \le K} A_{y=Y,m=M-j}^{g=IR}} \qquad \begin{array}{l} Y \text{ specific year} \\ M \text{ specific month} \\ K \text{ aggregated number of months} \end{array} \qquad (1)$$

For instance, $R^2_{2010,11}$ refers to the ratio of all cash flows from September to November 2010, while a ratio above one indicates the presence of more revenues than expenses (positive return). The notation $R(A_{Y,M})$ omits the superscript K, if $K = M - 1$, and is an aggregation of all realized cash flows in year $y = Y$ up to (and including) month $m = M$. Since ratios are specific for an entity e, for reasons of comparability, this work focuses on normalized ratios with values between zero and one per entity. Normalized ratios are obtained by subtracting the minimum ratio within an entity from R and dividing by the difference of its maximum and minimum ratio. For the readers convenience, the notation $_tR$ will refer to the normalized ratio instead of the entity specific ratio.

The suggested annual return margin target $(target)$ that an entity E has to reach at the end of the year is defined as $T(A_{y=Y})$ in year $y = Y$. As targets are unknown (to us), but business development measured with EBITDA figures seem rather stable over the years, the target in $y = Y$ is estimated by averaging the December actual ratios of the three preceding years $(R(A_{y=Y-j,m=12})$, with $j \in \{1, 2, 3\})$. The *revision* for ratios is defined as $_{12}R = R(_1F) - R(_2F)$, and describes the adjustment from the second last forecast to the last forecast before the actual. The last revision is used because generally the latest judgmental forecast incorporates the most information, and is the most accurate [19]. The *difference from target* is defined as $TargetDiff = T(A) - R(_1F)$. Finally, the *error* is $Err = R(A) - R(_1F)$. Ratios and revisions are not stored in the database but derived from the invoice items as shown in Eq. (1). Table 2 gives a brief overview of the metrics.

Table 2. Notation used in the empirical analyses.

Notation	Metric
$R(_1F)$	Forecast ratio (normalized)
$R(A_{y,m})$	Actual ratio (normalized)
$T(A_y)$	Target (normalized)
$TargetDiff$	Difference from target
$_{12}R$	Revision (normalized)
Err	Error (normalized)

4   Research Design

One of the primary KPI for corporate performance is the EBITDA [16]. For hypotheses, it is assumed that the derived margin targets are linked to the percentage EBITDA margins of the company. This seems plausible, as derived EBITDA margins result from revenues (by invoices issued) and expenses (by invoices received). The assumption is underpinned by the percentage EBITDA margin figures reported in the annual report, which are inline with the ranks of division ratios for December values (not reported in this paper). This allows retaining a substitute for percentage EBITDA margins.

To align the risk management to current and future business development, the corporation requires financial information by the subsidiaries. Based on the communication within the subsidiary, the forecaster might not be aware of the preconditions of the managers in planning and operations [6]. Additionally, when organizational structure motivates one manager (e.g., incentivization payment for managers in planning or operations), but not the other ones (e.g., the forecaster), the subsidiaries view might be organizationally biased by provided targets. These biases can result from the concealment of information – when managers with the different tasks do not have access to the same information [18].

To overcome this organizational bias, if the information retrieval processes can not be changed (the way how subsidiaries gather and transfer the forecast data to the corporation), the subsidiaries' forecasts need to be enhanced with information of planning and operations. The first hypothesis is that a statistical correction model will have higher explanatory power when additional key figures of planning and operations are provided, namely the *Target* listed in Table 2. The explanatory power of additional information can be shown, by using R^2-value for the explained variance. The second hypothesis is that re-adjustment of the subsidiaries' forecasts, utilizing the previously defined model, will provide meaningful accuracy improvement within the correction process. Therefore, the following hypotheses are proposed:

Hypothesis 1 (H1). *Incorporation of business key figures (that organizationally biases forecasts) is beneficial for the explanatory power of forecast correction models on corporate internal forecasts.*

Hypothesis 2 (H2). *Utilization of business key figures in a forecast correction model can improve the predictive results on corporate internal forecasts.*

To support the hypotheses two analyses are conducted in the empirical study. Thereby, the out-of-sample test period of the analyses covers the last year (2013) and the years before provide training period data (2008–2012).

First, in a linear regression model the R^2-values will be compared within the training period. The linear regression provides a first approach for a model to correct forecasts. Given the descriptive statistics provide evidence that a higher part of the variance can be explained with an organizational model compared to a basic statistic model, it will reason that more advanced models (e.g., random forests) can utilize this beneficial information of the key figures too. The R^2-value of the model using the business key figures must be greater than the model without that information in order to support (H1).

Second, the models defined for (H1) are compared in an out-of-sample test period of one year. Considering the seasonality in the business data, a model is trained for each month independently. Therefore, data is split into 12 subsets that are each accessed to train one specific model for each month. To support (H2), the monthly predictions of the models will be compared to the expert forecasts in absolute error numbers. In addition, percentage improvement in comparison to the expert forecast is analyzed for each month separately. To show empirically that differences of models are significant, the results on the percentage improvements are validated with a t-test.

5   Empirical Analysis

This section presents the empirical results. These consist of a linear regression analysis, analysis of the out-of-sample results, t-test evaluation, followed by the interpretation of the results. For the linear regression and hypothesis testing the experiments use the tools provided in [20].

5.1   Explanatory Power in Correction Models

As noted before, reaching margin targets is an important strategic goal. The influence of the integration of these figures with a key variable can be measured in two ways: by a significant estimate of the variable of the key figure, and by the R^2-value of the model. Significance of the estimate will provide support that the key figure has a non-random influence, while R^2-value describes the part of variance being explained by the model. The effect of an increased R^2-value is an important indication for the beneficial information measured by the additional amount of the variance explained by adding the information.

For (H1), two models are trained in the in-sample period using given data. The *Model 1* only uses basic statistical information: a constant for regression intercept, forecast ratio, and revision. The *Model 2* utilizes additionally the information of business key figures (difference from target) for the regression model. The two regression models are trained with the Eqs. (2) and (3).

$$\text{Model 1:} \qquad Err \sim \beta_0 + \beta_1(R(_1F)) + \beta_2(_{12}R) \qquad\qquad (2)$$

$$\text{Model 2:} \qquad Err \sim \beta_0 + \beta_1(R(_1F)) + \beta_2(_{12}R) + \beta_3(\textit{TargetDiff}) \qquad (3)$$

The resulting regression models are shown in Table 3. The integration of *TargetDiff* shows clearly strong influence on the response variable (ratio error), while the estimates' magnitude of the other dependent variables are reduced. Analysis of the R^2-values of the models supports (H1) (R^2-value of 0.608 for Model 2 compared to 0.310 for Model 1).

Table 3. Error dependencies for Model 1 and Model 2. Estimate of TargetDiff strongly influences the error in ratio. The explanatory power (R^2-value) nearly doubles by incorporating TargetDiff information. Note: $^*p < 0.1$; $^{**}p < 0.05$; $^{***}p < 0.01$; number of observations: 2,355.

Dep. variables for Err	Estimates Model 1	Estimates Model 2
Constant	0.178^{***}	-0.027^{***}
$R(_1F)$	-0.541^{***}	-0.001
$_{12}R$	-0.125^{***}	-0.039^{**}
TargetDiff	(Not utilized)	0.735^{***}
R^2-*value*	0.310^{***}	0.608^{***}

5.2 Predictive Results for Decision Support

Using the results of (H1), one model per month is trained to consider the seasonality of the year. The 2×12 models use Eqs. (2) and (3). To compare out of sample predictions with the expert error, $R(_1F)$ needs to be added to the model prediction to receive an prediction of the actual ratio $R(A)$. Forecast correction utilizes the results of Model 1 ($\widehat{x}_1$) and Model 2 ($\widehat{x}_2$) in the out-of-sample test period.

The results are presented in Table 4. The table shows the date of month in 2013. Further the table describes the aggregated forecast error of the expert forecast ($\widehat{x}_0 = R(_1F)$) or model prediction ($\widehat{x}_1, \widehat{x}_2$) by $\sum |Err(\widehat{x}_i)|$. The last two columns show the percentage improvement of the models, which is defined in Eq. (4). The improvements measure the model error by the forecast error of the expert.

$$Improvement\ (in\ \%) = \frac{\sum |Err(R(_1F))| - \sum |Err(\widehat{x}_i)|}{\sum |Err(R(_1F))|} \cdot 100\ \% \qquad (4)$$

Table 4. Out of sample test for the forecasts in 2013. Cumulative absolute forecast error for expert forecast, Model 1, and Model 2 are shown. Percentage improvement compares the specific model error to the expert error. Model 2 predictions have the lowest forecast error (7.05 in December) compared to expert and Model 1 predictions. The improvements at the end of the year show that Model 2 reduces error by 59.9 %, which is better than Model 1 with 10.4 %.

Data descriptive	Forecast error			Improvement (in %)	
Date	Expert	Model 1	Model 2	Model 1	Model 2
01/2013	18.77	18.01	18.40	4.0 %	2.0 %
02/2013	18.11	15.11	14.38	16.6 %	20.6 %
03/2013	16.79	14.74	10.86	12.2 %	35.3 %
04/2013	16.15	14.82	9.66	8.2 %	40.2 %
05/2013	16.59	15.42	9.25	7.1 %	44.3 %
06/2013	16.30	15.75	8.34	3.4 %	48.8 %
07/2013	16.84	15.38	7.91	8.7 %	53.0 %
08/2013	17.30	15.41	7.39	10.9 %	57.3 %
09/2013	18.51	16.58	7.35	10.4 %	60.3 %
10/2013	17.41	16.67	7.28	4.3 %	58.2 %
11/2013	17.79	15.95	7.10	10.3 %	60.1 %
12/2013	17.61	15.77	7.05	10.4 %	59.9 %

The organizational information to reach the assumed target clearly drives the forecast error of Model 2, as comparison for column three and four shows. Approaching the end of the fiscal year the Model 2 reduces the aggregated forecast error to 7.05, while expert and Model 1 error is stable. Considering that 100 % improvement is the maximum possible, the result shows the advantage of the organizational information. The results show that percentage improvement of the models with solely statistical information reduces the expert error up to 16.6 % in February, while the models that considers business key figures performs with up to 60.1 % improvement in November.

The paired t-test based on the monthly improvements of Model 1 and Model 2 (comparison of the last two columns in Table 4) provides the following statistics: $t = 6.7299$, df $= 11$, p-value $= 3.243e{-}05$. The statistics evidence that the out of sample period results are significant at the 0.01 % level, supporting (H2).

5.3   Service Implications in Corporate Information Systems

The aggregation of the forecasts to a subsidiary-specific level has several benefits. First, correction techniques on aggregates result in fewer issues recommended for manual review, and therefore fewer workload for managers, as the number of data points to inspect reduces compared to the validation of thousands of issues for individual forecasts, improving the number of issues for managers.

Second, organizational biases can inflict the accuracy of many underlying forecasts in different ways (for example, earnings management often affect a set of items), but at the aggregate level they can directly relate to a specific forecast. Utilizing the information at the aggregate level can provide beneficial insights to better identify the concernable issues at a level where meaningful decisions are made – the business level. And, as the issues concerns these business level the solved issues can improve forecast accuracy. For the sample corporation with tasks in risk management, obtaining an accurate forecasting basis is important as it can help to reduce the costs for the corporation by avoiding unnecessary currency hedges.

Finally, forecast correction techniques on a higher level can easily integrate key figures for these biases. Current predictive approaches in the literature often account for statistical information only, leaving important business information out of focus. Forecast support systems should therefore account for information on all relevant organizational levels to provide reliable managerial recommendations. While business organizations and scientific journals are aware of various biases, it seems that few effort was put into the combination of correction techniques and organizational biases to improve decision support services. Identifying the meaningful issues in a reliable manner, more accurate model predictions become more important as the aggregates combine many forecasts. Otherwise, falsely predicted issues potentially increase workload as it leads the managerial attention on unimportant work scopes. Therefore, the improved predictive value for decision support systems can also reduce the workload of forecast inspections in accounting information systems.

Information systems with services in business analytics benefit from the insights provided by this paper. The empirical study in this paper shows that business information provides meaningful key figures for predictive purposes.

Explicitly, forecast correction approaches that incorporate organizational dependencies (EBITDA information) can highly improve forecast accuracy. In sum, this paper argues with provided evidence and methodology to analyze fewer issues, but those that really matter.

6   Conclusions and Outlook

Analyses of forecast correction techniques in financial forecasting processes are very sparse in the service analytics community. This research addresses two research gaps: first, organizational business information that could increase the explanatory power has been left unattended for correction approaches. Second, current studies not performed comparative analyses for statistical debiasing and organizational debiasing – based on the predictive results of forecast correction techniques.

Further, this paper provides the first empirical analysis on this topic, based on cash flow forecast data of a large multinational corporation. The analysis reveals that forecast correction techniques can provide better predictions when organizational key figures are integrated. Specifically, this paper shows that forecast difference from target is the key to explain results in corporate invoice forecasting Empowering forecast correction models with these business figures result in highly improved forecasts when compared to basic statistic models. Circumstances at the organizational level might bias the managers in planning and operations systematically. Especially, the corporate invoice forecast and revision processes, that deliver data for risk management, must consider the systematic effects on the aggregate level. Disclosing biases at a level where hedging takes place, can show a way to reduce currency risks or hedging costs. Therefore, both corporations and researchers need to understand how the organizational environment can affect forecasting accuracy.

From a managerial perspective, the results provide unconsidered insights on how organizational biases pave the way of forecasting. The endeavor to improve service analytics and the decision support tools can reduce the managerial workload to identify forecasting items to revise. For example, based on this research, the research partner plans to utilize the model predictions with organizational debiasing to automatically check the validity of aggregated forecasts within the forecast support system. The model debiases at an aggregate level, where the business reasons matter. Overall, consideration of business key figures makes it easier to identify the most beneficial work packages.

In sum, provided findings for information systems that support managerial decisions have implications for future research. First, the monthly analysis revealed that recurrent characteristics in time should also be considered for service analytics, i.e., non-seasonal developments or fiscal business cycles that do not start in January. Second, analysis of aspects beyond corporate financial responsibilities can provide insights to understand interlinks to organizational structures. For example, self-governing departments can have serious implications for corporations, as departments' independence provides no obvious reason to align to corporate goals – besides benevolence. The incorporation of appropriate proxies for business key figures into prediction models surely leads to further improvements in information systems.

References

1. Almeida, H., Campello, M., Weisbach, M.S.: The cash flow sensitivity of cash. J. Financ. **59**(4), 1777–1804 (2004)
2. Blanc, S.M., Setzer, T.: Improving forecast accuracy by guided manual overwrite in forecast debiasing. In: Proceedings of the European Conference on Information Systems (2015)
3. Bromiley, P.: Do forecasts produced by organizations reflect anchoring and adjustment? J. Forecast. **6**(3), 201–210 (1987)
4. Burgstahler, D., Eames, M.: Management of earnings and analysts' forecasts to achieve zero and small positive earnings surprises. J. Bus. Financ. Account. **33**(5–6), 633–652 (2006)
5. Daniel, N.D., Denis, D.J., Naveen, L.: Do firms manage earnings to meet dividend thresholds? J. Acc. Econ. **45**(1), 2–26 (2008)
6. Fildes, R., Hastings, R.: The organization and improvement of market forecasting. J. Oper. Res. Soc. **45**(1), 1–16 (1994)
7. Graham, J.R., Harvey, C.R.: The theory and practice of corporate finance: evidence from the field. J. Financ. Econ. **60**(2), 187–243 (2001)
8. Hogarth, R.M., Makridakis, S.: Forecasting and planning: an evaluation. Manag. Sci. **27**(2), 115–138 (1981)
9. Kim, C., Mauer, D., Sherman, A.: The determinants of corporate liquidity: theory and evidence. J. Financ. Quant. Anal. **33**(3), 335–359 (1998)
10. Knöll, F.: The goal of forecasting – Predictability of cash flow revisions in corporate finance. In: Proceedings of the 26th European Conference on Information Systems: Beyond Digitization - Facets of Socio-Technical Change (ECIS), (2018)
11. Knöll, F., Setzer, T., Laubis, K.: On predictability of revisioning in corporate cash flow forecasting. In: Proceedings of the 51st Hawaii International Conference on System Sciences (HICSS), pp. 1583–1589 (2018)
12. Knöll, F., Simko, V.: Organizational information improves forecast efficiency of correction techniques. In: Proceedings of the 17th Conference on Information Technologies – Applications and Theory (ITAT), Computational Intelligence and Data Mining (WCIDM), vol. 1885, pp. 86–92 (2017)
13. Lawrence, M., Goodwin, P., O'Connor, M., Önkal, D.: Judgmental forecasting: a review of progress over the last 25 years. Int. J. Forecast. **22**, 493–618 (2006)
14. Leitner, J., Leopold-Wildburger, U.: Experiments on forecasting behavior with several sources of information – A review of the literature. Eur. J. Oper. Res. **213**(3), 459–469 (2011)
15. Lim, S.S., Wang, H.: The effect of financial hedging on the incentives for corporate diversification: the role of stakeholder firm-specific investments. J. Econ. Behav. Organ. **62**(4), 640–656 (2007)
16. Marr, B.: Key Performance Indicators (KPI): The 75 Measures Every Manager Needs to Know. Pearson, London (2012)
17. Martin, J.D., Morgan, G.E.: Financial planning where the firm's demand for funds is nonstationary and stochastic. Manag. Sci. **34**(9), 1054–1066 (1988)

18. McCarthy, T.M., Davis, D.F., Golicic, S.L., Mentzer, J.T.: The evolution of sales forecasting management: A 20-year longitudinal study of forecasting practices. J. Forecast. **25**(5), 303 (2006)

19. McNees, S.K.: The role of judgment in macroeconomic forecasting accuracy. Int. J. Forecast. **6**(3), 287–299 (1990)

20. R Core Team: R: A Language and Environment for Statistical Computing. Vienna, Austria: R Foundation for Statistical Computing (2013). http://www.R-project. org

21. Sanders, N.R., Manrodt, K.B.: The efficacy of using judgmental versus quantitative forecasting methods in practice. Omega **31**(6), 511–522 (2003)

22. Schweitzer, M.E., Ordóñez, L., Douma, B.: Goal setting as a motivator of unethical behavior. Acad. Manag. J. **47**(3), 422–432 (2004)

23. Stulz, R.M.: Managerial discretion and optimal financing policies. J. Financ. Econ. **26**(1), 3–27 (1990)

A Framework for the Simulation-Based Estimation of Downtime Costs

Clemens Wolff$^{(\boxtimes)}$ and Michael Voessing

Karlsruhe Institute of Technology, Kaiserstr. 89, 76133 Karlsruhe, Germany
{clemens.wolff,michael.voessing}@kit.edu

Abstract. Currently, industrial maintenance experiences a shift from traditional on-call business towards long-term maintenance contracts, as, for example, availability-based maintenance contracts. Recently, researchers presented Service Level Engineering (SLE) as an approach for customers to determine their cost-optimal long-term availability levels for their production equipment within one production system. However—in order to apply SLE—customers must know their costs of downtime, i.e. costs associated to the unavailability of production assets.

This work presents a generic approach for customers to determine their costs of downtime function, a function reflecting the costs arising for a manufacturer due to unavailable production equipment. The approach uses simulation studies to incorporate different downtime patterns that result in the same overall availability. Consequently, this work contributes to the successful application of availability-based maintenance contract in practice. In detail, this work addresses the question of determining the industrial costs of downtime, a necessary prerequisite for the successful application of SLE in industrial maintenance.

Keywords: Industrial cost of downtime · Service Level Engineering Simulation

1 Introduction

Today, industrial maintenance is still the backbone of industrial services. However, recently—driven by servitization of the manufacturing business [1,26]—industrial maintenance has become more and more a long-term engagement business. Instead of an on-call business, maintenance customers demand contracts that are closely aligned with their business goals [25]—the production of a predefined production output. However, for manufacturers to reach their goal, the production equipment must be available and performing as intended. As an answer to those new customer demands, maintenance providers come up with full service (e.g. [8,9,24]), availability-based (e.g. [2,5]), and performance-based (e.g. [10,17,18]) maintenance contracts. Given those contracts, maintenance providers now sell equipment availability or performance for a predefined, periodic fixed price instead of selling single maintenance jobs.

© Springer Nature Switzerland AG 2018
G. Satzger et al. (Eds.): IESS 2018, LNBIP 331, pp. 247–260, 2018.
https://doi.org/10.1007/978-3-030-00713-3_19

The new maintenance offerings have brought many advantages for customers. However, they also introduced new challenges, as, for example, the customer now needs to decide prior to manufacturing which service level (i.e. how much availability) he wants to purchase for a given production equipment unit. Wolff and Schmitz [27] propose the application of Service Level Engineering (SLE) for the given problem. The fundamental idea of SLE is the trade-off between additional costs for production equipment availability and the costs of equipment unavailability. Simplified, the costs of equipment unavailability are costs due to lost production output, one component of an industrial manufacturer's cost of downtime [23]. In order to calculate the lost production output, the manufacturer must know his production output for different combinations of equipment availability levels. However, in complex industrial production systems, the estimation of production output is a difficult task. First—as production systems usually rely on multiple heterogeneous production equipment units in series—the overall production output does not depend on the availability of an individual production equipment unit. Instead, the output depends on the availability combination of all production equipment operated within the given production system [27]. Second, given *availability* as a Service Level Indicator (SLI), Patti and Watson [19] note that the same combination of individual equipment availability may result in different production output quantities.

In order to successfully apply SLE in industrial maintenance contracting, manufacturers are in need for a generic approach to determine the lost production output. In the remainder of this work, we refer to the loss in production output as the reduction in production throughput. This research aims at providing guidelines for the estimation of the throughput loss function over a range of production equipment availability such that SLE can be applied in industrial maintenance.

The remainder of this work is structured as follows: In Sect. 2, we elaborate on fundamentals and related work on the given problem. In Sect. 3, we explain the Design Science Research approach used in this work. The proposed framework for the simulation-based estimation of downtime costs is introduced and explained in Sect. 4. The applicability and utility of the proposed framework for the given task is shown in Sect. 5 in an illustrative scenario. Finally, in Sect. 6, the work is summarized and limitations and future work are pointed out.

2   Fundamentals and Related Work

In this subsection, we elaborate on fundamentals for this work. Additionally, we distinguish related work and highlight differences between those and this work.

2.1   Fundamentals

First, we elaborate on fundamentals regarding industrial production, before, second, explaining the core idea behind Service Level Engineering (SLE). Third, basics of simulation studies are briefly introduced. Finally, we briefly present current states in the simulation-based estimation of production system throughput.

Manufacturing and Industrial Maintenance. For the remainder of this work, we use availability as SLI. In this work, availability is defined as shown in Eq. 1 and refers to the percentage of time a manufacturer has productive access to the asset, regardless of its actual usage due to, for example, planned downtime. Furthermore, availability during a given time period can only be determined ex-post.

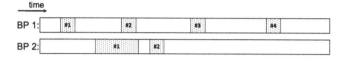

Fig. 1. Two different downtime patterns both resulting in the same overall availability of 80%

$$a = \frac{utilizable\ time}{total\ time} \tag{1}$$

However, the usage of availability as SLI has one major drawback. In order to calculate the true availability of a production equipment, one must know the downtime pattern of that equipment unit. The downtime pattern, however, depends on two stochastic variables, namely the breakdown rate (BR) and the breakdown duration (BD). BR refers to the quantity of breakdowns, whereas BD refers to the duration of equipment unavailability after a breakdown occurred. Ahead of time, both variables are stochastic, as we do not know when, how often, and how long equipment units fail. Consequently, one level of equipment availability can be reached by an unlimited number of downtime patterns. This is displayed in Fig. 1, where we see 2 downtime patterns resulting in the same overall equipment availability. As the downtime pattern is composed of the BR and BD, any overall availability can be met with an unlimited number of BR and BD combinations [3, pp. 23–32]. In practice, the BD greatly depends on the response time of the service provider, which, again, is mainly determined by its priority compared to other service demand. The determination of repair priority itself is a challenging task (e.g. [28]).

The dependency of equipment availability on BD and BR also effects the production system's throughput. In fact, researchers only presented models to calculate production system throughput for "[...] two-machine –one-buffer system and systems with infinite buffer capacity or without buffers" [14]. There are no exact throughput calculation models for more complex production systems. Hence, in practice, throughput analysis of entire manufacturing plants is estimated by using simulation. An recent overview over simulation in manufacturing is provided by Negahban and Smith [16].

Service Level Engineering in Industrial Maintenance. Service Level Engineering (SLE)—initially introduced by Kieninger et al. [11] for IT outsourcing scenarios—is a systematic engineering approach to determine a cost-optimal service level objective for the customer. It's application to industrial maintenance contracting has recently been proposed by Wolff and Schmitz [27].

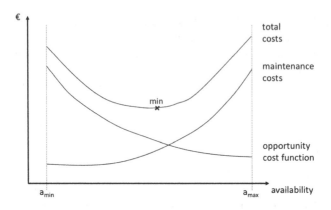

Fig. 2. Principle of SLE in industrial maintenance, adopted from [11]

The core idea behind SLE is the trade-off between the costs of additional equipment availability (maintenance costs) and the opportunity costs of unavailability. In industrial maintenance, the costs for additional availability refer to the periodic service fee the customer pays the provider. Furthermore, opportunity costs of equipment unavailability refer to opportunity costs introduced by the loss of production output. Both cost components are assumed to be non-linear with the availability level. As the customer faces the sum of both cost functions, the customer should—economically speaking—choose the production equipment availability such that his total costs are minimal. Furthermore, availability can only be within a range of possible availability levels $[a_{min}; a_{max}]$. This is depicted in Fig. 2.

From a customer perspective, the maintenance cost function can be derived by requesting the maintenance service costs for a range of available availability levels. Thus, the estimation of the maintenance cost function is rather a challenge for the provider—which is not addressed in this work. However, the determination of the opportunity cost function is essential for the application of SLE and must be derived by the customer himself. Wolff and Schmitz [27] define the opportunity cost function as shown in Eq. 2, in which $\boldsymbol{a}_{act}$ and $\boldsymbol{a}_{max}$ represents the combination of actual and maximum combination of asset availability (one availability for each production asset), c_u a product-value multiplier, and $n(\boldsymbol{a})$ the production throughput given availability $\boldsymbol{a} = (a_1, a_2, ..., a_n)$, with a_i corresponding to the availability of production asset i. In other words, the opportunity cost function reflects opportunity costs due to lost production output. The costs are derived by first determining the production throughput lost, and second, monetizing the loss with a certain cost component c_u. Unfortunately, Wolff and Schmitz [27] do not provide an approach on how to determine the manufacturing output function n for a given combination of asset availability $\boldsymbol{a}$. In the following, r refers to the function indicating the loss of production throughput due to the given availability levels $\boldsymbol{a}_{act}$.

$$c_d(\boldsymbol{a}_{act}) = c_u \cdot r(\boldsymbol{a}_{act})$$
$$= c_u \cdot (n(\boldsymbol{a}_{max}) - n(\boldsymbol{a}_{act})) \qquad (2)$$

Therefore, the goal of this work is the introduction of a generic simulation-based framework that enables manufacturers to determine their production throughput loss function r due to equipment unavailability. The framework must use simulation in order to be applicable to complex production systems. Furthermore, the introduction of simulation—coupled with multiple runs—allows for a higher consideration of the unlimited number of possible downtime patterns.

Simulation Studies. Under simulation, researchers understand the imitation of "[...] operations of various kinds of real-world facilities or processes" [13, p. 1]. Simulation models are a simplified representation of the real world and are usually designed to answer a question of interest that cannot be answered analytically [13, p. 5]. A simulation run underlies stochastic influences, thus, the output of an individual simulation run may result in unreliable results [13, p. 549]. To increase the reliability of knowledge gain, researchers run simulation models multiple times and use aggregated information to answer the question of interest. Given the set of responses, experts draw conclusion that are more reliable than conclusions drawn from an individual simulation run.

In practice, simulation runs are often conducted as part of *simulation experiments*—virtual experiments, in which multiple different scenarios are compared with respect to performance indicators. For example, a manufacturer might want to build a new production plant but cannot decide between two possible layouts. Using simulation experiments, a model is built for both scenarios and simulated multiple times, resulting in a set of values for each performance indicator. Experts refer to the variable simulation input as factors, whereas its output is referred to as response [13, p. 619].

2.2   Related Work

In this subsection, we highlight related work on the estimation of production system throughput. Furthermore, this section points out related work in the field of industrial cost of downtime estimation. Consequently, differences between this and related work are indicated.

Industrial Costs of Downtime. Most methods for estimating cost of downtime consider direct—costs that are directly associated with the repair of an asset (e.g. labour and material costs)—as well as indirect—consecutive costs of an asset failure (e.g. costs for production recovery, opportunity costs due to lost production output)—costs [23]. In the context of SLE, direct downtime costs are paid by the maintenance provider as part of his service offering. In return, the provider receives the pre-defined fixed maintenance fee periodically from the maintenance customer. Thus, direct downtime costs no longer influence the maintenance customer's profit and are not of further interest. However, indirect downtime costs, are still experienced by the maintenance customer. Therefore, the estimation of indirect downtime costs is of great relevance for this work.

So far, research focused on the estimation of (indirect) downtime costs for individual industrial cases (e.g. [6]) from a post-perspective. Furthermore, Edwards et al. [4] develop a method for predicting downtime costs of tracked hydraulic excavator in the mining industry. Predictions are based on historic data. The only exception found is the work by Liu et al. [15] and Wolff and Schmitz [27], who present methods to determine production losses due to downtime for simple production systems. Furthermore, the latter also notes that researches so far do not address the question on how manufacturers can estimate the costs of downtime prior to an actual downtime event for discrete manufacturing systems.

Related work on downtime cost estimation differs from this work such that for the application of SLE we do not need an estimate for an individual downtime event. Instead, we require an estimate of downtime costs for a—at the point of contract making—unknown series of downtime events. Therefore, the current rare work on downtime cost estimation does not yield sufficient insights for the given task.

Simulation-Based Estimation of Production Throughput. Today, simulation plays "[...] a significant role in evaluating the design and operational performance of manufacturing systems." [16]. Negahban and Smith [16] provide a comprehensive review on recent publications on simulation studies within the manufacturing sector. Most commonly, simulation studies are used to quantify production throughput under the variation of, for example, production layout, individual production equipment throughput, or scheduling decisions [16]. Simulation studies are used as no analytic throughput calculation exists for complex production systems [14]. In detail, Negahban and Smith [16] differentiate simulation studies in *manufacturing system design* and *manufacturing system operation* related studies. The proposed application of simulation in this work belongs to the manufacturing system design. Those simulation studies are furthermore categorized as *general system design and facility design/layout, material handling system design, cellular manufacturing system design,* and *flexible manufacturing system design.* However, those categories are not of interest for the given task. To the best of our knowledge, there is no work using simulation in order to estimate the production of throughout reduction in dependence of the individual production equipment availability. Therefore, this work is well distinguishable from previous simulation related work within industrial manufacturing.

3 Methodology

The following simulation framework was developed following a Design Science Research (DSR) approach with one design cycle as proposed by Kuechler and Vaishnavi [12]. The artifact of the DSR approach is a *method*, as the simulation framework consists of conceptual instructions [20]. Furthermore, the contribution of the simulation framework is an *improvement*, as we propose the new solution of simulation studies for the known problem of determining long-term production losses due to unavailable production equipment [7]. According to Peffers et al.

[21], an *improvement* can be evaluated by showing its suitability and utility on an illustrative scenario. The design cycle activities are depicted in Fig. 3.

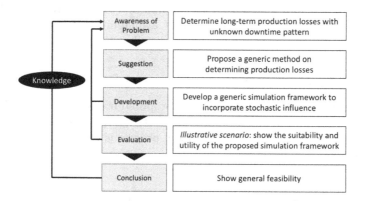

Fig. 3. Design cycle activities

4 Simulation-Based Framework for the Estimation of Downtime Costs

First, we explain how availability is modelled during an individual simulation run. Second, we present the developed simulation framework.

4.1 Availability Modelling Within the Simulation Framework

Previously, we stated that the production throughput of a production line is a random variable influenced by the BD and BR. Jointly—from a post perspective—BR and BD result in a downtime pattern that can be used to calculate overall equipment availability. Assuming a specific downtime pattern and starting buffer capacities, one is able to determine the throughput of a production line (determination may include simulation if no mathematical throughput calculation is possible) deterministically. However, this approach requires many different downtime patterns given in order to account for the many possible downtime cases. Furthermore, this approach is complicated by the influence of buffer capacity. Another approach, however, is to take the BR and BD distributions as the base unit. Given those two distributions, we are able to create many downtime patterns through repeated simulations. Consequently, as for the previous approach, this approach results in many simulation runs. However, this approach has the advantage of automatically creating downtime patterns and thus reducing the required user input for the simulation model.

Given those thoughts, we decided to model availability through the provision of a BR and BD distribution. Jointly, the two variables define the downtime pattern for each individual asset. Therefore, with the given purpose of this work, the BR and BD are seen as simulation experiment factors. Hence, those jointly define a simulation scenario and are varied across a predefined set of possible distributions during the experiment. Due to the stochastic influence, each scenario is simulated multiple times. Each simulation run will result in different downtime patterns, and thus also different availability levels for the different production assets—even though the factor distributions remain the same.

4.2   Simulation Framework

The prerequisite for the application of the proposed framework is a valid simulation model of the production system of interest. Overall, the framework follows a five step process, as shown in Fig. 4. Each process is explained individually in the following subsections.

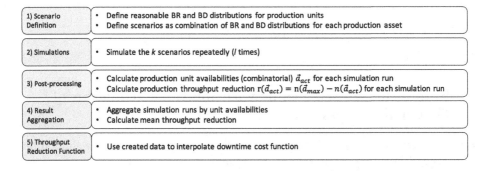

Fig. 4. Developed framework for the simulation-based estimation of downtime costs

Scenario Definition. First, the user must define possible experiment scenarios. Under such, we understand the range of possible input factors. As the breakdown behaviour of an asset can be modelled using BR and BD those are seen as the input factors of relevance. Consequently, one scenario consists of one BR and BD distribution for each production asset. Therefore, after this step, the user has a set S of o different scenarios with $S = \{s_1, s_2, \ldots, s_o\}$. Each scenario is defined by a combination of BD and BR distribution for the n production assets (e.g. $s_1 = ((BR_1, BD_1), (BR_2, BD_2), (BR_3, BD_3))$ for three production assets).

Simulations. For each scenario, the simulation model is executed p times. As explained previously, models must be executed repeatedly to guarantee generic results. Therefore, a total of $o \times p$ simulations are executed. As the determination of a suitable number of simulation runs has already been addresses by Law [13] it is not included in this work.

Postprocessing. At this point, the user has—simplified speaking—the results of $o \times p$ simulation runs. For each simulation run, we know the downtime pattern for each asset, thus allowing for the calculation of asset availability. Furthermore,

Table 1. Results after simulation runs have been executed and throughput reduction r has been computed

Scenario	Iteration	Availability a_{act}				$r(a_{act})$
		a_1	a_2	$\ldots$	a_n	
s_1	1	a_{111}	a_{112}	$\ldots$	a_{11n}	r_{11}
	2	a_{121}	a_{122}	$\ldots$	a_{12n}	r_{12}
	$\ldots$	$\ldots$	$\ldots$	$\ldots$	$\ldots$	$\ldots$
	p	a_{1p1}	a_{1p2}	$\ldots$	a_{1pn}	r_{1p}
$\ldots$	$\ldots$	$\ldots$	$\ldots$	$\ldots$	$\ldots$	$\ldots$
s_o	1	a_{o11}	a_{o12}	$\ldots$	a_{o1n}	r_{o1}
	2	a_{o21}	a_{o22}	$\ldots$	a_{o2n}	r_{o2}
	$\ldots$	$\ldots$	$\ldots$	$\ldots$	$\ldots$	$\ldots$
	p	a_{op1}	a_{op2}	$\ldots$	a_{opn}	r_{op}

we know the production throughput $n(a_{act})$ for the combination of individual asset availability $a_{act} = (a_1, a_2, ..., a_n)$ for n production assets. Availability levels should be rounded to a suitable level (e.g. only full percentage points). Given this data, we are able to calculate the production throughput reduction r for each simulation run using Eq. 3. Exemplary, the results after this step are shown in Table 1.

$$r(a_{act}) = n(a_{max}) - n(a_{act}) \tag{3}$$

Result Aggregation. Given the results from the previous step, we are able to aggregate the results depending on the combination of availability levels. Due to the repeated simulations and the dependence on the stochastic BD and BR, it is likely that multiple simulation runs result in the same combination of availability levels, yet result in different production throughput reductions. For each occurring combination of availability level, the mean reduction of production throughput $\overline{r(a)}$ is calculated.

Throughput Reduction Function. Given the data provided above, we are able to use interpolation to provide a production throughput reduction for each combination of availability levels. According to Eq. 2, the downtime costs reflect the throughput reduction multiplied with a cost factor c_u.

5   Framework Application on an Illustrative Scenario

As noted, evaluation is done by an illustrative scenario, as proposed by Peffers et al. [21]. An illustrative scenario describes the application of an artifact (i.e. the framework) to a synthetic situation in order to illustrate the suitability or utility of the designed artifact [20]. First, we explain the production system used, before applying the presented simulation framework. Finally, the resulting data points for interpolating the downtime cost function are shown.

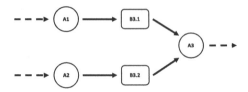

Fig. 5. Exemplary production system used for illustrative scenario evaluation

5.1 Introduction of the Illustrative Scenario

In order to illustrate the application of the presented framework a simple production system is used. The layout and material flow of the production system is shown in Fig. 5, in which circles and rectangles indicated production assets (i.e. equipment) and production buffers. Material flow is shown by directed arrows. Furthermore, dotted arrows indicate a non-starving or non-blocking material flow.

Therefore, the example contains three production assets as well as two production buffers. Each buffer has a capacity of 20 production units and asset *A3* can be seen as an integration, meaning that it produces by jointly using one production input unit coming from assets *A1* and *A2*. The machines A1, A2, and A3 have a lead time (in minutes) following a N(15, 5), N(16, 5), and N(14, 5) normal distribution, respectively. Therefore, the production system reflects a production system for which no analytic method for throughput determination exists. Under ideal conditions without any downtime, the production system's throughout is limited by the bottleneck [22], which is—in this case—machine A2. The unit price c_u is given with $c_u = 10$€.

5.2 Application of the Framework

In this subsection, we walk through the application of the simulation framework. Relevant steps and intermediate results are pointed out.

Scenario Definition. As previously introduced, a scenario i s_i is defined as $s_i = ((BR_1, BD_1), (BR_2, BD_2), (BR_3, BD_3))$, where BR_i and BD_i stands for the BR and BD distributions for production asset $j \in \{1, 2, 3\}$. For this illustrative scenario, the set of simulation scenarios is created by the permutation of the following individual distribution sets $BR_1, BD_1, BR_2, BD_2, BR_3$, and BD_3, reflecting the BR and BD distribution of the three production assets. Consequently, a total of $o = 729$ scenarios are used. BR follows an exponential distribution with $BR = Exp(\lambda)$ and $E(BR) = \frac{1}{\lambda}$ in days. Furthermore, BD (in hours) follows a normal distribution $N(\mu, \sigma)$. In detail, the following sets of distributions are used:

$$BR_i = \{Exp(5), Exp(10), Exp(15)\}, \forall i \in \{1, 2, 3\}$$
$$BD_i = \{N(6, 2), N(8, 2), N(10, 2)\}, \forall i \in \{1, 2, 3\}$$

Simulations. Each scenario is simulated $p = 10$ times. Therefore, a total of $o \times p = 7290$ simulation runs are executed. Each simulation run simulated a working period of 90 days. Given the simulation period of 90 days, the maximum production throughput is 8,100 units (assuming no lead time deviation).

Postprocessing. The simulation model automatically calculates the production throughput and the availability of the production equipment at the end of each simulation run. Therefore, for each simulation run r, we know the availability levels $\boldsymbol{a}_r = (a_1, a_2, a_3)$ and the throughput reduction r_r.

Result Aggregation. The results of the individual simulation runs are aggregated by the availability levels of the production equipment. Table 2 shows an extract of the aggregated data. In total, a set of 1,002 combinations of equipment availability levels were calculated.

Table 2. Extract of framework results: aggregated simulation results and opportunity costs

Availability $\boldsymbol{a}$			Mean throughput $n(\boldsymbol{a})$	Mean throughput reduction $r(\boldsymbol{a})$	Opportunity costs $c_d(\boldsymbol{a})$
a_1	a_2	a_3			
			...		
0.93	0.96	0.91	7,197	903	9,030 €
0.93	0.96	0.92	7,218	882	8,820 €
0.93	0.96	0.93	7,251	849	8,490 €
0.93	0.96	0.95	7,425.3	674.7	6,747 €
0.93	0.96	0.96	7,422.2	677.8	6,778 €
0.93	0.96	0.97	7,446.6	653.4	6,534 €
			...		

Throughput Reduction Function. Given the aggregated data, the manufacturer is able to interpolate the throughput reduction for a arbitrary availability levels of the production equipment. The throughput reduction r can be monetized in order to interpolate the opportunity cost function required for the successful application of SLE in industrial maintenance. For the given illustrative scenario, the opportunity costs for the combination of availability levels are given in Table 2. This data is now used to interpolate opportunity costs for any given combination of production equipment availability.

6   Conclusion

In this section, we summarize this work, show its contribution, highlight limitations, and point out directions for further work.

6.1   Summary and Contribution

This work presents a generic method to determine the production throughput losses due to unavailable production equipment. In this work, unavailability is used in a long-term context, thus unavailability is not limited to a single downtime event. Instead, a series of unknown downtime events result in a downtime pattern whose overall costs need to be determined. By applying simulation techniques, the framework is able to incorporate many possible downtime patterns in order to achieve an estimate of throughput reduction. The throughput reduction estimate can be monetized in order to gain a downtime cost estimate.

This work contributes to the application of Service Level Engineering (SLE) in industrial maintenance. In detail, the presented framework can be used to derive the required downtime cost function for the application of SLE, an approach to determine cost-optimal availability levels for production equipment from a customer perspective. The presented framework—and SLE—address the challenge of industrial manufacturers to determine long-term availability levels in new maintenance offerings, as, for example, availability-based maintenance contracts.

The framework was developed following a one-cycle Design Science Research (DSR) approach as proposed by Kuechler and Vaishnavi [12]. Following DSR terminology, the proposed framework is an artifact of the type *method* [20] and the contribution an *improvement* [7]. The suitability and utility of the developed artifact has been shown on an illustrative scenario [21].

6.2   Limitations and Future Work

First, as this work relies on simulation studies, known limitations of simulation studies apply to this work. For example, the results will only be as good as the simulation model itself. The user must decide which level of detail is required to receive accurate results. Furthermore, the execution of the entire simulation experiment is complex and time demanding, as the simulation model is executed repeatedly per scenario, resulting in overall $o \times p$ simulation runs.

Second, downtime costs are limited to opportunity costs of lost production output only, which is—in the context of SLE—a valid assumption. However, SLE being based on opportunity costs also implies that the market always demands the manufactured goods which allows the manufacturer to produce his goods without quantitative barriers.

Third, the computation of production throughput reduction r is based on simply averaging the results from the multiple simulation runs. In future work, the calculations could be extended to reflect a range of possible throughput reduction to reflect uncertainty. For example, one could calculate a worst-, average-, and best case scenario in terms of throughput reduction for a certain combination of availability levels due to different downtime patterns.

Fourth, more sophisticated evaluation of the proposed approach is required. So far, we evaluated the simulation framework by demonstrating its applicability on an illustrative scenario [20,21]. However, for further evaluation, we want to apply the framework on a real world use case. Therefore, a promising field of research lies ahead.

References

1. Baines, T., Lightfoot, H., Benedettini, O., Kay, J.: The servitization of manufacturing. J. Manuf. Technol. Manag. **20**(5), 547–567 (2009)
2. Datta, P.P., Roy, R.: Cost modelling techniques for availability type service support contracts: a literature review and empirical study. CIRP J. Manuf. Sci. Technol. **3**(2), 142–157 (2010)
3. Ebeling, C.E.: An Introduction to Reliability and Maintainability Engineering. Waveland Press, Inc., Long Grove (2010)
4. Edwards, D.J., Holt, G.D., Harris, F.C.: Predicting downtime costs of tracked hydraulic excavators operating in the UK opencast mining industry. Constr. Manag. Econ. **20**(7), 581–591 (2002)
5. Erkoyuncu, J.A., Roy, R., Shehab, E., Kutsch, E.: An innovative uncertainty management framework to support contracting for product-service availability. J. Serv. Manag. **25**(5), 603–638 (2014)
6. Fox, J.P., Brammall, J.R., Yarlagadda, P.K.: Determination of the financial impact of machine downtime on the Australia post large letters sorting process. In: 9th Global Congress on Manufacturing and Management (GCMM), pp. 732–738 (2008)
7. Gregor, S., Hevner, A.R.: Positioning and presenting design science research for maximum impact. MIS Q. **37**(2), 337–355 (2013)
8. Huber, S., Spinler, S.: Pricing of full-service repair contracts. Eur. J. Oper. Res. **222**(1), 113–121 (2012)
9. Huber, S., Spinler, S.: Pricing of full-service repair contracts with learning, optimized maintenance, and information asymmetry: pricing of full-service repair contracts. Decis. Sci. **45**(4), 791–815 (2014)
10. Hypko, P., Tilebein, M., Gleich, R.: Clarifying the concept of performance-based contracting in manufacturing industries. J. Serv. Manag. **21**(5), 625–655 (2010)
11. Kieninger, A., Westernhagen, J., Satzger, G.: The economics of service level engineering. In: Proceedings of the 2011 44th Hawaii International Conference on System Sciences, pp. 1–10 (2011)
12. Kuechler, B., Vaishnavi, V.: On theory development in design science research: anatomy of a research project. Eur. J. Inf. Syst. **17**(5), 489–504 (2008)
13. Law, A.M.: Simulation Modeling and Analysis, 5 edn. McGraw-Hill Education, New York (2015)
14. Li, L., Chang, Q., Ni, J.: Data driven bottleneck detection of manufacturing systems. Int. J. Prod. Res. **47**(18), 5019–5036 (2009)
15. Liu, J., Chang, Q., Xiao, G., Biller, S.: The costs of downtime incidents in serial multistage manufacturing systems. J. Manuf. Sci. Eng. **134**(2), 021016 (2012)
16. Negahban, A., Smith, J.S.: Simulation for manufacturing system design and operation: literature review and analysis. J. Manuf. Syst. **33**(2), 241–261 (2014)
17. Ng, I., Maull, R., Yip, N.: Outcome-based contracts as a driver for systems thinking and service-dominant logic in service science: evidence from the defence industry. Eur. Manag. J. **27**(6), 377–387 (2009)

18. Ng, I., Yip, N.: Identifying risk and its impact on contracting through a benefit based-model framework in business to business contracting: case of the defence industry. In: Proceedings of the 1st CIRP Industrial Product-Service Systems (IPS2) Conference, pp. 207–215 (2009)
19. Patti, A.L., Watson, K.J.: Downtime variability: the impact of durationfrequency on the performance of serial production systems. Int. J. Prod. Res. **48**(19), 5831–5841 (2010)
20. Peffers, K., Rothenberger, M., Tuunanen, T., Vaezi, R.: Design science research evaluation. In: Peffers, K., Rothenberger, M., Kuechler, B. (eds.) DESRIST 2012. LNCS, vol. 7286, pp. 398–410. Springer, Heidelberg (2012). https://doi.org/10.1007/978-3-642-29863-9_29
21. Peffers, K., Tuunanen, T., Rothenberger, M.A., Chatterjee, S.: A design science research methodology for information systems research. J. Manag. Inf. Syst. **24**(3), 45–77 (2007)
22. Roser, C., Nakano, M., Tanaka, M.: Productivity improvement: shifting bottleneck detection. In: Proceedings of the 34th Winter Simulation Conference (WSC), San Diego, California, pp. 1079–1086 (2002)
23. Salonen, A., Tabikh, M.: Downtime costing—attitudes in Swedish manufacturing industry. In: Proceedings of the 10th World Congress on Engineering Asset Management (WCEAM 2015), Tampere, pp. 539–544 (2016)
24. Schmitz, B., Duffort, F., Satzger, G.: Managing uncertainty in industrial full service contracts: digital support for design and delivery. In: Proceedings of the 2016 IEEE 18th Conference on Business Informatics (CBI), pp. 123–132 (2016)
25. Stremersch, S., Wuyts, S., Frambach, R.T.: The purchasing of full-service contracts: an exploratory study within the industrial maintenance market. Indus. Market. Manag. **30**(1), 1–12 (2001)
26. Vandermerwe, S., Rada, J.: Servitization of business: adding value by adding services. Eur. Manag. J. **6**(4), 314–324 (1988)
27. Wolff, C., Schmitz, B.: Determining cost-optimal availability for production equipment using service level engineering. In: Proceedings of the 2017 IEEE 19th Conference on Business Informatics (CBI), pp. 176–185 (2017)
28. Wolff, C., Vssing, M., Schmitz, B., Fromm, H.: Towards a technician marketplace using capacity-based pricing. In: Proceedings of the 51st Hawaii International Conference on System Sciences, pp. 1553–1562 (2018)

Combining Machine Learning and Domain Experience: A Hybrid-Learning Monitor Approach for Industrial Machines

Daniel Olivotti[1]([⊠]), Jens Passlick[1], Alexander Axjonow[2], Dennis Eilers[1], and Michael H. Breitner[1]

[1] Information Systems Institute, Leibniz Universität Hannover, Königsworther Platz 1, 30167 Hannover, Germany
{olivotti,passlick,eilers,breitner}@iwi.uni-hannover.de
[2] Lenze SE, Hans-Lenze-Straße 1, 31855 Aerzen, Germany
alexander.axjonow@lenze.com

Abstract. To ensure availability of industrial machines and reducing breakdown times, a machine monitoring can be an essential help. Unexpected machine downtimes are typically accompanied by high costs. Machine builders as well as component suppliers can use their detailed knowledge about their products to counteract this. One possibility to face the challenge is to offer a product-service system with machine monitoring services to their customers. An implementation approach for such a machine monitoring service is presented in this article. In contrast to previous research, we focus on the integration and interaction of machine learning tools and human domain experts, e.g. for an early anomaly detection and fault classification. First, Long Short-Term Memory Neural Networks are trained and applied to identify unusual behavior in operation time series data of a machine. We describe first results of the implementation of this anomaly detection. Second, domain experts are confronted with related monitoring data, e.g. temperature, vibration, video, audio etc., from different sources to assess and classify anomaly types. With an increasing knowledge base, a classifier module automatically suggests possible causes for an anomaly automatically in advance to support machine operators in the anomaly identification process. Feedback loops ensure continuous learning of the anomaly detector and classifier modules. Hence, we combine the knowledge of machine builders/component suppliers with application specific experience of the customers in the business value stream network.

Keywords: Machine monitoring · Hybrid learning
Long Short-Term Memory Neural Networks · Product-service systems

1 Introduction

Availability of industrial machines and plants in the manufacturing industry is essential because a breakdown can result in high costs for the breakdown itself

© Springer Nature Switzerland AG 2018
G. Satzger et al. (Eds.): IESS 2018, LNBIP 331, pp. 261–273, 2018.
https://doi.org/10.1007/978-3-030-00713-3_20

as well as loss of production. The monitoring of machines and the prediction of possible failures can reduce the possibility of a breakdown. Reliable machine monitoring requires a huge amount of domain knowledge and experience about the interaction of the machine components. Machine builders as well as component suppliers have detailed product knowledge and can offer this knowledge to their customers in form of services. It must be taken into account that every machine can be individual. Each components can be installed in very different environments. Therefore, the domain knowledge about the components has to be combined with information about the particular application of the individual component. In so called product-service-systems (PSS) physical assets are combined along with services to gain more value [19, 20]. Possible services which contributes to machine availability and aim to reduce breakdowns are preventive maintenance, machine monitoring or predictive maintenance services. Suppliers of these services can be component suppliers as well as machine builders or a third party, based on the individual value stream network. But according to what concept the implementation of such a service can deliver valuable and reliable results? McArthur et al. [16] present a condition monitoring system model but not in a PSS context. We think it is important to combine the competences of all participants of a PSS in the best promising way. Machine builders and component suppliers have a lot of experience in using their machines and components in very different scenarios or applications. The service provider, on the other hand, has access to very detailed lifetime data on every individual applications of a machine.

In this article we show how the experiences of each party of a PSS can be combined to realize a hybrid-learning machine monitoring approach. Such an approach should use different algorithms to process various sensor data, enable predictive maintenance services and a better fault diagnosis. In this article we present a first approach. The developed approach consists of three modules: anomaly detector, monitor and classifier. In Sect. 3 an overview of the developed approach is given. Before that we describe the research background. Herein the focus is on smart services in accordance to product-service systems and the application of condition monitoring and predictive maintenance in the industrial context. The anomaly detection with Long Short-Term Memory Neural Networks (LSTM) and its application is explained in Sect. 4. This also includes the application with a real-world machine. A description of the monitoring and the classifier for anomalies follows in Sect. 5. Finally, discussion of results, limitations and conclusions are presented in Sect. 6.

2   Research Background

Service science enters more and more into manufacturing industries [14, 19]. In general offering (smart) services comes along with the satisfaction of individual consumer needs or new revenue channels [19]. This interaction enables co-creating value between several partners in value stream networks [5]. Machine learning and intelligent machines support this value co-creation of smart services

[5]. Different authors state out that the usage of information and communication technologies, sensors and context information enable smart services [3,4]. A continuous feedback loop between different partners [24] as well as including different data sources [1] are also important factors.

The combination of products with services and supporting infrastructure is known as product services systems [17]. In manufacturing industries it plays an essential role since component suppliers and machine builders build physical products and a trend towards offering PSS is seen [14]. Possible smart services in the manufacturing industries are condition monitoring or predictive maintenance as a service. As stated in the introduction high availability of machines and plants is a major challenge. This availability is reduced by failures or planned maintenance. With better knowledge of the actual condition of a machine, the possibility of a breakdown can be reduced and the machine only needs to be maintained when it is necessary (based on the condition). Therefore, actual sensor data need to be combined with prediction models [10]. The implementation of monitoring systems for machines has already been discussed. Teti et al. [21] give an overview of the different implementation options. [25] show in a case study the added value that condition monitoring can bring to maintenance scheduling. In their example, they improve the maintenance cost rate by 32%. In this way, their work underlines the usefulness of condition monitoring. One way to realize condition monitoring is the use of neural networks. Wu et al. [22] use a neural network to predict the lifetime distribution. They optimize the threshold value for the probability of failure, which corresponds to the lowest maintenance costs. A more complex structure is presented by Wu et al. [23]. They develop a three-stage system to minimize the operating costs. On the first level, vibration data is stored in a database. In the second stage, the data is used to predict the life expectancy with a neural network. In the last step, the prediction is then used to optimize the expected costs per unit. This shows that condition monitoring approaches can consist of several levels. A possible visualization of condition monitoring data has already been presented. For example, a patent describes a tool for visualizing sensor data of a power plant [18]. In the event of a failure, all abnormal sensors are displayed on a single screen. In addition, the technician can navigate through the details of each individual sensor. From our point of view, this is a good way to provide a technician with all sorts of data, but without overloading him or her. Besides displaying and recognizing errors, another challenge is the classification of errors. Axinte [2] describes a tool for error detection and classification during broaching. He uses a neural network to classify the errors. McArthur et al. [16] describe the idea of using different systems for the monitoring of machines. They describe a condition monitoring system for an electrical transformer. To do this, they use a neural network and the k-means algorithm. To detect and classify faults in the transformer they use "ultra high frequency measurement" [16]. They use data that is generated during operation. However, these classifiers are often for very specific applications [13]. According to Lee et al. [13], they are often not flexible enough to process more complex information. They give the following reasons for this: no close human-machine interaction, no use of all available information and no adaptive learning [13]. To mitigate these problems, they propose cluster algorithms with feedback loops. Further research shows that classification can be further improved by using completely different sensors (e.g. image and sound data) [8]. Hirt and Kühl [8] show how classifications can be further improved by using several different data sets. They use texts as well as the name and the profile picture of Twitter users to decide which gender the user has.

The previous literature shows that neural networks were very often used for the analysis and prediction of deterioration processes. Combinations of different mathematical methods and algorithms have already been presented. However, to our knowledge, there is still no combination of error detection in combination with a classification of the detected error using additional (rich) sensor data. In this context, rich means that these sensor data require additional methods for their analyses and the automated processing is still complex (e.g. image and sound data). It has also not yet been described how the classification can be improved by the inclusion of expert knowledge. In this context, we also want to point out the opportunities that arise from working together in a PSS. Based on the discussed literature, we would like to present such an approach in the following.

3   A Hybrid-Learning Machine Monitoring Approach

The hybrid-learning machine monitoring approach proposed in this section is based on three inter-playing modules. Figure 1 provides an overview of the implementation.

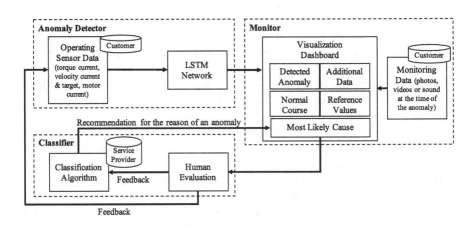

Fig. 1. The developed machine monitoring approach

For the first module we developed a procedure for detecting anomalies with the operation data of a machine. In this example we define operating data as a time series of the motor torque which provides the necessary information about the current operating status of the machine. During normal operation, similar repetitive patterns can be observed which represents for example the transport of a material box on a conveyor belt. Anomalies are defined as deviations of the observed time series from the expectation. In Sect. 4 we show how an LSTM is trained on large time series data sets and subsequently applied to identify deviations from a normal operation behavior.

In case of a detected anomaly, the monitor module is triggered. The goal of this module is the identification of causes and possible countermeasures as quickly and reliable as possible to prevent high breakdown cost of the machine. The core of this module is a visualization dashboard which contains a visual representation of the detected time series anomaly in addition to snapshots from different monitoring data sources. Monitoring data are defined as additional observations which can provide further clues to the cause of an anomaly like short video and audio sequence from the point in time of the detected event as well as sensor data from different component of the machine, like vibration and temperature. In addition, values of normal operation sequences are provided for comparison.

The actual identification of causes is then supported by the classifier module. In case of an anomaly, the classifier receives all available operating and monitoring data as inputs to calculate the likelihood of predefined anomaly scenarios. The recommendation for a possible cause is returned and displayed in the monitor module. By investigating the actual cause, the machine operator or other domain experts now provide an indirect feedback to the model about the quality of the general anomaly detection and classification. If the detected anomaly is indicated as a false positive after a screening by human experts, the respective case is added to the training database for normal operating examples. Therefore, the model continuously learns and adapts to the normal operation behavior of the machine. In case of a true positive in the anomaly detector model but a wrong classification of the cause, two cases can occur. (1) The actual anomaly is known but mismatched or (2) the anomaly was detected for the first time. In the first case, again the database is updated by adding the new pattern with the correct label which leads to a continuous improvement of the model. In the second case, a new class must be defined by the human experts to update the models. True positive anomalies which are also correctly classified, are simply approved by human experts and also added to the database to reinforce right decisions of the models and further advance the automation process.

As a machine builder, component supplier and service provider of such a hybrid-learning machine monitoring approach, the question of how to initialize the system and thus to ensure a direct value co-creation with the customers arise. To this end, we define machine specific anomalies and application specific anomalies. The anomaly detector module completely depends on application specific data since the application scenarios differ from customer to customer. Therefore, to initialize the anomaly detector module, the data must be collected individually for each customer during a normal operation behavior. The necessary effort, however, is also manageable since no labeling process is necessary for time series anomaly detection. In contrast to this, the classifier requires labeled data to calculate the likelihood for the predefined anomaly classes. To realize this approach, the classifier must be initialized by the machine builders or component suppliers who have the soundest knowledge about the internal operation of the product. Data about possible anomalies that affect the machine itself (e.g. friction causes abrasion) must be recorded and continuously updated according to experiences in different application scenarios with changing conditions. The service provider can fall back on a large pool of experience with various customers who provide the machine specific data for example in return for the provided service. To cover application specific anomalies like a misplaced load, the approach explicitly introduces the feedback loops, which allows the customers to easily build up their own labeled data pools. If no machine specific anomaly matches the detected anomaly, human experts from each customer are able to label these new cases by simply introducing a new class of anomalies to the model and provide a possible solution approach for this case. The approach therefore allows to continuously update the knowledge base of the classifier module which was initialized and continuously updated only with machine specific anomalies from the service provider. Hence the module relies on the experience of the different partners in the value stream network about the machine itself and continuously adapts to the needs of each individual customers. The more machines are in use, the faster the adaption can be realized due to more application specific anomaly detections.

4   Anomaly Detection with Long Short-Term Memory Neural Networks

Different cases were realized with the industry partner Lenze, a German manufacturer of automation solutions. Their products and automation solutions are used in automotive, robotics, packaging and material handling applications. Their main components are motors, gearboxes, inverters and controllers.

The implementation of the complete proposed approach is currently not realized. Although the module for anomaly detection is already implemented and evaluated. For this purpose, a specially designed demonstration machine is used which represents a typical application of the Lenze automation solutions and components. This machine consists of three conveyor belt modules in a row to move various goods. Typical applications could be the movement of cargo, baggage or parcels. To imitate the data gathering process, application specific time series data for normal operations are collected by simulating nearly 600,000 time series windows of different load patterns on a conveyor belt. In this case we focus on the torque time series which reflects the operation status of a machine in each point in time. The initialization of the module is now realized by the following procedure.

1. The idea is that anomalies are defined in this context as deviation of measured operating data and the expectations. The expectation about a normal operation behavior is based on a machine learning model which investigates the recent data history and provides a prediction for the expected operation status in the next time step.

2. To realize this prediction, the machine learning model must be trained with many usual operation behavior patterns to learn an approximation of a usual time series. A method which is able to handle ever-changing load patterns in addition to complex, noisy and non-linear torque time series patterns is required. One particularly suitable model type for this purpose are LSTMs. The focus of our work is not to benchmark the optimal model for such a time series analysis. Statistical quality control techniques like CUSUM (cumulative sum), exponentially weighted moving average analyses, other autoregressive models or machine learning techniques can also be applied. Since [6] establish LSTM models as a well functioning standard method for similar task, we implement this technique in our anomaly detector module as well. A complete introduction to LSTMs is beyond the scope of this article, hence we refer to [9].

3. The LSTM is trained on 64% of the generated data. Before the training process the values are scaled between 0 and 1. To prevent the LSTM from overfitting, an early stopping approach is used. Therefore 16% of the data are used for continuously validation during training. A hyperparameter tuning is conducted by a grid search approach. The resulting LSTMs consist of two hidden layer with 20 hidden neurons, a batch-size of 64 and a windows size of 50.

4. Based on the remaining 20% out-of-sample data sets which again contain usual patterns of the time series, the absolute values of the residuals for each period are summed up to generate a maximal threshold value for normal deviations of the operation behaviour.

5. In a further step, actual anomalies are simulated. Simulated anomalies contain for example a motor blockade or a misplaced load.

6. Test data of normal cases as well as simulated anomalies are now evaluated by the LSTM model which compares the actual measurement with the expected measurement. If the absolute deviations lay above the maximal threshold value for a normal operation behaviour, the respective time series window is labeled as a potential anomaly.

The presented procedure is a standard approach in time series anomaly detection, regardless of the model type used [15]. Figure 2 represents an example for a normal operation of a machine (left) and an illustration of a detected anomaly (right). This type of anomaly occurs due to insufficient motor bias. This creates a slip between the toothed belt and the drive shaft of the asynchronous motor. To evaluate the implemented approach, we perform 208 test runs on the demonstration machine. In 153 test runs, we simulate an anomaly at a certain point in time caused, for example, by loading a wrong mass, or setting a wrong acceleration. 55 test runs are carried out without any anomaly. Table 1 shows the respective confusion matrix. The results show an accuracy of 90.9%, a true positive rate (correctly identified anomalies) of 91.5% and a false positive rate (false alarm) of 10.9% in the current setting.

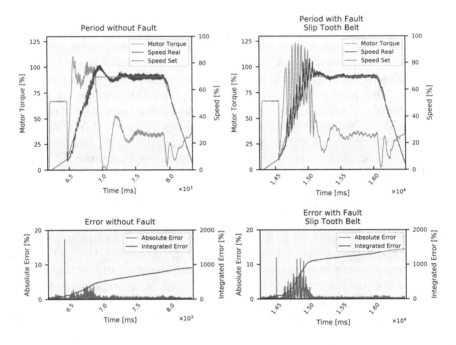

Fig. 2. Normal operation scenario vs. detected anomaly

Table 1. Confusion matrix

	No anomaly	Anomaly
Predict no anomaly	49	13
Predict anomaly	6	140

In a real application, the customer of the proposed solution is able to generate the necessary training data either within a controlled environment with own demonstration machines or during manually monitored real live operation scenarios to ensure the quality and representativeness of the normal training examples.

5   Future Integration of the Monitor and Classifier Module

In cooperation with our industry partner Lenze, the anomaly detector module is the part of the approach which is already implemented, trained on generated data and tested in a realistic setting. The monitor and classifier module are developed conceptually and not implemented so far. This section provides the conceptual ideas about the future integration into the presented approach.

The classifier aims to support domain experts in identifying the cause of an anomaly reliably and quickly. Therefore, all available operating and monitoring data sources should be used to train a classification model. As Hirt et al. [8] show, the overall classification performance can be improved by implementing a cognitive meta model which combines individual classification algorithms trained on different data sources. We follow the idea of constructing a cognitive meta model and currently develop individualized machine learning models for different available data pools. To this end, in the first place we build a database of all detected anomalies and the reported causes which serve as the necessary labels. In the phase of a first initialization of the module, we currently focus on a small number of four classes. For each of the classes, the current effort is to simulate the respective anomalies and relate the respective data to the anomaly. The available monitoring data sources like temperature and displacement of a bearing are again available in the form of sensor time series. Here, we currently test again LSTMs for multiclass time series classification. Audio and video data sets are currently not yet available. Our current research effort in this area is to better understand if and how audio and video can provide valuable information for the classification task. We therefore investigate optimal positions of audio and video sensors on the production line. We currently also experiment with audio data from other domains to better perform the necessary multiclass classification task. For such classification tasks with visual data sources like pictures and videos, Convolutional Neural Networks are widely discussed as a promising approach in the literature [11,12] which will therefore also be implemented and tested in our setting. As suggested by Hirt et al. [8] the idea is now to train these specialized models on different data pools and then combine the predictions within a meta model. The implementation and validation of this promising approach are future research tasks. The involvement of the user also plays a key role in this process since the knowledge of the service provider is also continuously improved by the possibility of correction by the user. In a value stream network, the proposed feedback loops which generate new and improved training patterns, are essential for a collaborative value co-creation. This in turn benefits all customers of the service provider, as the classification improves for all. Machine builders and component suppliers can benefit from this data since product enhancement can be achieved. It is also important to create a high level of acceptance of the tool among users. Dashboards can help to increase acceptance and usability for different users. This is to be achieved by displaying all relevant information about the problem classification to the user in the right form which will be investigated in our ongoing research to realize a mutual improvement of the models and experts within our hybrid-learning approach. The proposed approach raises the question where the individual calculations and algorithms should be carried out best. The training of the networks will probably take place in the cloud but the calculation of the trained nets can maybe better done by the respective customer. Hence, a concept for fog, edge and cloud computing need to be discussed.

6   Discussion, Limitations and Conclusions

In this article we propose a hybrid-learning machine monitoring approach which is based on three interplaying modules for anomaly detection, monitoring and classification. With our research we try to address different challenges which arise for providers and customers of such services and aim to initiate a broader discussion about hybrid-learning systems for an improved integration of human experts in analytics based service tasks. A major issue in today's solutions, is still the integration of machine learning techniques and human domain knowledge. One can argue, that with the huge amount of available data and machine learning techniques like deep learning, no domain expertise is necessary anymore to understand the dependencies between a certain operation behavior and different monitoring data sources. But especially application specific anomalies have often only small data histories which still require human judgement. Therefore, in a first step, we implement a fully automated anomaly detection procedure which only analyses deviations from normal operation time series data. The second step provides a clearly prepared dashboard solution for the machine operator or other experts which allow them to manually analyze the current situation of the machine in case of a detected anomaly. The overview contains different sources of monitoring data like temperature of the machine but also corresponding video and audio material at the point in time of the anomaly detection. This setup should support the human expert to assess the current state of the machine. By identifying the reason for the anomaly, or the assessment of a false positive detection, a feedback is provided which reinforce the modules to continuously improve and adapt to new situations. The machine learning parts learn from human feedback and therefore provide better suggestions in future anomaly situations. But the idea of such a hybrid intelligence interaction is difficult to realize. Questions, for example about the necessary number of available patterns, arise. To continuously improve the machine learning parts of the approach requires lots of labeled data which is difficult to generate only during a normal operation without provoking certain error types. The whole approach is therefore most likely applicable in a large value stream network with many different application domains of a machine type and the willingness of all partners involved to participate in the knowledge sharing process. As in many big data projects, the question arises who owns which data and who is allowed to use it for what. In the presented approach, it is very important that the service provider can access the classification data and the user feedback. To achieve that it requires a certain amount of confidence at the part of the customers that they provide the data to the service provider and that security as well privacy is ensured.

Two types of anomalies need to be distinguished. Application specific anomalies and machine specific anomalies. An initialization of machine specific anomalies can be provided from the service provider in advance by provoking many anomaly patterns for example on a demonstration machine. To further improve the quality for the whole value stream network, these anomalies can be anonymized but used for training better models which can be applied in the whole network again. The main challenge are application specific anomalies which only arise in certain applications of the machine or component. To retrieve the necessary amount of data for these cases is challenging. Here, maybe other learning approaches need to be invented for much smaller data sets. The goal is still to implement a classification module which is able to adapt to new and domain specific situations, but the questions arise if this is possible based on sparse training data sets. Of course at this point the integration of knowledge from domain experts is essential. Currently we only show the applicability of the anomaly detection module. It can further be argued that this way is not appropriate to evaluate such an approach. A single case study is only a weak form of underpinning the proposed idea. Besides the final implementation of the modules for monitoring and classification, especially the evaluation of the presented ideas in different contexts is the main focus of our ongoing research. To ensure generalizable insights we will follow a rigorous Design Science Research approach. For our research suitable evaluation methods are additional multi case/field studies and architecture analyses as proposed by [7].

The presented approach shows how practitioners can implement a reliable machine monitoring solution. This approach consists of three modules: anomaly detector, monitor and classifier. Operative sensor data (e.g. the current torque) is connected to more complex data (e.g. images and sound). With an LSTM network, the operative data are examined for problematic deviations at a certain frequency in the anomaly detector module. If an error is detected, it is displayed to a technician on a dashboard (monitor module). In addition, the more complex data is presented in comparison to data from normal machine operation. Based on the additional data, the classifier proposes a reason for the cause to the technician. This can be confirmed or improved, which also leads to an improvement of the classification algorithm. This means that a large amount of sensor data is used together with the experience and the expert knowledge of humans to classify faults. The longer the classifier runs, the less human intervention is required in the classifier. At the current state the anomaly detector module is implemented and tested in a real-world demonstration machine. The monitor and classifier module are described conceptually and will be implemented in further research.

References

1. Alahmadi, M., Qureshi, M.: Improved model to test applications using smart services. Sci. Int. **27**(3), 2275–2280 (2015)
2. Axinte, D.: Approach into the use of probabilistic neural networks for automated classification of tool malfunctions in broaching. Int. J. Mach. Tools Manuf. **46**(12–13), 1445–1448 (2006). https://doi.org/10.1016/j.ijmachtools.2005.09.017
3. Byun, J., Park, S.: Development of a self-adapting intelligent system for building energy saving and context-aware smart services. IEEE Trans. Consum. Electron. **57**(1), 90–98 (2011). https://doi.org/10.1109/tce.2011.5735486
4. Calza, F., et al.: Fuzzy consensus model for governance in smart service systems. Procedia Manuf. **3**, 3567–3574 (2015). https://doi.org/10.1016/j.promfg.2015.07.715

5. Gavrilova, T., Kokoulina, L.: Smart services classification framework. In: Proceedings of the Federated Conference on Computer Science and Information Systems, 13–16 September, Lodz, Poland, pp. 203–207 (2015). https://doi.org/10.15439/2015f324

6. Gers, F.A., Eck, D., Schmidhuber, J.: Applying LSTM to time series predictable through time-window approaches. In: Dorffner, G., Bischof, H., Hornik, K. (eds.) ICANN 2001. LNCS, vol. 2130, pp. 669–676. Springer, Heidelberg (2001). https://doi.org/10.1007/3-540-44668-0_93

7. Hevner, A.R., March, S.T., Park, J., Ram, S.: Design science in information systems research. MIS Q. **28**(1), 75–105 (2004)

8. Hirt, R., Kühl, N.: Abbildung kognitiver Fähigkeiten mit Metamodellen. In: Eibl, M., Gaedke, M. (eds.) Proceedings of the INFORMATIK 2017, 25–29 September, Chemnitz, Germany, pp. 2301–2307. Gesellschaft für Informatik, Bonn (2017). https://doi.org/10.18420/in2017_231

9. Hochreiter, S., Schmidhuber, J.: Long short-term memory. Neural Comput. **9**(8), 1735–1780 (1997). https://doi.org/10.1162/neco.1997.9.8.1735

10. Kaiser, K., Gebraeel, N.: Predictive maintenance management using sensor-based degradation models. IEEE Trans. Syst. Man Cybern. - Part A: Syst. Hum. **39**(4), 840–849 (2009). https://doi.org/10.1109/tsmca.2009.2016429

11. Karpathy, A., Toderici, G., Shetty, S., Leung, T., Sukthankar, R., Fei-Fei, L.: Large-scale video classification with convolutional neural networks. In: Proceedings of the 2014 IEEE Conference on Computer Vision and Pattern Recognition, CVPR 2014, 23–28 June, Columbus, OH, USA, pp. 1725–1732. IEEE Computer Society, Washington, DC (2014). https://doi.org/10.1109/CVPR.2014.223

12. Krizhevsky, A., Sutskever, I., Hinton, G.E.: Imagenet classification with deep convolutional neural networks. In: Proceedings of the 25th International Conference on Neural Information Processing Systems, NIPS 2012, 03–06 December, Lake Tahoe, NV, USA, vol. 1, pp. 1097–1105. Curran Associates Inc., USA (2012)

13. Lee, J., Kao, H.A., Yang, S.: Service innovation and smart analytics for industry 4.0 and big data environment. Procedia CIRP **16**, 3–8 (2014). https://doi.org/10.1016/j.procir.2014.02.001

14. Lerch, C., Gotsch, M.: Digitalized product-service systems in manufacturing firms: a case study analysis. Res.-Technol. Manag. **58**(5), 45–52 (2015). https://doi.org/10.5437/08956308x5805357

15. Malhotra, P., Vig, L., Shroff, G., Agarwal, P.: Long short term memory networks for anomaly detection in time series. In: Proceedings of the European Symposium on Artificial Neural Networks, Computational Intelligence and Machine Learning, 22–24 April, Bruges, Belgium, p. 89. Presses universitaires de Louvain (2015)

16. McArthur, S., Strachan, S.M., Jahn, G.: The design of a multi-agent transformer condition monitoring system. IEEE Trans. Power Syst. **19**(4), 1845–1852 (2004). https://doi.org/10.1109/tpwrs.2004.835667

17. Mont, O.: Product-service systems: panacea or myth? Ph.D. thesis, Lund University (2004)

18. Neubauer, C., Cataltepe, Z., Yuan, C., Cheng, J., Fang, M., McCorkle, W.: Tool for sensor management and fault visualization in machine condition monitoring, US Patent 7,183,905 (2007)

19. Oliva, R., Kallenberg, R.: Managing the transition from products to services. Int. J. Serv. Ind. Manag. **14**(2), 160–172 (2003). https://doi.org/10.1108/09564230310474138

20. Schrödl, H., Bensch, S.: E-procurement of cloud-based information systems - a product service system approach. In: Proceedings of the Thirty Fourth International Conference on Information Systems (ICIS), 15–18 December, Milan, Italy (2013)

21. Teti, R., Jemielniak, K., ODonnell, G., Dornfeld, D.: Advanced monitoring of machining operations. CIRP Ann.-Manuf. Technol. **59**(2), 717–739 (2010). https://doi.org/10.1016/j.cirp.2010.05.010

22. Wu, B., Tian, Z., Chen, M.: Condition-based maintenance optimization using neural network-based health condition prediction. Qual. Reliab. Eng. Int. **29**(8), 1151–1163 (2013). https://doi.org/10.1002/qre.1466

23. Wu, S.j., Gebraeel, N., Lawley, M.A., Yih, Y.: A neural network integrated decision support system for condition-based optimal predictive maintenance policy. IEEE Trans. Syst. Man Cybern.-Part A: Syst. Hum. **37**(2), 226–236 (2007). https://doi.org/10.1109/tsmca.2006.886368

24. Wünderlich, N.V., Wangenheim, F.V., Bitner, M.J.: High tech and high touch: a framework for understanding user attitudes and behaviors related to smart interactive services. J. Serv. Res. **16**(1), 3–20 (2013). https://doi.org/10.1177/1094670512448413

25. You, M., Liu, F., Meng, G.: Benefits from condition monitoring techniques: a case study on maintenance scheduling of ball grid array solder joints. Proc. Inst. Mech. Eng. Part E: J. Process Mech. Eng. **225**(3), 205–215 (2011). https://doi.org/10.1177/2041300910393426

Exploring the Value of Data – A Research Agenda

Tobias Enders[(✉)]

Karlsruhe Institute of Technology, Karlsruhe, Germany
`tobias.enders@kit.edu`

Abstract. Big data has been a technological quantum leap in recent years. Organizations are provided the opportunity to leverage this data by applying analytics to derive competitive advantages and increase operational efficiencies. However, the amount of value that is hidden within a set of data can often only be determined when used in a particular context. Being able to determine the value of their data assets as such is an even greater challenge for organizations. We conducted a structured literature review and identified three clusters of discussion in IS literature that address the value of data form different perspectives. Based on this review and the literature gap identified, we propose a research agenda to (1) identify the factors that influence the value of data, (2) cluster data according to value, (3) develop value-based data governance guidelines, and (4) quantify the value contribution of a single data source.

Keywords: Big data · Data value · Analytics

1 Introduction

When Microsoft acquired LinkedIn for 26.2 bn USD in 2016 [1], the purchase price went far beyond the sum of tangible and intangible assets as stated on the balance sheet. With approximately 100 million active users per month during the time of acquisition, Microsoft gained access to massive amounts of user data. As part of the due diligence, Microsoft's analysts have made assumptions on the value of that data. The question therefore arises, how the value of data – and its potential business impact – can be determined. While data resources are commonly viewed as an important IT asset within an organization [2], their significance cannot be overstated. Precision farming in agriculture, for example, enabled thought the use of high resolution satellite imaginary and weather data, allows farmers to increase their yield by up to 2% while saving up to 10% on input materials [3]. In a survey, 40% of companies reported that losing access to their data center for 72 h would put their survival at risk and 93% even filed for bankruptcy within one year after losing access to their data center for 10 days [4].

The need for being able to determine the value of data is manifold. While the acquisition of an organization or part thereof may be one of the most obvious external triggers for data evaluation, there are business benefits for companies that have a data valuation process in place. Akred and Samani [5] summarize the reasons why organizations would want to evaluate their data assets into three categories: direct data monetization, internal investment, and mergers & acquisitions. They argue that

© Springer Nature Switzerland AG 2018
G. Satzger et al. (Eds.): IESS 2018, LNBIP 331, pp. 274–286, 2018.
https://doi.org/10.1007/978-3-030-00713-3_21

knowing the worth of the data assets allows the organization to make smarter decisions when monetizing data by selling it directly or offering products based thereof. Knowing the value of data can guide the management when it comes to making decisions on investments into data assets. This includes make vs. buy decisions or knowing how much to spend on a particular data set.

With an ever-increasing amount of data being created through, for instance, social media and the Internet of Things (IoT), the need for companies to evaluate their data assets increases. While most data value frameworks focus on the business benefit derived from the use of data, little focus has been put on the data source itself. For example, Douglas Laney [6], Chief Data Office at Gartner, developed a comprehensive framework consisting of six key performance indicators. Three focusing on improving an organization's information management discipline and three aiming at economic benefits derived from information. However, little attention is given to the data source itself. By looking at data as an intangible asset of the organization, we might want to consider its value as a "raw material" instead of focusing on the business impact through its use. With this way of thinking in mind, we open up a new field of research and, in addition, respond to a call for research to understand the implications of big data use by organizations [7, 8].

The remainder of this paper is structured as follows: in Sect. 2, we provide an overview of the fundamentals and related work related on data and value. Section 3 describes the methodology applied for the literature review with a summary of the results in Sect. 4. To further explore the value of data, we propose a research agenda in Sect. 5. Finally, Sect. 6 concludes this paper with closing remarks.

2   Fundamentals and Related Work

In the following, a brief introduction to data and value is given. It describes the fundamentals, frameworks and schools of thought that remainder of the paper is based upon.

2.1   Data

The extant literature does not give a clear definition of what data is. Zins [9] argues that the understanding of data depends on the angle and approach. Definitions of data reach from being "the coded invariance", "an abstraction" to "a set of symbols representing a perception of raw facts" [9, p. 480]. With a focus on IS, Levitin and Redman [10] refer to data as an immaterial and necessary asset of an organization. In contrast to rival goods, such as cars, data is considered a non-rival good, which allows its use by several people at the same time in different places [11].

There is a general consensus among researchers that data by itself is to be considered a raw material, whereas information is processed data and knowledge a form of actionable insights [12, 13]. Being an integral part of the IT within an organization [2], the importance of data management and data governance has been increasing over the past years [14]. A well-defined and executed data governance model helps maximizing the value that an organization can derive from a data asset [15] and it supports the alignment of data-related programs with company objectives [16]. The confidence in making data-based management decisions is dependent on the quality of the data [17], which triggers the need for a thorough data management plan. Aaltonen and Tempini [18], for example, argue that business opportunities can be lost if data quality and granularity does not meet the requirements and expectations of the client or project scenario. The consequences of poor data management have been addressed in a number of recent studies [e.g. 19–22].

While the basic definition of data, as outlined above, is rather broad, it would not be complete without a brief introduction of the term *big data*. The creation of massive amounts of data is often referred to as *big data* and generally characterized by five V's: value, variability, variety, velocity, and veracity [23]. Powerful computational power is needed by organizations to process big data and to unveil trends and other insights [7]. However, organizations show different maturity levels in the use of big data and therefore have a diverse understanding of it [24].

2.2   Value

Discussions about the term value and its meaning date back to Aristotele (4th century B.C.) and have been part of many fields of research since [25]. In its central theory, economics describes two central meanings of value: "value-in-exchange" and "value-in-use". The traditional perspective on value is referred to as a goods-dominant view, which finds its basis in "value-in-exchange". In contrast, the "value-in-use" approach has strong parallels to the service-dominant logic as proposed by Vargo and Lusch [26].

While Aristotele was the first to distinguish between use-value and exchange-value [25], it was not until the scholars of scholasticism that extended this theory by arguing that the value-exchange process is based on the needs of the consumers [27]. Extant literature provides multiple approaches and frameworks on how to determine value [e.g. 28, 29]. Drawing on the field of financial accounting, several methods have been established for the valuation of intangible assets, such as patents. While cost-based, market-based, and income-based models are frequently used for balance sheet calculations, option-based models have a far smaller application space [30].

3   Methodology of the Literature Review

We conducted a systematic literature review [31] in order to identify the research streams around the value of data in IS. The review process consists of a search and selection process to identify the relevant papers and a description of the research streams, which have been identified in the process.

3.1   Search and Selection

The goal of our literature research is to identify work that is relevant in the field of IS. It shall help identifying research streams that have formed around data value in recent years.

We started our search by querying multiple scientific databases – namely AISeL, Scopus, ScienceDirect, and EBSCO – with a focus on leading IS journals. In order to account for the fast-moving pace of the IS practice and the review cycles of journals, we decided to also include the leading IS conference (ICIS) in our search results. Initially, we did not limit the search time frame for relevant papers to ensure a generalizability. A brief analysis of the papers shows, however, that the majority of relevant publications (85%) have been published in 2010 or after. This indicates that the relevance of data value in research and practice has spiked in recent years.

We limited our initial search to the title, abstract, and key words. The search string used was "value" and "data" or "big data". The result of all databases – excluding duplicates – yielded 245 papers. From here, we excluded articles that did not meet the focus of our search – especially papers that have a strong technical focus, deal with general investments into IT infrastructure, focus on the creation of social value and such that put non-profit purposes (e.g. healthcare) into the focus of their research. Papers were excluded after a thorough review of the abstracts. In line with Webster and Watson [31], we performed a forward and backward search of the remaining 23 papers, which added an additional 5 articles to the total amount of relevant papers. A summary of the search process is depicted in Fig. 1.

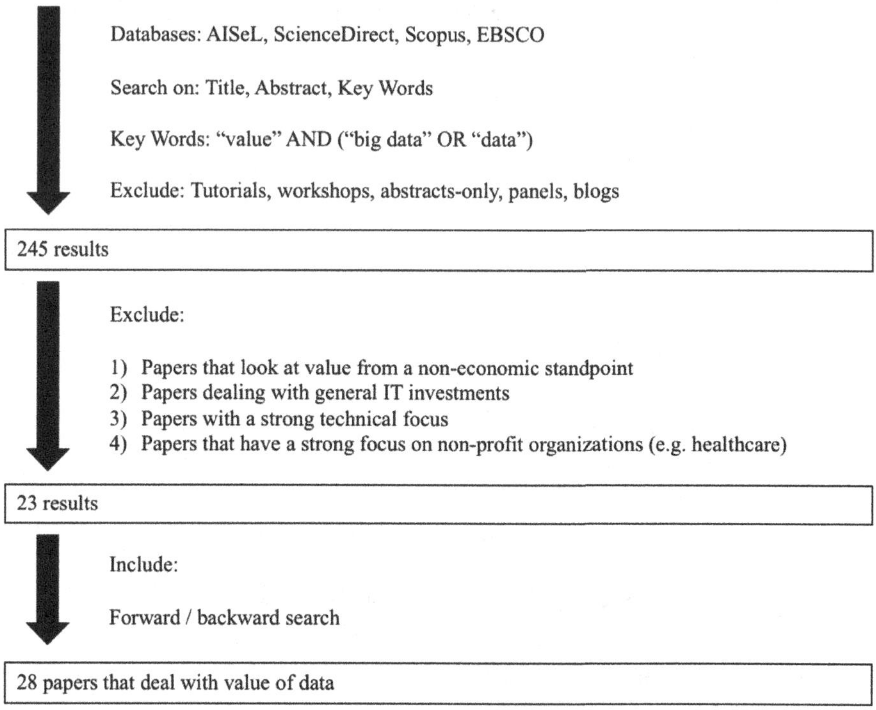

Databases: AISeL, ScienceDirect, Scopus, EBSCO

Search on: Title, Abstract, Key Words

Key Words: "value" AND ("big data" OR "data")

Exclude: Tutorials, workshops, abstracts-only, panels, blogs

245 results

Exclude:

1) Papers that look at value from a non-economic standpoint
2) Papers dealing with general IT investments
3) Papers with a strong technical focus
4) Papers that have a strong focus on non-profit organizations (e.g. healthcare)

23 results

Include:

Forward / backward search

28 papers that deal with value of data

Fig. 1. Search and selection process

3.2   Identification of Clusters

In order to identify and describe research streams covered by the selected papers, we summarized the information for each relevant article in a separate document. The information extracted included: research methodology applied, underlying theoretical assumptions and frameworks, main findings, suggested future work and limitations as well as the broader topic (e.g. organizational, process, technical) covered. In a sub-sequent step, we grouped the papers that had a strong overlap in their topics; leading to a total of three research streams. We shaped groups that are as homogenous as possible within and heterogeneous in-between. In the following, each of the streams is described and gaps in research are pointed out.

4   Results of the Literature Review

The structured literature review helped us in understanding and clustering the state-of-the-art research being conducted in the field of data value. Eventually, we were able to identify three research streams that broach the issue of value of data: (I) business value & organization, (II) data management, and (III) data value chain. Figure 2 introduces the general framework of the data value chain consisting of data (collection), information and insight building, and business value creation [32–35]. The marks (I – III) depict the stage of each identified research stream within the data value chain.

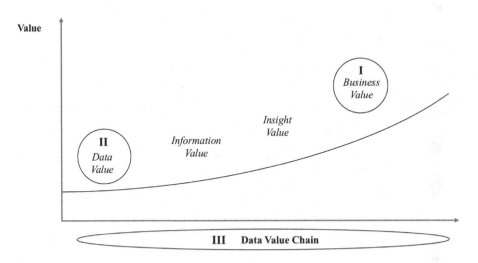

Fig. 2. Research streams on data value

4.1   Business Value and Organization

There is a general agreement among scholars that value of data is mostly realized when it is used for a specific purpose. A major part of the discussion around deriving value from data deals with the opportunities and challenges that organizations face when intending to realize value. Günther et al. [36] identify six debates that are central to how firms realize value from big data – pointing out that the discussion happens on multiple levels: individual, organizational and supra-organizational. While most organizations still struggle deriving value from data [37], some of them have been able to identify it as a source of competitive advantage [38] and to develop an holistic view of their customer base [39]. Furthermore, the source of the data is playing an increasingly important role for realizing business value. While many organizations still explore the possibilities of using data generated by themselves (e.g. transaction or product information), some have moved on to also leveraging external open data sources [40, 41]. To this end, the value contribution of a single data source, in contrast to the value contribution of big data analytics leveraging multiple sources, is being discussed in a number of papers [e.g. 38, 42]. Looking at business value derived from data and IT from a resource-based view, Melville et al. [43] argue that certain internal (e.g. complementary organizational resources) and external factors (e.g. trading partners) need to be in place to derive a benefit for the organization as a whole.

4.2 Data Management

The amount of value that can be extracted from data depends on the organization's capabilities to derive insights and to turn them into business value. However, the value creation process starts even sooner. The way that organizations manage their data has a strong impact on data value and, in a subsequent step, on business value. While data sets that contain non-sensitive information can be easily exploited by organizations, those that contain sensitive data, such as personal information, often add a layer of complexity and influence the value of the data [37, 42]. The more granular the data, the easier it is to track it back to an individual, which is why current research calls for new processes and technical mechanisms to protect the privacy of individuals [44] and still be able to realize value from that data. The ability of an organization to safely store and process data, described in data governance procedures, becomes more and more important when value is to be realized from granular sensitive data [15].

4.3 Data Value Chain

Going back to the original definition of what a value chain is, Rayport and Sviokla [45] describe it as a model of value-adding activities and processes that connect the organization's supply and demand side. With the raise of big data in the past decade, the value chain framework has been adapted for this use case. Abbasi et al. [32] have developed a comprehensive IS research agenda, which calls for additional research to further explore the impact of big data analytics on the value chain and the economic value that can be derived. Following this call, Lim et al. [46] have developed a nine-factor framework that covers the stages of the data value chain and explains how data-based value is created. While data generation, data acquisition, data storage and data analytics are key activities along the value chain, data visualization and data-based decision-making are often neglected as value-generating activities. Saggi and Jain [47] point out that all steps are relevant to e.g. get a better understanding of the customer or to prepare for M&A activities. Already at an early stage of the development of data value chains, Barua et al. [48] highlight the opportunities, which emerge when connecting a traditional value chain with data derived from Internet transactions.

4.4 Quo Vadis?

The literature review has shown that most of the research carried out in the context of data value focuses on realizing business value; e.g. by executing analytics projects. Figure 3 summarizes the relationship between the research streams. More than 50 per cent of the relevant papers identified discuss this particular view on data value. However, little attention is given to the source of the value: the data itself. While some papers discuss how to "treat" data from a data management perspective, to the best of our knowledge, no research has been looking at the data source itself – independent of the potential use cases of this data. While there is a common agreement among scholars that the true value of data is derived during its use, we argue that by looking at the data source itself, we can already make statements about its potential value. In the following, we present a research agenda to further explore this field of research.

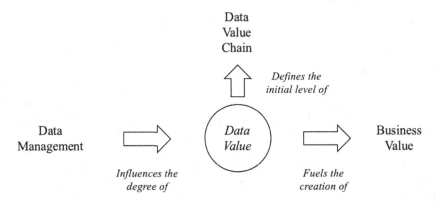

Fig. 3. Relationship of data value and research streams

5 Proposed Research Agenda

Based on the literature gap identified in the previous section, we propose a research agenda to further explore the value of data. This agenda constitutes four clusters that build on each other (cf. Fig. 4). Therefore, we have defined an overarching research question:

> *"How can organizations determine the value of their data assets and how does that impact their data management activities?"*

In the following, we describe the streams of research that we propose to open up for further exploration by scholars.

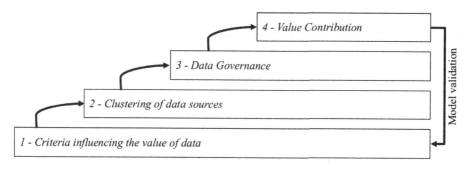

Fig. 4. Research agenda overview

5.1 Influencing Criteria

The need for determining factors that influence the value of data is not new: several models have been proposed by consulting agencies and market research institutes [6, 49]. While these models may be suitable for certain industries or business scenarios, little research has been carried out looking at these factors conclusively. By opening up this field of research, we will be able to make qualitative statements about the factors that influence the value of data and it lays the foundation for quantitative considerations in the future. While no in-depth research has been conducted on the factors influencing data value, a handful of researchers have mentioned that there are certain characteristics that may influence its value. Malgieri and Custers [44], for instance, argue that the reusability and the value decay over time may have an influence on the data's overall value. Furthermore, the source of data, e.g. freely available, purchased or proprietary, could influence the value [38]. Eventually, the quality of data may impact its value [15].

Looking at the quality of a good to determine its value is an approach where IS can leverage the knowledge of other disciplines. Materials management, for example, uses this approach to divide raw material into different groups. As such, steel has certain grades, which defines its quality, its purpose and therefore its value. Steel with a "higher" grade can be used for advanced purposes and it therefore creates a higher value since customer's willingness-to-pay for the end product is higher. In consequence, we may be able to leverage existing knowledge and transfer it onto the data use case. Nevertheless, we need to account for the specifics of data, e.g. its ability to be reused by multiple parties at the same time.

5.2 Clustering

Every organization has to make decisions on how to orchestrate its investment activities. This is true for traditional assets such as buildings and vehicles, however, this also applies to intangible assets such as data. Investments into data can be manifold; for example, the acquisition of data sets and the management of existing data sources.

Since not all data have the same level of significance in contributing to the firm's economic success, there is a need to be able to cluster data according to its value.

Determining the factors that influence the value of data lays the groundwork for clustering it into multiple categories. By using these clusters, organizations have the opportunity to focus their investments on data that contributes most to their success and competitive advantage in the marketplace.

The need to cluster items into categories by value is not new. Similar as to the first point of the research agenda, materials management uses methods to assign parts into categories. One of the most commonly used ones is the ABC-analysis, where the parts with the highest value are assigned the A category, the ones with medium value the B category and the ones with the least value the C category. Oftentimes, this analysis is used in connection with the Pareto principle. This approach could be a starting point to introduce value-based data clustering.

5.3   Data Governance

Our literature review has shown that data management, and data governance in particular, is one of the streams in today's research on data value. While most of the research conducted today focuses on the way data is managed on a firm level, we argue that data governance should also adopt a value-based view.

Having shaped value-based clusters of data enables the definition data governance rules and guidelines for each of these clusters. This allows organizations to treat data that has a high value to the firm differently from such that only has a lower value. Examples of differentiation in data management could relate to encryption, storage, and access management. Also, data governance based on data value could guide organizations in terms of their openness to share data with business partners and customers.

The implications of data governance go far beyond just managing data. The rules and guidelines that organization define for themselves imply key strategic decisions on a business level such as platform strategies and IT investments. By adding a value-based view, we argue that data can be better exploited as a strategic asset.

5.4   Quantifying the Value

While the first three research clusters – influencing criteria, clustering, data governance – shed light on the qualitative side of data value, this research block addresses the quantitative side. Understanding the quantitative value of data enables organizations to compare it to non-data assets and to better estimate its strategic value. As outlined before, current research has been addressing the business value derived from data. Therefore, we suggest leveraging current research in order to determine the value of a single data source. I.e. instead of choosing a bottom-up approach to determine the business value, we follow a top-down approach to determine the value of a single data source. Since these calculations are specific to every organization, we propose conducting a series of case studies. Furthermore, this approach allows a validation of the clustering approach described in Sect. 5.2: by determining a monetary value associated with a data source, we can check if the data source had been assigned to the correct cluster. If needed, a revision of the initial model can be performed.

5.5   Link to Related Academic Fields

The need for collecting, processing and making data available is not new. Related disciplines such as information management and decision support systems have done extensive research associated with the management of data. Information management (IM), for instance, provides guidance for managing information – a structured form of data - over its lifecycle and is associated with activities such as sensing, collecting, organizing, processing, and maintaining of information [50, 51]. Therefore, information management touches on data, processes, and systems within an organization, which establishes a link to the proposed research agenda. Scholars in the field of IM have conducted studies around data value [e.g. 38, 46], however, their work focuses on the value derived from data; no conclusive research has been targeting the data asset and its value-based management itself.

In decision support systems (DSS), data along with the users and models/applications shape the fundamentals of every BI system [52]. While BI systems are designed to process big data to derive insights, a key success factor remains the same regardless of the quantity of data processed: the higher the data quality, the greater the value that can be derived [53]. Data quality, as the literature review suggests, is also likely to be one of the influencing criterion for the value of a data asset.

The research agenda proposes exploring data and its management from a value perspective while IM addresses the data lifecycle management and DSS discusses data quality as an important input factor. Therefore, the availability of extant research in related disciplines shapes a solid foundation for the proposed research agenda and new studies can be built upon existing insights and research results.

6   Conclusion

Our research provides an overview of the current discussions around the value of data. A structured literature review serves as basis to identify a literature gap and to propose a research agenda to address this gap. We identified three research streams that address the value of data from different perspectives. The majority of articles identified focuses on the *business value* that can be derived from data and how different organizational levels influence this process. The second stream puts the data itself into the spot light and discusses *data management* as an influencing factor to data value. The third stream explains how data value is embedded into the overall *data value chain*.

Based on the discussions identified through the literature review, we propose a research agenda to further explore the value of data by looking at the data source itself – independent of knowing all of its potential use cases within the organization. We argue that data by itself has certain characteristics – e.g. re-usability, shelf life – that may influence its value. Finally, knowing the data value allows clustering it according to value and deriving value-based data governance guidelines.

With the proposed research agenda, we enable further research into the field of data value and to build a deeper understanding of the strategic value of data assets for an organization. By answering the questions raised through the research agenda, organizations will be able to manage their data assets from a value perspective. This enables, for instance, a more profound decision-making on which data to keep within the organization and which to share externally to foster collaboration and product innovation.

References

1. Microsoft: Microsoft to acquire LinkedIn. https://news.microsoft.com/2016/06/13/microsoft-to-acquire-linkedin/. Accessed 11 Mar 2018
2. Fisher, T.: The Data Asset: How Smart Companies Govern Their Data for Business Success. Wiley, London (2009)
3. Claas: Precision Farming. http://www.claas.co.uk/blueprint/servlet/blob/1079246/bff071e5cbfe11eb09d672adb3d933ab/282575-dataRaw.pdf. Accessed 30 Mar 2018
4. Atkinson, K., McGaughey, R.: Accounting for data: a shortcoming in accounting for intangible assets. Acad. Account. Financ. Stud. J. **10**, 85–95 (2006)
5. Akred, J., Samani, A.: Your Data is Worth More Than You Think. https://sloanreview.mit.edu/article/your-data-is-worth-more-than-you-think/. Accessed 20 Feb 2018
6. Laney, D.: Why and how to measure the value of your information assets (2015). https://www.gartner.com/doc/3106719/measure-value-informationassets. Accessed 15 Feb 2018
7. George, G., Haas, M.: From the editors: big data and management. Acad. Manag. J. **57**, 321–326 (2014)
8. Markus, M.L., Topi, H.: Big data, big decisions for science, society, and business. In: Report on a Research Agenda Setting Workshop. Bentley University, September 2015
9. Zins, C.: Conceptual approaches for defining data, information, and knowledge. J. Assoc. Inf. Sci. Technol. **58**, 479–493 (2007)
10. Levitin, A., Redman, T.: Quality dimensions of a conceptual. Inf. Process. Manag. **31**, 81–88 (1995)
11. Pantelis, K., Aija, L.: Understanding the value of (big) data. In: Proceedings of IEEE International Conference on Big Data, pp. 38–42 (2013)
12. Alavi, M., Leidner, D.E.: Knowledge management and knowledge management systems: conceptual foundations and research issues. MIS Q. **25**, 107–136 (2001)
13. Liew, A.: Understanding data, information, knowledge and their inter-relationships. J. Knowl. Manag. Pract. **7**, 1–16 (2007)
14. DalleMule, L., Davenport, T.H.: What's your data strategy? Harv. Bus. Rev. **2017**, 112–121 (2017)
15. Otto, B.: Data governance. Bus. Inf. Syst. Eng. **3**, 241–244 (2011)
16. Cheong, L., Chang, V.: The need for data governance: a case study. Proc. ACIS **2007**, 999–1008 (2007)
17. Madnick, S.E., Wang, R.Y., Lee, Y.W., Zhu, H.: Overview and framework for data and information quality research. ACM J. Data Inf. Qual. **1**, 1–22 (2009)
18. Aaltonen, A., Tempini, N.: Everything counts in large amounts: a critical realist case study on data-based production. J. Inf. Technol. **29**, 97–110 (2014)
19. Kokemueller, J.: An empirical investigation of factors influencing data quality improvement success. In: Proceedings of AMCIS 2011 (2011)
20. Wechsler, A., Even, A., Weiss-Meilik, A.: A model for setting optimal data-acquisition policy and its application with clinical data. In: Proceedings of 34th International Conference on Information Systems (2013)
21. Shankaranarayanan, G., Blake, R.: Data and information quality: research themes and evolving patterns. In: Proceedings of AMCIS 2015 (2015)

22. Ghasemaghaei, M.: The impact of big data on firm data diagnosticity : mediating role of data quality. In: Proceedings of 38th International Conference on Information Systems (2017)
23. Philip Chen, C.L., Zhang, C.Y.: Data-intensive applications, challenges, techniques and technologies: a survey on big data. Inf. Sci. (Ny) **275**, 314–347 (2014)
24. Demirkan, H., Bess, C., Spohrer, J., Rayes, A., Allen, D., Moghaddam, Y.: Innovations with smart service systems: analytics, big data, cognitive assistance, and the internet of everything. Commun. Assoc. Inf. Syst. **37**, 733–752 (2015)
25. Fleetwood, S.: Aristotle in the 21st century. Camb. J. Econ. **21**, 729–744 (1997)
26. Vargo, S.L., Lusch, R.F.: Evolving to a new dominant logic for marketing. J. Mark. **68**, 1–17 (2004)
27. Dixon, D.: Marketing as production: the devlopment of a concept. J. Acad. Mark. Sci. **18**, 337–343 (1990)
28. Short, J.E., Todd, S.: What's Your Data Worth? https://sloanreview.mit.edu/article/whats-your-data-worth/. Accessed 10 Feb 2018
29. Nagle, T., Sammon, D.: The data value map : a framework for developing shared. In: Twenty-Fifth European Conference on Information Systems (ECIS), Guimarães, Port. 2017, pp 1439–1452 (2017)
30. Matsuura, J.H.: An overview of intellectual property and intangible asset valuation models. Res. Manag. Rev. **14**, 33–42 (2004)
31. Webster, J., Watson, R.T.: Analyzing the past to prepare for the future : writing a literature review. MIS Q. **26**, 13–23 (2002)
32. Abbasi, A., Sarker, S., Chiang, R.: Big data research in information systems: toward an inclusive research agenda. J. Assoc. Inf. Syst. **17**, 1–32 (2016)
33. Gao, J., Koronios, A., Selle, S.: Towards a process view on critical success factors in big data analytics projects. In: Proceedings of 21st Americas Conference on Information Systems (2015)
34. Sharma, R., Mithas, S., Kankanhalli, A.: Transforming decision-making processes: a research agenda for understanding the impact of business analytics on organisations. Eur. J. Inf. Syst. **23**, 433–441 (2014)
35. Seddon, P.B., Constantinidis, D., Tamm, T., Dod, H.: How does business analytics contribute to business value? Inf. Syst. J. **27**, 237–269 (2017)
36. Günther, W., Mehrizi, M.H.R., Huysman, M., Feldberg, F.: Debating big data: a literature review on realizing value from big data. J. Strateg. Inf. Syst. **26**, 191–209 (2017)
37. Günther, W., Hosein, M., Huysman, M., Feldberg, F.: Rushing for gold : tensions in creating and appropriating value from big data. In: Proceedings of 38th International Conference on Information Systems (2017)
38. Mamonov, S., Triantoro, T.M.: The strategic value of data resources in emergent industries. Int. J. Inf. Manag. **39**, 146–155 (2018)
39. Padmanabhan, B., Zheng, Z., Kimbrough, S.: An empirical analysis of the value of complete information for eCRM models. MIS Q. **30**, 247–267 (2006)
40. Jetzek, T., Avital, M., Bjørn-Andersen, N.: Generating value from open government data. In: Proceedings of 34th International Conference on Information Systems, pp. 1737–1756 (2013)
41. Zeleti, F.A., Ojo, A.: Open data value capability architecture. Inf. Syst. Front. **19**, 337–360 (2017)
42. Kugler, L.: The war over the value of personal data. Commun. ACM **61**, 17–19 (2018)
43. Melville, N., Kraemer, K., Gurbaxani, V.: Review: information technology and organizational performance: an integrative model of IT business value. MIS Q. **28**, 283–322 (2004)
44. Malgieri, G., Custers, B.: Pricing privacy - the right to know the value of your personal data. Comput. Law Secur. Rev. **34**, 289–303 (2018)

45. Rayport, J., Sviokla, J.: Exploiting the virtual value chain. Harv. Bus. Rev. **73**, 75–85 (1995)
46. Lim, C., Kim, K.H., Kim, M.J., Heo, J.Y., Kim, K.J., Maglio, P.P.: From data to value: a nine-factor framework for data-based value creation in information-intensive services. Int. J. Inf. Manag. **39**, 121–135 (2018)
47. Saggi, M.K., Jain, S.: A survey towards an integration of big data analytics to big insights for value-creation. Inf. Process. Manag. **54**, 758–790 (2018)
48. Barua, A., Konana, P., Whinston, A., Yin, F.: An empirical investigation of net-enabled business value. MIS Q. **28**, 585–620 (2004)
49. Brown, B., Kanagasabai, K., Pant, P., Pinto, G.: Capturing value from your customer data. https://www.mckinsey.com/business-functions/mckinsey-analytics/our-insights/capturing-value-from-your-customer-data. Last accessed 10 Mar 2018
50. Marchand, D.A., Kettinger, W.J., Rollins, J.D.: Information orientation: people, technology and the bottom line. MIT Sloan Manag. Rev. **41**, 69–80 (2000)
51. Mithas, S., Ramasubbu, N., Sambamurthy, V.: How information management capability influences firm performance. MIS Q. **35**, 237–256 (2011)
52. Watson, H.: Revisiting Ralph Sprague's framework for developing decision support systems. Commun. Assoc. Inf. Syst. **42**, 363–385 (2018)
53. Watson, H.: Preparing for the cognitive generation of decision support. MIS Q. Exec. **16**, 153–169 (2017)

Service Topics Open Exploration

Investigating the Alignment Between Web and Social Media Efforts and Effectiveness: The Case of Science Centres

Marlene Amorim[1,2(✉)], Fatemeh Bashashi Saghezchi[1,2],
Maria João Rosa[1,2,3], and Pedro Pombo[1,4]

[1] University of Aveiro, 3810-193 Aveiro, Portugal
{mamorim,fatemeh,m.joao,ppombo}@ua.pt
[2] The Research Unit on Governance,
Competitiveness and Public Policies (GOVCOPP), Aveiro, Portugal
[3] Centre for Research of Policies of Higher Education (CIPES),
Matosinhos, Portugal
[4] Fábrica Ciência Viva Science Center, 3810-171 Aveiro, Portugal

Abstract. The adoption of Internet technologies has led to important transformations in the way in which organizations reach and interact with their audiences. The context of museums and exhibitions is also witnessing a consolidation of the usage of Web 2.0 for disseminating information and for supporting the interactions to complement, improve and augment visitors' experiences. Despite the proliferation of online evidence about the efforts that these organizations are making to build a strong Web presence and a vibrant existence in Social Media, there is far less knowledge about the effectives of such investments from the perspective of the users. The purpose of this study is to propose a framework that allows the assessment and the discussion about the alignment between online presence of an organization and its effectiveness towards its target audience and service promise. The study builds on the analysis of Web and Social Media presence of Science Centers and addresses the whole network of Centers in Portugal. It offers an overview of different types of online efforts and resource allocation, while discussing it towards recent results about its importance for users' awareness about the Centers.

Keywords: Web presence · Social media usage · Museum services
Museum visitors perceptions

1 Introduction

In recent years we have been witnessing the growth and diversification of the online and social media presence of museums, and exhibition providers in general. The incorporation of such elements in museum service experiences reflects the tendency of involving audiences beyond the physical onsite visits. These trends have been happening together with the adoption of approaches that support the development of museum experiences as to informal learning environments, building on visitors' engagement, interactivity, participation and fun [1]. The effective setup of such

© Springer Nature Switzerland AG 2018
G. Satzger et al. (Eds.): IESS 2018, LNBIP 331, pp. 289–302, 2018.
https://doi.org/10.1007/978-3-030-00713-3_22

experiential environments relies on enabling visitors with an ease and agile access to the providers' information and resources, both during the visit and in the pre and post phases of the service. Despite of the proliferation and diversity of web and social presence of museums and organizations alike, there is still a lack of structured knowledge to inform the allocation of resources for the development of effective online presence, as well as to assess its effectiveness.

A great number of museums, exhibitions centers and alike have been consistently deploying resources to build and make available digital content about their collections and content. Moreover, general information on opening hours, ticket prices, calendars of events, maps, and directions has become part of the expected online portfolio of information. Adding to this, institutions have also embraced the adoption of web 2.0 tools on their sites, as well as the nurturing of a vibrant and very visual presence in social media, as a response to the emerging patterns of consumption and service expectations [2]. The efforts placed by museums for creating a presence on the Web and Social Media have, to some extent, emerged as a complement to the visits experiences, and much as a response to "a perceived demand for different forms of engagement" [3]. Nevertheless, the existing evidence suggests that such investments are not informed by structured approaches to develop and assess an adequate online service offering, but rather that there is missing knowledge on how to articulate, organize and understand such multiple presences. This study offers a contribution to this debate building on the analysis of the online and social media presence of a particular segment of museum services: Science centers. These organizations offer portfolios of services oriented to promote the public engagement with science, by means of adopting an exhibition approach that is very hands-on, relying on interactive exhibits that call for visitors' participation and interaction. For the particular nature of the service experience proposed by Science centers the study of online presence is of key importance, given the important role played by online resources to enable good preparation for the experiential, on site visits.

2 Conceptual Background

2.1 The Service Experiences of Museums and Science Centers

Science centers and museums offer a diverse portfolio of service experiences that are designed to promote public engagement with science. Many science centers were developed after governmental initiative and have emerged as informal learning environments able to attract much diversified audiences. Often, Science centers are also adopted as complementary learning contexts, that amplify the services provided by schools and educators in general, who promote class visits to the Centers' exhibitions and activities, while aligning its content with the topics on the existing school curricula [4]. Science centers offer a vivid context to learn about the management of the service activities of museums and exhibitions, as they have been pioneers in the adoption of new visitors' service approaches that privilege service experience attributes such as visitors' participation, emotional engagement, fun for the purpose of improving awareness and access to knowledge by means of informal learning among children and adults alike.

Overall, science centers subscribe to a contemporary approach to the development of visitors' experiences that does not focus on the assembly and passive exhibition of the content, but rather on the development of service experiences that enable the engagement of visitors, by means of participation, interactivity and rich flows of emotion and communication among visitors, centers' staff and the exhibition artefacts [5]. Digital technologies can play a substantial role in achieving these service experiences. Science centers, in particular, have done remarkable advances in the adoption of technologies to enrich the interactive nature of the visits, resorting to elements such as personal digital assistants, digital information kiosks among others [6].

Overall, museums and Science Centers alike have been increasingly making use of digital technologies resources to improve the service experiences offered. The adoption of digital technologies has transformed the nature of the museum experiences, changing the way visitors and staff interact with each other and with the exhibitions' resources, both during the visit, before and after it. It is now common practice for visitors to search for information and resources to learn about and prepare a visit, and afterwards, either by providing online feedback, revisiting the available online resources or resorting to the web and social web to share their experience with other visitors [7]. The proliferation of online information and resources, that results from these evolutions, is adding layers of complexity to the management of information resources for these organizations while also expanding the expectations of visitors who increasingly expect access to the data and resources beyond the hours of the visit. Moreover, there is a, expanding, variety of means to support a disseminate digital information and resources, ranging from an institutional web presence to social media and networks, that make the task of defining and maintaining a relevant and updated online presence a very demanding task. While museums, Science centers and exhibitors in general face increasing calls to deal with multiple uses and demands for information resources, there is a paucity in what regards studies about the how such resources are effectively used and perceived by visitors, and how they contribute to the building of the visit experience.

2.2   Web and Social Presence

Currently most of the existing museums and Science Centers have embraced the opportunities offered by websites to improve the attractiveness of their offerings and services to potential visitors. In general, companies, including the ones active in the service industry, need to keep up with emerging information and communication technologies to convey their knowledge and message to the customers. In fact, Various authors, namely Pettigrew and Reber [8] have already identified that web presence can remarkably improve a company's success and visibility. Day [9] argued that web presence is even more important that the service itself. Bonsón and Ratkai [10] addressed undesirable effects of poor web presence, which can lead to decreasing the press coverage of a company. Different studies have consistently supported that most museums and similar institutions are investing in the development of online presences, and are engaged in learning on how to make the most of these new resources [11]. The nature of the online functionalities that museums aim to implement go beyond functional elements, as in the business websites, for the reason that museums aim to setup systems that are able to deliver emotions and promote engagement and fun. This perspective makes the study of the online presence for the case of museums a partic-ularly new and largely unaddressed topic. The purpose of service systems of museums, Science centers and alike is to offer enjoyable and informative experience, through the combination of on line and on-site resources and information. Adding to this, the ability to digitalize content and service elements serves also the museums purpose of reserving knowledge and reaching wider audiences [12].

Museums and exhibit services in general are adopting practices to promote a pres-ence and activity in social media for fostering communication purposes with customers via virtual communities (i.e., by facilitating the creation and sharing of information, etc.), so customers can easily become a fan and follow their brand's fan page [13]. Research findings suggested that museums should implement adequate strategies to attract young people. In that regard, using social media (i.e., Facebook, Google+, Instagram, Twitter, etc.) can be effective as it is a quite popular activity, especially among youngsters [14]. Moreover, museum managers hold to the belief that online social presence is important to promote awareness, release information and reach wider audiences. Evidence suggests however that the usage of such means, is, in general still incipient as they mostly rely on one-way communication [15]. Social media has also demonstrated a big impact on museum sector, as a quick way for prompting its activities, providing additional channels to spread museum's knowledge and receive the feedback from visitors [16, 17]. Social media analytics are increasingly incorporated by museums to evaluate the impact of organized events explored the feasibility of adopting a com-putational model to automate the information extraction from the visitors' social media, allowing larger amount of data to be analyzed in close to real-time [18]. Overall, social media offer museums and science centers a very disperse network of interactions that can cover social networking sites, blogs, wikis, podcasts, photo and video sharing, etc., contemplating an array of combinable and customizable means to interact with their audiences. However, dedicated resources are required in the organization to enable a continuous nurturing of communications and relationships.

3 Research Approach, Data and Methods

3.1 Research Strategy

This paper presents and discusses the results from an exploratory work addressing the use of new digital media in museum contexts, particularly for the case of Science Centers. The purpose of the study was to contribute to the understanding the variety of efforts placed by these organizations for developing a web and social media presence, while assessing its effectiveness in the eyes of the visitors. The study involved a combination of methods to examine and categorize different types of online efforts for the various units from the whole network of Science Centers in Portugal, including: (i) a qualitative approach, including a literature review for advancing our understanding about how Internet and particularly the Web and social media presence can enhance the users' awareness of museums, Science centers and the service experiences of exhibi-tions in general; and (ii) a quantitative approach, building on data collection from the science centers websites and social media presences, complemented with interviews, when necessary, to validate and triangulate data.

3.2   The Units of Analysis

The study addressed the network of Science centers in Portugal - Science Centers Network (SCN) – whose units are listed in Table 1. The SCN was established in 1997 with a mission of disseminating science and promoting scientific and technological culture. Over the years the network consolidated a wide national presence and recognition as a trusted and innovative provider of scientific culture. The data collection scope included the whole network of existing centers, labelled as *Centros Ciência Viva (CCV)*, whose literal translation to English is "Living Science Centers". A Science center is in fact a place where science and technology fracture the walls of the laboratories, to their visitors, emphasizing on the hands-on approach, a way to promote learning through experience and enjoyment, where you can interact with what is exposed; touch, experiment, discover, etc. [19, 20].

At present there are 19 CCVs in Portugal (including the center in Guimarães and excluding the closed one in Amadora). However, because this study builds on data from the latest report made available by CCV, published in 2016 [20], which reports 18 CCVs, the analysis reported is restricted to the available data. Table 1 lists all active CCVs at the time this report was prepared, summarizing information about their location, foundation year, web address and number of visitors. Note that the table contains 18 centers since the Amadora center was already close when the report was prepared and there was also no information available about the new center in Guimarães. The number of visitors per year, according to the data provided by the official website of the SCN. The average number of visitors per center, taking into account the last four years, varies between 7,000 and 70,000 annual visitors per center, depending on its size and location. The only exception is the Pavilhão do Conhecimento in Lisbon, with an average of about 220,000 visitors per year over the same period. The CCV Network, as a whole, has an average of 550,000 annual visitors in the last ten years [20, 21].

Table 1. Analyzed national science centers of the Portugal.

List of science centers (location)	Year	Websites	Total visitors
1. Centro Ciência Viva de Sintra (Sintra)	2006	oficinacienciasintra.pt	9.782
2. Centro Ciência Viva da Floresta (Proença-a-Nova)	2007	ccvfloresta.com	81.3
3. Centro Ciência Viva de Bragança (Bragança)	2007	braganca.cienciaviva.pt	11.095
4. Centro Ciência Viva de Constância (Constância)	2004	constancia.cienciaviva.pt	36.613
5. Centro Ciência Viva de Estremoz (Estremoz)	2005	ccvestremoz.uevora.pt	17.059
6. Centro Ciência Viva de Porto Moniz (Porto Moniz, Madeira)	2004	portomoniz.cienciaviva.pt	16.648

(continued)

Table 1. (*continued*)

List of science centers (location)	Year	Websites	Total visitors
7. Centro Ciência Viva de Tavira (Tavira)	2005	cvtavira.pt	27.704
8. Centro Ciência Viva de Vila do Conde (Vila do Conde)	2002	viladoconde.cienciaviva.pt	82.963
9. Centro Ciência Viva do Algarve (Faro)	1997	ccvalg.pt	215.034
10. Centro Ciência Viva do Alviela - Carsoscópio (Alcanena)	2007	alviela.cienciaviva.pt	44.241
11. Exploratório - Centro Ciência Viva de Coimbra (Coimbra)	1998	exploratorio.pt	13.328
12. Fábrica – Centro Ciência Viva de Aveiro (Aveiro)	2004	www.ua.pt/fabrica	12.656
13. Pavilhão do Conhecimento (Lisboa)	1999	pavconhecimento.pt	15.435
14. Planetário Calouste Gulbenkian (Lisboa)	1965	ccm.marinha.pt/pt/planetario	22.237
15. Centro Ciência Viva de Lagos (Lagos)	2009	lagos.cienciaviva.pt/home	34.265
16. Planetário do Porto - Centro Ciência Viva (Porto)	1998	planetario.up.pt	14.49
17. Expolab (Lagoa)	2009	expolab.centrosciencia.azores.gov.pt	13.376
18. Centro Ciência Viva do Lousal-"Mina de Ciência" (Grândola)	2010	lousal.cienciaviva.pt	16.968

Given that the context of the study was a coherent network of Science centers that share a common purpose and national approach it could be hypothesized that all Science center units would have substantially similar web and social media presence. Former studies have proposed that online resources should reflect an organizations' mission, for which they should exhibit functional comparable web presences, related to the essential characteristics of their mission [22].

4   Analysis and Discussion

To assess the presence of CCV Science centres in social media, we study their presence in the following media, following Shang et al. [23]: Facebook, Google+, Instagram, LinkedIn, Twitter, YouTube, Viemo, Issuu and Trip Advisor. Table 3 summarises the results for all 18 CCV science centres, collected till 28 March 2018 as our reference date. In the following, we briefly define the metrics that we introduce in this table. For the case of Facebook, '*people like*' and '*people share*' are the number of people that like and follow a Facebook page, respectively. '*FB Likes*' indicates the number of likes that the most liked post posted during March 2018 has received, and the parameter '*share*' indicates the number of times that the most shared post had been shared until the reference date. Finally, *FB comments* indicates the number of comments that the most commented post received until the reference date. On the other hand, for the case of Instagram, the table reports the number of people that follow the target science centre on this medium, where the parameters '*likes*' and '*comments*' represent respectively the number of likes that the most liked post and the number of comments that the post with the highest number of comments had received, amongst the nine posts that had been posted most recently, had received till the reference date. For Google+, LinkedIn and Twitter, we simply consider the number of followers. For YouTube, we consider the number of people subscribing for the channel created by the science centre itself. For Trip Advisor, we simply count the number of reviewers, i.e., people who posted any review on the considered science centre. The 'Other Media' includes the Vimeo and Issuu media, where the numbers report the aggregate number of followers for these two media. Finally, at the bottom of the table, the parameter 'Total' indicates the sum of all abovementioned parameters for each science centre. This is indeed a measure of science centres' presence in the social media, and from now on, we refer to it as *social media presence*.

Table 2 ranks all science centres, according to their presence in the social media. We observe that C13 (Pavilhão) with total social media presence of 103,257 is by far the most present science centre in the social media, more than five times greater than its following science centre, C11 (Coimbra). The presence in social media for the rest of science centres, varies between 4,787 and 19,960, with C6 (Moniz) and C4 (Constância) having the least presence.

Table 2. Ranking of social media presence of different science centres in Portugal.

Ranking	Centros Ciência Viva	Social media presence	Science center
1	Pavilhão	103,257	C13
2	Coimbra	19,960	C11
3	Planetário Calouste Gulbenkian	14,837	C14
4	Estremoz	14,254	C5
5	Floresta (Proença-a-Nova)	13,997	C2
6	Planetário do Porto	13,850	C16
7	Algarve (Faro)	13,818	C9
8	Lagos	13,059	C15
9	Sintra	11,925	C1
10	Alviela - Carsoscópio	11,616	C10
11	Fábrica	10,997	C12
12	Lousal-"Mina de Ciência"	10,006	C18
13	Bragança	8,367	C3
14	Tavira	6,950	C7
15	Expolab Lagoa Açores	6,068	C17
16	Vila do Conde	5,757	C8
17	Moniz	4,928	C6
18	Constância	4,787	C4

Similar to the approach followed by [23, 24], a Web content analysis of these science centers was also conducted. Table 3 summarizes the results.

In this table, all elements are normally either 1 or 0, indicating the existence or lack of a respected item, respectively. The list of items that we use in this table is suggested by previous research. [23, 25]. Padilla-Meléndez and Del Águila-Obra [24] suggest a comprehensive list of items for analyzing the web and social media usage, without clearly defining these items. Therefore, we adopted the definitions provided by Shang et al. [23]. Finally, for assessing Facebook usage, we employed the metrics (namely, likes, comments and shares) proposed by Bonsón and Ratkai [10]. In the sequel, we briefly define each of these items:

- *Website,* whether the centre has a website or not.
- *Blog* is a website used for socialising and narrating life stories.
- *Online Forum* is a website where the visitors can discuss about specific topics.
- *Downloadable files* indicate whether the centre provides any PDF or PPT material that can be downloaded from its website. The score for this item indicates how many of these file types is available on the website. For example, the score for C5 is two, which means that it has both PDF and PPT files on its website.
- *Email* indicates whether the science centre's website provide email service.
- *Newsletter subscription* means whether it is possible to subscribe for receiving a regular newsletter on the website or not.
- *e-service* item includes four different components. Among other services, the CCV centres also offer birthday celebration for visitors with a 'flavour of science and technology'. For reservation, part of the total cost is normally required to be paid in advance, say 50%. This payment can be done either through a bank transfer or by paying on the centre's desk. For the first component of the e-service item, we give one score for each centre that accepts bank transfer for the reservation. Some centres also have an online shop; we consider one additional score for these centres – the second component.
- For the third component, we give one more score to those centres that also sell tickets online. Finally, for the last component, we give one additional score to those centres that offer booking for a visit through their website.
- *Google maps, earth, street* indicates whether the website provides an integrated Google map along with street address on its website.
- *Geographic coordinate* indicates whether the website provides the GPS coordinates of the centre's premises.
- *Photo* indicates whether the centre provides photos of its activities, lab facilities or organized events such as exhibitions, workshops, etc.
- *iTunes* indicates whether the centre has published any multimedia content in iTunes or not.
- *Online chat* means that if there is any assistance provided online by the website.
- *Online database* means whether the website provides any statistic or archived information about the visitors over a certain period of time, e.g., monthly, yearly, in the last ten years, etc.
- *Podcasts* indicates whether the centre provides any archived podcast, an episodic series of digital audio or video files which a user can download and listen to.

Table 3. Social media presence of *"Rede Nacional de Centros Ciência Viva"* National Science Centres.

	C1	C2	C3	C4	C5	C6	C7	C8	C9	C10	C11	C12	C13	C14	C15	C16	C17	C18
Facebook																		
People like	5,775	6,503	3,884	2,361	7,070	2,455	3,314	2,768	6,678	5,181	9,477	4,931	47,685	7,232	6,127	6,325	2,992	4,761
People follow	5,684	6,429	3,821	2,356	7,042	2,417	3,264	2,687	6,494	5,100	9,417	4,892	46,837	7,198	6,036	6,343	2,975	4,697
FB likes	23	57	60	26	86	14	22	24	107	176	69	25	53	176	18	316	25	20
share	14	7	8	16	16	5	15	48	28	163	23	12	16	60	34	222	5	6
FB comments	0	2	2	0	0	2	3	0	4	3	17	2	3	23	0	15	0	1
Google+	0	147	0	0	0	0	0	0	0	74	0	0	0	0	0	46	57	0
Instagram	190	74	130	0	0	0	0	220	0	572	718	147	896	0	283	206	0	166
Most likes & comments from the last 9 posts	130	151	210	00	00	00	00	100	00	362	1893	100	704	00	112	220	00	281
LinkedIn	0	0	36	0	0	0	0	0	43	9	0	0	0	0	0	0	0	0
TripAdvisor Reviewers	30	34	37	28	40	25	15	0	82	14	28	1	742	113	127	2	1	62
Tweets Followers	196	716	368	0	0	10	308	0	381	285	0	0	6,755	0	401	353	0	214
Vimeo	0	0	0	0	0	0	0	0	0	0	0	0	31	0	0	0	0	0
Video	0	0	1	0	1	0	0	0	0	1	0	1	1	0	1	0	1	0
Virtual visit	0	1	0	0	0	0	0	0	0	0	0	0	0	0	0	0	1	0
Language	1	1	3	2	1	0	2	2	1	2	1	0	1	1	2	1	1	1
Total	11,925	13,997	8,367	4,787	14,254	4,928	6,950	5,757	13,818	11,616	19,960	10,997	103,257	14,837	13,059	13,850	6,068	10,006

- *Video* indicates whether the centre provides any video integrated in its website or not. That is, the video can be played and watched on the host website. The video can introduce the centre itself or can be some parts of their activities, exhibitions, science talks, etc.
- *Virtual visit* means whether the centre provides any 360° virtual tour from its facilities on its website or not.
- *Language* indicates the website is in how many different languages. For instance, C3's website is in three different languages, namely Portuguese, Spanish and English.

In Table 4, the last row indicates the overall Web presence score of different science centres. Each element is simply the sum of all elements in the respected column. Table 5 lists all science centres according to their Web usage rank.

Note that among all 18 centres one of them (C6) surprisingly does not have any website; in fact, it previously had a website, but now it is only active in social media. The web presence of different science centres varies between 8 and 14, with C13 having the first rank. Interestingly, C13 has also by far the first rank in the social media presence.

Table 4. Web presence of different science centres.

	C1	C2	C3	C4	C5	C6	C7	C8	C9	C10	C11	C12	C13	C14	C15	C16	C17	C18
Website	1	1	1	1	1	0	1	1	1	1	1	1	1	1	1	1	1	1
Blog	0	0	0	0	0	0	0	0	0	0	0	0	0	0	0	0	1	0
Online Forum	0	0	0	0	0	0	0	1	0	0	0	0	1	0	0	0	0	0
Downloadable files	1	1	1	1	2	0	1	1	1	1	2	1	1	2	0	1	1	1
Email	1	1	1	1	1	0	1	1	1	1	1	1	1	1	1	1	1	1
Newsletter subscription	0	1	1	1	1	0	1	1	1	1	0	1	1	1	1	1	1	0
e-services	2	1	3	1	2	0	3	1	3	3	2	1	3	4	0	2	1	0
Google maps, earth, street	1	1	1	1	1	0	1	1	1	1	1	1	1	1	1	1	1	1
Geographic coordinate	0	1	0	1	1	0	1	1	1	1	1	1	1	1	1	1	1	1
Photo	1	1	1	1	1	0	1	1	1	1	1	1	1	1	1	1	1	1
iTunes	0	1	0	0	0	0	0	0	0	0	0	1	1	0	0	0	1	0
Online chat	0	0	0	1	0	0	1	1	0	0	0	0	0	0	0	0	0	0
Online database	0	0	1	1	0	0	0	1	1	1	1	0	1	1	0	1	1	1
Podcasts	0	0	0	0	0	0	0	0	0	0	0	0	0	0	1	0	0	0
Video	0	0	1	0	1	0	0	0	0	1	0	1	1	0	1	0	1	0
Virtual visit	0	1	0	0	0	0	0	0	0	0	0	0	0	0	0	0	1	0
Language	1	1	3	2	1	0	2	2	1	2	1	1	1	1	2	1	1	1
Total	8	11	14	12	12	0	13	13	12	14	11	11	15	14	10	11	13	8

The overall scenario for the web and social media for the network of Portuguese Science Centers offers unexpected patterns. A first rank of comments regards the relatively intense heterogeneity in the patterns of web and social media presences exhibited by all the units, despite their common goals and frame of activities. All the Science Centers have similar target audiences and host similar type of content to display, and resort to a shared vision of a participatory and engaging service experience. Data analysis did not reveal any relevant association of the observed differences to variables such as age of the centers, or the population in surrounding areas or the

Table 5. Web presence ranking based on the web items listed in Table 4.

Ranking	Centros Ciência Viva	Total web items	Science center
1	Pavilhão	15	C13
2	Bragança	14	C3
3	Alviela - Carsoscópio	14	C10
4	Planetário Calouste Gulbenkian	14	C14
5	Tavira	13	C7
6	Vila do Conde	13	C8
7	Expolab Lagoa Açores	13	C17
8	Constância	12	C4
9	Estremoz	12	C5
10	Algarve (Faro)	12	C9
11	Floresta	11	C2
12	Coimbra	11	C11
13	Fábrica	11	C12
14	Planetário do Porto	11	C16
15	Lagos	10	C15
16	Sintra	8	C1
17	Lousal-"Mina de Ciência"	8	C18
18	Moniz	0	C6

number of average visitors per year. Such differences could explain different approaches in selecting different intensities and target presences in social media channels, but it was not the case evidenced by the data. Therefore, the existing differences in web and social media presence are either related to different strategic approaches by each science center or rather likely by a lack of a structured frame of action to develop and monitor such efforts for online presence. Beyond the observation of the heterogeneity in the choices for the online and social media presence, the study involved two further analysis: (i) the alignment between the web and social media presence; (ii) the alignment between the online efforts and its impact on visitors' awareness about the science centers. The purpose of (i) was to investigate the existence of distinct online positioning strategies, according to which centers would privilege web or rather the social media efforts. The purpose of (ii) was to get insights about the relevance of the online efforts on the eyes of the visitors. To this end the study resorted to secondary data from studies with the centers' visitors, conducted by national authorities and entities and that were publicly available.

The analysis of the relative positions, or rankings of the science centers for web presence and for social media evidenced the existence of important gaps, suggesting that centers don't invest equally in the development of a web and the maintenance of a social media presence. For example, the center from Bragança (C3) ranked in the 13rd position regarding the social media effort, while it was ranked 3rd in the density and richness of the features and functions of its web presence. This was however an unusual positioning given that most of the centers exhibited a stronger bet on the construction of social media resources. Also, in the case of the center C6 (Moniz) the data supported a stronger presence on social media, given that the center even decided to withdraw its web presence closing the webpage it maintained for several years in the past.

The analysis of the effectiveness or impact of the online presence efforts was done by looking at recent data released by the network of science centers about the importance of the web for the awareness about the center, and its information. Data from a survey from 2016 revealed that the proportion of visitors who declared to be acquainted with the online pages of the centers was substantially lower than those referring to be aware of the presence of the center in social media. Following a similar approach to the other analysis in the study we used the survey data to rank the centers according to their visitors' "Web Presence Awareness" and "Social Media Awareness", as follows in Table 6. Results evidenced misalignments between what is relevant and visible to the visitors, and the amount of effort evidenced by the web and social media presence.

Table 6. Web presence and social media awareness about the Centers.

Centers																	
C1	C2	C3	C4	C5	C6	C7	C8	C9	C10	C11	C12	C13	C14	C15	C16	C17	C18
Social media awareness																	
15	8	17	4	9	7	12	1	18	5	11	2	13	10	14	6	3	16
Web presence awareness																	
5	12	9	11	16	3	14	7	15	13	8	10	1	4	17	2	6	18

Overall the study brings forward the need to monitor the evidence of the efforts placed by organizations in building an online presence towards several dimensions: (i) the alignment between the efforts placed on the web and in social media; (ii) the alignment between efforts placed in online means and channels and visitors online uses; and iii) the comparative assessments of the alignment of the online efforts towards those of similar organizations.

5   Conclusions

The exploitation of Internet technologies for communication and interaction with customers have resulted in remarkable changes in the way service businesses are operated. In this paper, we examined the employment of social media and web presence by science centres, addressing the alignment and importance of their presence in these media on the user's awareness. In particular, we focused on the network of science centres in Portugal and studied the richness of their websites, adopting acknowledged items from the literature (e.g., Blog, Online Forum, Online Shop, Booking for a visit, Email, Virtual visit, etc.), and their popularity in several social media, namely Facebook, YouTube, Instagram, Google+, Twitter, LinkedIn, TripAdvisor, Vimeo and Issuu, using metrics such as number of likes, shares, and comments. The results revealed that the centres do not invest equally on their web and social media presence, and overall they invest more on social media than on web presence. The rationale behind this can be the widespread use of social media by their target audience, which are mostly young people. Similar to Bonsón and Ratkai [10], our quantitative study reveals that Facebook posts are more liked, than commented on or shared. However, unlike this work where the comments are more than shares, our study shows that the number comments are in the minimum. Therefore, it is important to intrigue visitors to leave more comments since receiving their feedback (either positive or negative) is helpful for the improvement. As for the limitations of this study, we addressed the problem using a quantitative approach. It is also important to approach it from a qualitative aspect, e.g., by studying the visitors' mood, which can be positive, negative and neutral [10]. We have also gathered data on a certain date. For future work, it is also interesting to extend the research to monitor these centres over several months and check the variations in the visitor's interactions to help managers assess the effectiveness of their adopted strategies. We also suggest analysing the user's interactions in social media in further details. For instance, we considered only the number of YouTube subscribers, but it is also helpful for managers to understand the number of views for their shared content.

References

1. Soren, B.J.: Museum experiences that change visitors. Mus. Manag. Curatorship **24**(3), 233–251 (2009)
2. López, X., Margapoti, I., Maragliano, R., Bove, G.: The presence of web 2.0 tools on museum websites: a comparative study between England, France, Spain, Italy, and the USA. Mus. Manag. Curatorship **25**(2), 235–249 (2010)
3. Kidd, J.: Enacting engagement online: framing social media use for the museum. Inf. Technol. People **24**(1), 64–77 (2011)
4. Bamberger, Y., Tal, T.: An experience for the lifelong journey: the long-term effect of a class visit to a science center. Visit. Stud. **11**(2), 198–212 (2008)
5. Waltl, C.: Museums for Visitors: Audience Development-A Crucial Role for Successful Museum Management Strategies, pp. 1–7. Intercom, Chicago (2006)
6. Heath, C., Vom Lehn, D.: Configuring 'interactivity' enhancing engagement in science centres and museums. Soc. Stud. Sci. **38**(1), 63–91 (2008)
7. Marty, P.F.: Museum websites and museum visitors: before and after the museum visit. Mus. Manag. Curatorship **22**(4), 337–360 (2007)
8. Pettigrew, J.E., Reber, B.H.: The new dynamic in corporate media relations: how Fortune 500 companies are using virtual press rooms to engage the press. J. Pub. Relat. Res. **22**(4), 404–428 (2010)
9. Day, A.: A model for monitoring web site effectiveness. Internet Res. **7**(2), 109–115 (1997)
10. Bonsón, E., Ratkai, M.: A set of metrics to assess stakeholder engagement and social legitimacy on a corporate Facebook page. Online Inf. Rev. **37**(5), 787–803 (2013)
11. Kabassi, K.: Evaluating websites of museums: state of the art. J. Cult. Herit. **24**(March–April), 184–196 (2017)
12. Styliani, S., Fotis, L., Kostas, K., Petros, P.: Virtual museums, a survey and some issues for consideration. J. Cult. Herit. **10**(4), 520–528 (2009)

13. De Vries, L., Gensler, S., Leeflang, P.S.: Popularity of brand posts on brand fan pages: an investigation of the effects of social media marketing. J. Interact. Market. **26**(2), 83–91 (2012)
14. Manna, R., Palumbo, R.: What makes a museum attractive to young people? Evidence from Italy. Int. J. Tour. Res. **20**, 508–517 (2018)
15. Fletcher, A., Lee, M.J.: Current social media uses and evaluations in American museums. Mus. Manag. Curatorship **27**(5), 505–521 (2012)
16. Cameron, F.: Object-oriented democracies: conceptualising museum collections in networks. Mus. Manag. Curatorship **23**(3), 229–243 (2008)
17. Russo, A.: Transformations in cultural communication: social media, cultural exchange, and creative connections. Curator Mus. J. **54**(3), 327–346 (2011)
18. Gerrard, D., Sykora, M., Jackson, T.: Social media analytics in museums: extracting expressions of inspiration. Mus. Manag. Curatorship **32**(3), 232–250 (2017)
19. Pearce, S.M.: Exploring Science in Museums. Athlone Press, London (1996)
20. Garcia, J.L., Ramalho, J, Silva, P.S.: Os Públicos Da Rede Nacional De Centros Ciência Viva – Relatório Final. Instituto de Ciências Sociais, Universidade de Lisboa para a Rede Nacional de Centros Ciência Viva (2016). http://www.cienciaviva.pt/img/upload/estudo.PDF. Accessed 31 Mar 2018
21. Cienciaviva. http://www.cienciaviva.pt. Accessed 31 Mar 2018
22. Mason, D.D., McCarthy, C.: Museums and the culture of new media: an empirical model of New Zealand museum websites. Mus. Manag. Curatorship **23**(1), 63–80 (2008)
23. Shang, S., Li, E., Wu, Y., Hou, O.: Understanding Web 2.0 service models: a knowledge-creating perspective. Inf. Manag. **48**(4–5), 178–184 (2011)
24. Padilla-Meléndez, A., del Águila-Obra, A.: Web and social media usage by museums: online value creation. Int. J. Inf. Manag. **33**(5), 892–898 (2013)
25. Del Águila-Obra, A., Padilla-Meléndez, A., Serarols-Tarres, C.: Value creation and new intermediaries on Internet. An exploratory analysis of the online news industry and the web content aggregators. Int. J. Inf. Manag. **27**(3), 187–199 (2007)

Exploring Customers' Internal Response to the Service Experience: An Empirical Study in Healthcare

Gabriela Beirão[1]([✉])[ID] and Humberto Costa[2][ID]

[1] INESC TEC, Faculty of Engineering, University of Porto,
Rua Dr. Roberto Frias, 4200-243 Porto, Portugal
gbeirao@fe.up.pt
[2] Faculty of Engineering, University of Porto,
Rua Dr. Roberto Frias, 4200-243 Porto, Portugal
humbertoccosta@gmail.com

Abstract. Service organizations increasingly understand the importance of managing the customer experience to enhance customer satisfaction and loyalty. This study aims to develop a better understanding of the customer experience by investigating how the customer's internal mechanisms influence it. That is, how it is perceived and processed at three different levels (visceral, behavioral and reflective), which determines a person's cognitive and emotional state. To this purpose an exploratory multi-method ethnographic study was undertaken in a healthcare service. The results showed the emotions provoked by the service experience at each level. These levels are interconnected and impact each other working together to influence a person's cognitive and emotional state, and thus playing a critical role in the overall evaluation of a service. Results show that elements such as servicescape aesthetics, face-to-face and non-human interactions influence emotions and service evaluations. The service should be designed in a way that induces positive emotions, and a feeling of being in control. Especially in healthcare services there is a need to balance the conflicting responses of the emotional stages that may be triggered at the visceral and behavioral levels, while providing reassurance and calm at the reflective level that the health problem is going to be taken care. Using service design approaches this understanding of the customers' brain can be translated into improving the customer experience.

Keywords: Customer experience · Emotions · Service design Healthcare

1 Introduction

Companies have focused on improving the customer experience to create differentiation, customer satisfaction and loyalty [1]. Customer experience has been conceptualized as "a multidimensional construct focusing on a customer's cognitive, emotional, behavioral, sensorial, and social responses to a firm's offerings during the customer's entire purchase journey" [2 p. 71]. That is, customer experience is holistic, integrating

© Springer Nature Switzerland AG 2018
G. Satzger et al. (Eds.): IESS 2018, LNBIP 331, pp. 303–315, 2018.
https://doi.org/10.1007/978-3-030-00713-3_23

all contacts with a firm, as such all encounters need to be seamlessly designed [1, 3]. As such, customers co-create the experience when they integrate resources and interact with multiple service elements [4]. Although is important understanding dyadic interactions, research has shown that value is co-created in complex networks of interdependent many-to-many interactions among actors [5, 6].

Service design approaches may be used to enhance the customer experience and value co-creation process [7], through a human-centered, creative, and holistic approach to the creation of services [8, 9]. The design of the desired experience requires a coherent set of elements, or clues, along the customer journey [10]. These service elements are the context where the experience takes place, that is the physical and relational elements in the experience environment, namely physical environment, service employees, and fellow customers [1]. Context can intensify engagement and emotional connections [11, 12]. Thus, the experiential side of service is a very important aspect of service design [11]. However, research is needed to clarify which service elements produce the most compelling contexts, and how to use them to create customers' emotional connections to a given service [1].

Emotions assume an important role in the service experience [13], as such when designing servicescapes both the functional and the affective aspects of the service experience must be carefully considered [11]. Voss et al. [14] suggests that is useful to view experiential services in terms of the emotional journey performed by customers. That is, looking not only for the functional aspects required at each stage of interaction with the service, but also for the emotional impact of the interactions.

Norman [15] says that human emotions and behaviors are the result of three different levels of processing: visceral, behavioral and reflective. The visceral level is the response to sensory perception or immediate effect. Is about rapid judgments, such as what is good or bad, safe or dangerous, attractive or unattractive. This level may be positively or negatively influenced by the environment within which it is perceived, and by other levels of the cognitive process. The behavioral level interprets available sensory data controlling everyday behavior. The behavior level can enhance or inhibit the visceral and reflective layer. The reflective thought is the contemplative side and the most developed, involving conscious consideration and reflection on past experiences. While it does not have direct access either to sensory data or to the control of behavior, it watches over, reflects upon, and tries to influence the behavioral level. It should be noted that the three levels interact with each other in complex ways.

Norman's theory [15] provides a structure that may be used for understanding the brain's processing layers, applicable to both cognitive and emotional processing of the customer experience. This knowledge may be translated into the development of evidence useful for better designing the service experience. The visceral dimension is about the initial sensory stimuli, it aims to appeal to the senses, for example, through visual, audial, and tactile stimuli. The emotional response is usually quick. The behavioral regards the use and experience of a product or service and includes many aspects like functionality, performance, or usability. In the reflective dimension the experience is interpreted, understood, and reasoned. Only at this level the full impact of emotions and thought are experienced. That is, through reflection, each person is able to integrate their own experiences with, for example services, other people, artifacts, into their broader life experiences and, over time, associate meaning and value with them.

Enhancing service experiences has been considered a research priority [16]. In healthcare this is especially important since it is an essential service for societal well-being [17]. Research should aim to understand the perspective of the customers in determining their own experiences. More specifically, research is needed to understand which service elements creates the most compelling contexts and [1], and also understanding the impact of the aesthetic aspects [18].

This study aims to develop a better understanding of the customer experience by investigating how the customer's internal mechanisms influence the experience. That is, how it is perceived and processed at the three different levels (visceral, behavioral and reflective), which determines a person's cognitive and emotional state. This knowledge can be translated into valuable insights to support the service design process, thus enhancing the customer experience.

The paper is organized as follows. Next, the methods used and data collection are explained. Section 3 presents the findings for each processing level, which are then integrated to enable understanding the customer experience. The final section concludes the study and discusses the future work.

2 Method

To address the research aims the study builds on an exploratory case study using a multi-method ethnographic approach, combining field observation, shadowing, interviews, and focus groups. This approach enabled to elicit data on customer's brain processing at the three different levels (visceral, behavioral and reflective) proposed by Norman [15]. Data were collected in a hospital in Brazil. The Hospital preferably serves elderly people, but is open to everyone, and most patients are between ages 45 and 70. This Hospital provided a rich context since it was recently constructed and follows a service model aiming to give humanized care. The scope of analysis was limited to the reception room since it would be too complex to extend the analysis to the entire servicescape due the size of the hospital, and the number of different health care services provided (e.g. consultations, exams, surgeries).

Data collected from the different methods were analyzed individually using qualitative research approaches [19, 20], and then results were combined to allow a comprehensive understanding of the customer experience [21]. Next, the study research design and methods of data analysis are detailed. Data collected during the research involved different participants in each phase for two reasons. First, the hospital requested that the service was not disturbed during the research, and second because each phase involved different methods requiring different size samples. This allowed obtaining information from different participants, thus increasing results validity.

The study started with a field observation study at the hospital and analysis of documents. This enabled mapping the service system, creating personas, and drawing the customer journey using the personas previously identified. Personas are defined as "fictional profiles, often developed as a way of representing a particular group based on their shared interests" [22]. Personas provide information about a specific kind, or segment of customers, who will use the service.

These data framed the implementation and analysis of the qualitative methods undertaken for collecting data enabling understanding the levels of processing within the brain.

To collect data for understanding the behavioral dimension shadowing, contextual inquiry, and emotional scales were used [22, 23]. Shadowing was undertaken with 10 patients that were closely followed by the researcher to identify the actors and artifacts involved in the service, as well as the emotions, expectations and habits. After the service provision patients were interviewed to ascertain their perceptions about the experience. Following the interview patients were shown photographs representing the different touchpoints in the customer journey and asked to identify the emotion felt from eight given alternatives (enjoyed, happy, anguished, confused, scared, apathetic, sad, and nervous). Using several methods enabled data triangulation and increasing reliability through the use of use of multiple sources of evidence [24].

Data for the visceral dimension were collected using four emotional wheels developed, based on existing research [25–30], for this study for each of the four senses: sight, hearing, smell and touch. The emotional wheels were previously tested to verify the adequacy, legibility and time to complete. Data were collect from 100 patients waiting for a consultation or exam. The patients were asked to identify if they felt a given emotion during their experience with the service and its intensity, and if they would like to have felt another emotion and the expected intensity. Spearman rank-correlation coefficients were calculated to determine correlations between emotions.

The reflexive dimension was studied using a focus group with eight participants: one doctor, one nurse, two administrative staff, 4 patients, and 1 patient relative. This stage included both service employees and customers to understand each actor's perspective, and elicit both differences and similarities.

Participants were asked to talk about their health care service experience, the positive and negative points, the emotions felt, and recommendations for improvements. To increase group dynamics and help respondents evoking emotions issue cards with photographs portraying different emotions were showed. Data were analyzed following Miles, Huberman et al. approach [19]. Coding started with a first cycle, where data was initially coded around patterns and themes. The service quality attributes identified in extant literature were used for codification, but allowing the list to be expanded during the analysis to capture emerging themes. Then, in the second cycle, coding was refined and patterns grouped into a smaller number of categories composing the service experience factors. This process was done separately by two researchers to ensure results accuracy and validity [19].

3 Findings

3.1 Behavioral Dimension

The results emphasized the emotions at the behavioral level of processing facilitating or damaging the customer experience. Customers perceived different emotions as they walkthrough the necessary activities of the service, as shown in Fig. 1.

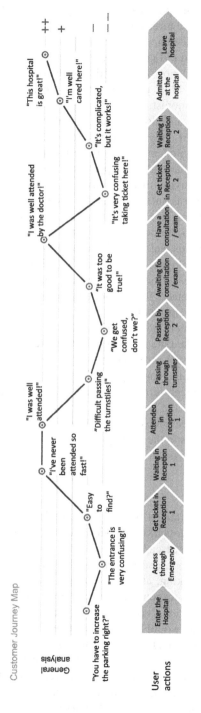

Fig. 1. Emotions evoked along the customer journey

The emotions range from happiness and joy to sadness, anxiety, and fear. Negative emotions were mostly caused when patients need to interact with artifacts such as the ticket dispenser, or pass through the turnstiles. This occurs especially with patients that were not familiar with the servicescape. For example, patients felt that the ticket dispensers were difficult to see and did not realize they need a take ticket before being served. Also, several patients did not understand how to open the turnstile, which provoked confusion and anxiety. Most patients experienced positive emotions when waiting to be called by the receptionist. This happens because the waiting time was usually small, since most exams and doctor appointment were previously schedule. Also, after consultations patients usually felt positive emotions.

The interviews results enable understanding more in-depth patients service experience and emotions felt. Both intangible (i.e. personal contact, quality of care) and tangible elements (i.e. physical environment) contributed to a positive and enjoyable experience. The patients used words such as 'attentive', 'caring', 'tranquility', 'polite, and 'patience' to characterize the service. The interactions with doctors and employees are very important contributing to the positive emotions. For example a patient mentioned that "the employees makes us feel relaxed". Patients liked "being given attention and care by doctors", "having employees that talk calmly and are patient", and "quality of healthcare". However, patients do not like when the receptionist focused on the computer not establishing eye contact. Thus, emotions showed by the healthcare staff have and influence on customers emotional state.

Overall the physical environment also contributed to a good experience eliciting positive emotions. The patients liked the hospital infrastructure, and nice spacious clear rooms. A well-designed and pleasant physical environment can reduce stress evoking positive emotions on patients. On the contrary, if it is poorly designed creates pressure and anxiety, as in the cases of using turnstiles or the ticket dispenser.

However, patients would also like improvements in some procedures such as calls to inform about appointments cancelations, being able to schedule consults or exams by phone, more entertainment options in the waiting room, and better location of the ticket dispenser to facilitate access and visualization.

3.2   Visceral Dimension

The visceral dimension was ascertained by analyzing the impact of each of the four senses (sight, audition, smell and touch) on emotions felt and expected by respondents in the hospital reception. The servicescape under analysis is a spacious reception, with neutral colors, a clear simple layout, and natural lighting. The space is mostly silent, only with sounds from people talking, and phones ringing occasionally.

Figure 2 shows the emotions respondents felt and expected on each of the four senses. The emotional reactions to the visual stimuli on each of the four senses were mostly positive, although patients would like to have felt even more positive emotions. Tranquility, happiness, relief and satisfaction are the positive emotions more evoked, and sadness, fear, anxiety and tension the negative ones. The level of negative emotions felt was low, however patients still wanted the to be lower. Interestingly, sight is the sense that evokes more positive and negative emotions. On the contrary hearing elicited the less positive emotions and more negative emotions.

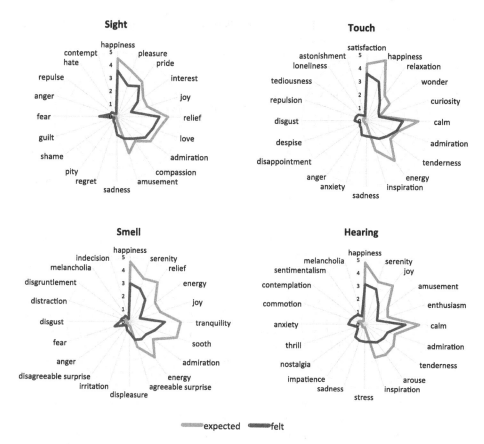

Fig. 2. Emotions felt and expected on the servicescape

3.3 Reflective Dimension

The results from the focus group enabled understanding respondents' perceptions of the service experience and emotions evoked in the servicescape. The most positive (Caring and Courtesy) and most negative aspects (Accessibility, availability, and Communication) for the participants are summarized in Table 1 (only factors with the highest and lowest perceptions are shown). Care and courtesy are very important for the respondents and they felt that they are deeply related and must be present in the service provision to evoke positive emotions. Respondents identified several problems with accessibility to the hospital, namely the installations layout, information access organization, and long time waiting in the reception, which provoked on the receptionists feelings of incapability to solve the problems, and on patients frustration.

Communication problems were also identified, thus damaging the customer experience because the necessary information was not provided. Interestingly, healthcare personnel and patients showed different views on improvement recommendations. While employees focused on improving the physical evidence such as the artifacts, the patients wanted more and deeper face-to-face interactions.

Table 1. Focus group results with examples and illustrative quotes summarizing most positive and most negative service experience factors

Service experience factors	Examples
Caring	• *Focusing on personal interactions* (e.g. "Technology is essential, has to be present, but nothing replaces doctor's care with the patient, being well received." [patient]) • *Sufficient time listening to patients* (e.g. "Caring for the patient was prioritized, and to achieve it we now have more time in the infirmary to be with patients." [nurse]) • *Being given complete information* (e.g. "They explain everything well, then they stop and give us attention." [patient])
Courtesy	• Being cared with courtesy (e.g. "Since the first time in the emergency I was well treated by everyone." [patient]) • Being served rapidly and calmly (e.g. "We have to be calm when the patient arrives, sometimes he or she is nervous, with fear, and anxiety. If we are calm the patient perceives that the health problem may be treated." [receptionist])
Accessibility and availability	• Accessing the facilities easily (e.g. "sometimes the reception is full of people and it takes to long to attend the patients" [receptionist]) • Longer visiting hours (e.g. "Only one time for visiting patients, and there is a lot of people." [patient]) • Having enough staff to facilitate access (e.g. "We have a good infrastructure, but since we don't have enough staff it doesn't work." [nurse]) • Accessing the information system (e.g. "The information system is slow, it makes things very difficult" [patient])
Communication	• Accessing patients healthcare information to provide information (e.g. "We don't have information where the patient is, if he or she went to the room, or other place." [receptionist]) • Providing information on how to use the service (e.g. "The first day I got to the hospital I was lost." [patient]) • Providing healthcare information (e.g. "Sometimes I have difficulties in understanding information. They explain well, also my daughter helps me understanding, it facilitates." [patient])

3.4   Integrating Results for Improving the Customer Experience

The results enabled better understanding customer's internal mechanism response forming the service experience. A summary of the most important positive factors and the ones needing improvement is presented in Table 2. The dimensions are interconnected, influencing each other, as such improvements in one may impact the others contributing to an overall better service experience.

Knowledge about the factors perceived by the customers as positive and the ones needing improvement are important for providing a good overall service experience, as the three dimensions are intertwined influencing each other. It is known that the customer experience has a holistic nature, as such all elements and interactions with the

Table 2. Summary of most important factors of the customer experience: positive and needing improvement

	Visceral dimension	Behavioral dimension	Reflexive dimension
Positive factors	Sigh – evokes the most positive emotions and less negative emotions	Accessibility • Easy access to reception • Easy interaction with artifacts Fast service Courtesy and empathy	Care • Face-to-face interactions • Time dedicated for service Courtesy • Service provided with serenity and courtesy
Improvements priorities	Sight, hearing, smell and touch • Improve visual elements, sounds, smells and better materials reinforcing positive mood on patients and facilitate the service provision	Physical evidence • Better signaling • Better tickets control • Easier to use turnstiles Interactions • Consistent service • Calling patients for consultations • Better information provision	Accessibility/availability • Longer visiting hours • Better design of physical layout • Faster information system • More employees Communication • Information transparency • Better communication methods • Usage of language patients can understand

service provider along the customer journey are important [10, 31]. The service provider needs to maintain the factors contributing to positive emotions and satisfaction. However, results showed that some factors require attention since they are damaging the service experience by provoking negative emotions. First, the factors in the visceral dimension should be improved since at this level rapid judgments are made, and is the start of affective processing [15]. Results showed that even in a healthcare service customers desire to experience more positive emotions and less negative emotions.

At the behavioral level, results showed that some elements in the servicescape inhibit behavioral control provoking negative emotions. Actions are usually associated with an expectation, and a positive or negative outcome results in a positive or negative affective response, respectively. Results showed the different emotions felts in the customer journey from confusion and frustration to satisfaction. In some situations just moving the ticket dispenser to a more visible place enhances the customer experience, because enables patients to feel more in control.

The reflective dimension, where deep understanding develops and highest level of emotions emerge, showed the need to provide information to customers about the flow of activities and healthcare information. Results evidence the difference in perspectives that patients and healthcare personnel had. While the first want more humanized and personal care, the second focused on using artifacts to improved the service provided. Results showed the need to design the service integrating the perspectives and needs of the different actors interacting in the service.

The results of the three dimensions should to be combined to enable understanding and improving the global experience, since they work together and all play a critical role in the overall evaluation of a service (see Fig. 3). For example, previous bad experience may damage future service evaluations. Results showed the confusion and frustration felt by patients for not being able to hearing information, or knowing that they have to take a ticket before an appointment, which in turn affects service comprehension and usage. The service should be designed in a way that induces positive emotions, and a feeling of being in control. Especially in healthcare services there is a need to balance the conflicting responses of the emotional stages that may be triggered at the visceral and behavioral levels, while providing reassurance and calm at the reflective level that the health problem is going to be taken care.

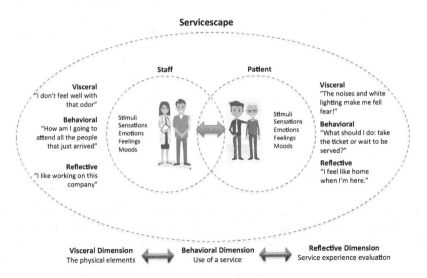

Fig. 3. Internal response mechanisms in the servicescape

4   Conclusion

The aim of this paper was to explore more in depth the internal customer's response that forms the customer experience. The results from the empirical study show that studying the customer experience at the different levels determining a person's cognitive and emotional state provides useful considerations for improving the customer experience. As such, this study contributes with a more comprehensive view of the customer experience, extending knowledge beyond functional elements. Using service design approaches this understanding of the customers mind can be translating into improvements on the service, which in turn affect service evaluations.

This has also important managerial implications because it shows how to improve customer experience by incorporating the internal customer responses. Hospital may redesign their service addressing the most important elements in the servicescape that determine customers' cognitive and emotional state, and thus perceptions of service quality. Tools such as service blueprinting could be used to portray the service and the interactions between frontstage and backstage facilitating service improvements and innovations [32, 33]. For example, procedures for receiving patients, such as displaying positive emotions, including greeting, speaking in a calm vocal tone, smiling, and making eye contact could be established.

This study has several limitations. First, it was applied to only one context of a healthcare service. Second, only one part of the customer journey was considered, so it does not offer an overall assessment of the customer experience. Third, results may be biased since respondents are mostly older patients.

Future research should extend to the entire customer journey, expanding to other actors involved in the customer experience, and adopting a network perspective. Healthcare specific context should be considered since it is complex and involve multiple dynamic interactions [6]. For example, a recent service design method (SD4VN) [34] developed for health care service systems may be useful in designing network-level services addressing actors' interrelated activities, interaction and objectives.

Acknowledgments. This research was funded by Conselho Nacional de Desenvolvimento Científico e tecnológico (CNPQ), through a post-doctoral scholarship (Humberto Costa).

References

1. Zomerdijk, L.G., Voss, C.A.: Service design for experience-centric services. J. Serv. Res. **13**(1), 67–82 (2009)
2. Lemon, K.N., Verhoef, P.C.: Understanding customer experience throughout the customer journey. J. Mark. **80**(6), 69–96 (2016)
3. Meyer, C., Schwager, A.: Understanding customer experience. Harvard Bus. Rev. **85**(2), 117–127 (2007)
4. Vargo, S.L., Lusch, R.F.: Institutions and axioms: an extension and update of service-dominant logic. J. Acad. Mark. Sci. **44**(5), 5–23 (2016)
5. Pinho, N., Beirão, G., Patrício, L., Fisk, R.P.: Understanding value co-creation in complex services with many actors. J. Serv. Manag. **25**(4), 470–493 (2014)
6. Beirão, G., Patrício, L., Fisk, R.P.: Value cocreation in service ecosystems: investigating health care at the micro, meso, and macro levels. J. Serv. Manag. **28**(2), 227–249 (2017)
7. Patrício, L., Gustafsson, A., Fisk, R.: Upframing service design and innovation for research impact. J. Serv. Res. **21**(1), 3–16 (2018)
8. Blomkvist, J., Holmlid, S., Segelström, F.: Service design research: yesterday, today and tomorrow. In: Stickdorn, M., Schneider, J. (eds.) This is Service Design Thinking: Basics - Tools-Cases, pp. 308–315. BIS Publishers, Amsterdam (2010)
9. Wetter-Edman, K., Sangiorgi, D., Edvardsson, B., Holmlid, S., Grönroos, C., Mattelmäki, T.: Design for value co-creation: exploring synergies between design for service and service logic. Serv. Sci. **6**(2), 106–121 (2014)

10. Berry, L.L., Carbone, L.P., Haeckel, S.H.: Managing the total customer experience. Sloan Manag. Rev. **43**(3), 85–89 (2002)
11. Pullman, M.E., Gross, M.A.: Ability of experience design elements to elicit emotions and loyalty behaviors. Decis. Sci. **35**(3), 551–578 (2004)
12. Costa, H.: Design para serviços e consistência estética: Proposição de um protocolo de avaliação estética em serviços. Doctoral thesis, Universidade Federal do Paraná, Curitiba (2017)
13. Mattila, A.S., Enz, C.A.: The role of emotions in service encounters. J. Serv. Res. **4**(4), 268–277 (2002)
14. Voss, C.A., Roth, A.V., Chase, R.B.: Experience, service operations strategy, and services as destinations: foundations and exploratory investigation. Prod. Oper. Manag. **17**(3), 247–266 (2008)
15. Norman, D.A.: Emotional Design: Why We Love (Or Hate) Everyday Things. Basic Books, New York (2004)
16. Ostrom, A.L., Parasuraman, A., Bowen, D.E., Patrício, L., Voss, C.A.: Service research priorities in a rapidly changing context. J. Serv. Res. **18**(2), 127–159 (2015)
17. Anderson, L., et al.: Transformative service research: an agenda for the future. J. Bus. Res. **66**(8), 1203–1210 (2013)
18. Kumar, D.S., Purani, K., Sahadev, S.: Visual service scape aesthetics and consumer response: a holistic model. J. Serv. Mark. **31**(6), 556–573 (2017)
19. Miles, M.B., Huberman, A.M., Saldaña, J.: Qualitative Data Analysis: A Methods Sourcebook, 3rd edn. SAGE Publications, Thousand Oaks (2014)
20. Corbin, J., Strauss, A.: Basics of Qualitative Research, 4th edn. Sage Publications, Thousand Oaks (2015)
21. Yin, R.K.: Case Study Research: Design and Methods, 4th edn. Sage, Los Angeles (2009)
22. Stickdorn, M., Schneider, J.: This is Service Design Thinking: Basics, Tools, Cases. BIS, Amsterdam (2012)
23. Martin, B., Hanington, B.: Universal Methods of Design: 100 Ways to Research Complex Problems, Develop Innovative Ideas, and Design Effective Solutions. Rockport, Beverly (2012)
24. Maxwell, J.: Understanding and validity in qualitative research. Harv. Educ. Rev. **62**(3), 279–301 (1992)
25. Scherer, K.R., Shuman, V., Fontaine, J.J.R., Soriano, C.: The grid meets the wheel: assessing emotional feeling via self-report. In: Fontaine, J.J.R., Scherer, K.R., Soriano, C. (eds.) Components of Emotional Meaning: A Sourcebook. Oxford University Press, Oxford (2013)
26. Scherer, K.R.: What are emotions? And how can they be measured? Soc. Sci. Inf. **44**(4), 695–729 (2005)
27. Zentner, M., Grandjean, D., Scherer, K.R.: Emotions evoked by the sound of music: characterization, classification, and measurement. Emotion **8**(4), 494–521 (2008)
28. Chrea, C., et al.: Mapping the semantic space for the subjective experience of emotional responses to odors. Chem. Senses **34**(1), 49–62 (2009)
29. Kjellerup, M.K., Larsen, A.C., Maier, A.: Communicating emotion through haptic design: a study using physical keys. In: Proceedings of the Kansei Engineering and Emotion Research International Conference (KEER 2014), Linköping, Sweden (2014)
30. Desmet, P., Hekkert, P.: Framework of product experience. Int. J. Des. **1**(1), 13–23 (2007)
31. Teixeira, J., Patrício, L., Nunes, N., Nóbrega, L., Fisk, R.P., Constantine, L.: Customer experience modeling: from customer experience to service design. J. Serv. Manag. **23**(3), 362–376 (2012)

32. Bitner, M.J., Ostrom, A.L., Morgan, F.N.: Service blueprinting: a practical technique for service innovation. Calif. Manag. Rev. **50**(3), 66–94 (2008)
33. Patrício, L., Fisk, R.P., Cunha, J.F.: Designing multi-interface service experiences: the service experience blueprint. J. Serv. Res. **10**(4), 318–334 (2008)
34. Patrício, L., Pinho, N.F., Teixeira, J.G., Fisk, R.P.: Service design for value networks: enabling value cocreation interactions in healthcare. Serv. Sci. **10**(1), 76–97 (2018)

Health Information Technology and Caregiver Interaction: Building Healthy Ecosystems

Nabil Georges Badr[1(✉)], Maddalena Sorrentino[2],
and Marco De Marco[3]

[1] Higher Institute for Public Health, USJ, Beirut, Lebanon
nabil@itvaluepartner.com
[2] Università degli Studi di Milano, Milan, Italy
maddalena.sorrentino@unimi.it
[3] Università Telematica Internazionale Uninettuno, Rome, Italy
marco.demarco@uninettunouniversity.net

Abstract. This qualitative study explores the widely recognized role of the informal caregivers (ICGs) as key co-producers in the delivery of effective and sustainable healthcare systems. The central argument is that to enhance the quality of care in non-clinical settings and the healthcare ecosystem as a whole, developers of Health Information Technology (HIT) need to harness the knowledge and experiences of the ICGs to better align their products to practice. The paper has two aims: to improve the understandability of informal caregivers' role in non-traditional healthcare settings, and to identify and formulate valuable guidelines for the development of "fit-for-use" HIT solutions that acknowledge the needs of the ICGs.

Keywords: Informal caregivers · Health IT · Co-production
HIT development guiding principles

1 Introduction

Health Information Technology is the application of information technology to health and healthcare. The fact that Health Information Technology ('HIT') brings together the management and computerization of health information with a wide range of stakeholders means it has far-reaching effects on the delivery and consumption of healthcare and health-related services [1: iv]. In this paper, HIT is defined as the use of information and communication technology (ICT) by formal (i.e., professional) and informal caregivers to deliver healthcare services in non-traditional settings. Basically, caregivers would combine the use of social media, collaborative platforms, online portals, bedside terminals, assistive technologies, handhelds, electronics, and electronic medical records into "effective means of accessing, communicating and storing information to improve patient care and population health, and reduce healthcare expenditures" [2: 476, 3, 4].

The sustained care of informal caregivers (ICGs) could enable people with chronic diseases, e.g., cancer [5], AIDS [6], Alzheimer's [7], stroke [8], Severe Multiple Sclerosis [9], and certainly the aging [10], to be assisted by family members either at

© Springer Nature Switzerland AG 2018
G. Satzger et al. (Eds.): IESS 2018, LNBIP 331, pp. 316–329, 2018.
https://doi.org/10.1007/978-3-030-00713-3_24

the hospital or at home [11]. Nevertheless, to provide an effective care package, healthcare providers who enlist informal caregivers as active patient care partners [12] need to pay careful attention to the interactions between the ICGs, the healthcare professionals and the patients [13]. HIT has done much to make the informal caregivers part of healthcare service delivery ecosystems [14] but the evidence so far suggests that, overall, it is not enough to improve outcomes across the care continuum [2, 15]. Prior studies have demonstrated significant advances in knowledge but uncertain improvements in skills [16] - in particular, in the training of disease management planning. The attitude of the healthcare provider to the ICG as a user of HIT is often the result of a mutual misunderstanding of needs [17, 18]. On the other hand, the perception of the ICGs and care recipients is that they are at the mercy of the healthcare professionals.

Motivation

The difference between an informal caregiver (or family carer or caregiver, informal carer or care provider or caregiver) and a healthcare professional (e.g. clinical staff, nurse, and physician) is that the former is usually a family member, a close member of the patient's societal context, or a non-clinical social worker. ICGs operate as patient's advocates providing a source of continuity of care for their care recipient during their transitions through care settings. They also perform the functions of *system navigators* (who locate, evaluate and integrate relevant knowledge and information on behalf of the care recipient); or *gatekeepers* of support and services (assist the care recipient in the navigation of the often complex healthcare system); and *coordinators of care* (scheduling medical appointments and coordinating care services) [21].

The use of HIT by informal caregivers raises crucial issues for service providers. Central among these is the need for the latter to work effectively with the former [1, 2]: "understanding caregivers' needs, their varied experiences and the complex interactions between caregivers, healthcare professionals and patient is important" [1: 154]. Hence, to fully comprehend what is required of this type of HIT it is vital is to analyze the practical everyday needs of the ICGs and their role as boundary spanners. "Significant developments in public sector services, where open standards and architecture are facilitating disintegration of services and their recombination around what has been termed service ecosystems" have yet to consider the growing role of ICGs [19: 136]. Hence, the inspiration for our research: *What insights can we glean from the experiences of the ICGs, and how could these translate into some basic guidelines for the design of more tailored HIT applications?*

The contribution of the qualitative analysis followed in this paper is twofold. It first seeks to consolidate our current understanding of the role of HIT in healthcare ecosystems with a review of the relevant literature. It then builds on the knowledge of the authors to elaborate a set of principles that factors in the ICG interaction to identify and formulate basic guidelines meant to productively inform HIT development.

2 Approach

Health information technology (HIT) is an umbrella-term that refers to a set of computer systems, devices and interfaces used in health management and healthcare. It supports the exchange of health information between patients, carers, consumers, providers, payers, and quality monitors, in order to improve the quality of care. [17]. Vargo and Lusch [20] define HIT ecosystems as "spatial and temporal structures of largely loosely coupled, value-proposing social and economic actors interacting through institutions, technology, and language to co-produce service offerings, engage in mutual service provision, and to co-create value".

This preliminary-descriptive study explores a set of focal characteristics of a HIT ecosystem, using - as the basic units of analysis - papers and statements that show the needs of the ICGs and their interaction with formal and informal providers who co-operate and co-ordinate their activities to deliver tailor-made care.

In light of the emerging nature of these notions, a qualitative approach was taken. The research path consisted of three basic steps.

The first step was to map the relevant literature on Information Systems, Healthcare Management Informatics, and Service Management to take stock of the current state of the research and to gain insights from the evidence presented so far. We searched electronic databases, including EBSCO, Scopus and Web of Science. The primary goal was to screen for papers focused on the use of technology for caregiver interaction, scanning for the types of software, interfaces and devices that facilitate interaction practices between the caregivers and the other stakeholders of the networked service ecosystem. We conducted a search for peer-reviewed publications written in the English language ["ICT" OR "Information Technology" OR "Health Technology" OR "Health Information Technology"] AND "Caregiver Interaction". After removing duplicates, references were screened on title and abstract and then on full text. A total of 68 articles were read in full, 47 of which were selected for review due to their depth analysis of caregiver needs [5, 16, 39, 41, 44] and their relevance to the research goals.

The second step was to perform a textual analysis. Each researcher applied their personal store of knowledge and experience to identify and select the most relevant parts for the purpose of this study, then independently codified each "meaning unit" (i.e., that portion of the text associated with an identifiable theme or issue) [22]. Scientific evidence was categorized according to the Dialogue, Access, Risk, and Transparency through the lens of the ('DART') model of interaction developed by Prahalad and Ramaswamy [23] (Sect. 3) and coded into Key Concepts (Sect. 4). DART assumes a service dominant logic, and takes into account the service ecosystem. The 'blocks (or pillars) of interactions' identified by the original DART model encompass two types of actors: "the firm" and "the consumer". The fact that dialogue, access, risk, and transparency are key variables in the patient's experiential sphere thus makes the DART model [23] a good analytical fit for the interactions between the service provider and the ICG [13, 24]. The coding results and interpretations were shared and discussed, focusing on cases in which the same meaning unit could be slotted into several categories. The key pieces of evidence were organized by one of the researchers via an iterative process.

The third and final part of the study was to formulate a set of basic principles that factor in ICG interaction as a preliminary response to our research question (Sect. 4, Table 1). The studies reviewed inform that the DART approach shifts the focus from the features and technicalities to the implications of the multiple experiences in which ICGs are engaged. In other words, the adoption of a service logic [4] enables the charting of an ontological path [25] to connect each DART pillar of interaction. The insights generated were then translated into some basic principles (or guidelines) to help the HIT designers and developers better align their products to the needs of the ICGs.

3   Applying the DART Lens: Pillars of Interaction

Previous research work has recognized informal caregivers as patient advocates, gatekeepers of support and services, and coordinators of care (Table 1). Sorrentino et al. [21] qualified the service roles of informal caregivers under the lens of building blocks of interaction.

Table 1. The Dialogue, Access, Risk, and Transparency (DART) map and the intersecting roles of informal caregivers [21]

	Patient's advocate	System navigator and gatekeeper	Coordinator of care
D	Dialog and information exchange between caregivers	Understanding of needs between formal and informal caregivers	Coordination supported by relationships of shared goals/shared knowledge
A	Data access issues can impede the effectiveness of informal caregivers	Access to multiple sources enables the exchange of information on patient condition	Technological access to useful data optimizes the coordination of care
R	Problem-solving approach to reduce patient health risks	Possible induced risk to the informal caregiver	Informal caregiver preparedness
T	Care-recipient spokespersons and intermediaries with service providers	Transparency of information flow impeded by fears over security and loss of privacy	Improved disease management through patient engagement and family caregiver assistance

The authors of this paper find it interesting to apply this conceptual approach to mapping the role of the informal caregivers to contextual implications of using HIT. It then extends prior research and offers a crisper look into how HIT could enable informal providers in each of their identified roles in the health service ecosystem.

Enhanced Dialog between Formal and Informal Caregivers (D)
ICGs use HIT tools to reinforce their communication within the healthcare organization [26]. The literature supports the consistent incorporation of novel ideas for patients and ICGs to manage their own conditions and to foster communication among the circle of

care [27]. Some solutions include secure messaging for communication among the patient care team, putting the patient in control of his/her health data privacy [28]. ICGs place critical value on continuous communication with primary care providers, especially to receive recommendations, guidance and endorsement to sources of caregiving information in the early stages of the caregiving journey [12]. Whether patients or their informal carers actually make use of and benefit from available HIT applications still hinges upon the attitude of their professional care provider towards such sources of information [17]. Nevertheless, the attitude of the professional care provider towards HIT use by informal caregiver may affect the understanding of needs between formal and informal caregivers. Hence, workflow clarity is a vital component for role definition and understanding between formal and informal caregivers: the potential for tension arises as the use of technology might result in a successful outcome for one party (professional caregiver) but not necessarily for another (family carers) [29]. This is a critical requirement for the coordination of care between formal and informal caregivers, which is supported by relationships of shared goals and shared knowledge [30].

Access to Data by Informal Caregivers (A)
HIT provides access to useful data known to optimize the coordination of care. Online availability of data sources have shortened physical distance [31], consolidated health information about the patient, and provided an opportunity to maintain up-to-date information about care professionals and care-related goals [28]. Access to multiple sources enables the exchange of information on patient condition. Subsequently, the ability to gather data and analyse it into essential feedback for monitoring and alerting (e.g. breathing monitors in premature infants), diagnosis and important treatment information known to improve the interaction between caregivers ('CGs') [32]. This enhances caregiver knowledge base but not necessarily caregiver skills, which still requires training on specific disease management plans. Moreover, extant pressure of technology usage skills may introduce data entry errors due to barriers of computer illiteracy. Computer illiteracy can be a barrier to online services [33]; however, access to online assisted literacy resources require little training if they are well-designed [34]. Such information may include the understanding of the course of the disease, the importance of caring for self, finding respite and long-distance care provision [35]. ICGs expect HIT tools to be easy to use whereby navigation ought to provide access to information quickly and easily [36]. The use of memory aids, visual aids, access to training programs [37]. Examples could be used with text and displays that are easy to read [38].

Patient Risk Mitigation and Lessening the Burden on ICG (R)
Extant pressure of technology usage skills may introduce data entry errors due to barriers of computer illiteracy [43]. Risks of patient health and privacy issues often offset efforts to increase caregiver preparedness [36]. Education on patient's health risks, ought to be imbedded in portals [20, 36, 57] with guidance on treatment provides potential improvement of possible induced risk to the informal caregiver [29, 43]. Technology-based interventions can reduce the caregiving burden, depression, anxiety and stress, and improve the ICG's coping ability [41], but not without unforeseen additional pressures, such as checking equipment, interpreting movements and having

additional care responsibilities [29]. Assistive technologies have been proposed to overcome elderly problems such as fall risk, chronic disease, dementia, social isolation and poor medication management. Assistive technologies help reduce the burden among ICGs of older adults [39]; these technologies were also found to reduce anxiety in informal carers of people with dementia [40]. Components of HIT such as computerized prescription tools for instance, are clear examples of how risk reduction occurs in the case of a patients' and the informal caregiver are reminded of the correct intake of the prescribed medicine. The benefits of HIT components, such as computerized prescription tools and medication dosage calculation devices which tell the ICG the correct intake of their care recipients' prescribed medicine, include a reduction in the risk of error [43] and higher rates of prescription adherence. However, for this to be effective, an already functioning adherence program needs to be in place [44] to ensure the ICG's preparedness. Care recipients are set up with health monitoring technologies at home, growing the ICG support base of non-family members to include close family members, thus enriching the ICG's social and behavioral support [26]. While monitoring [through technologies] may provide peace of mind for carers, it may be perceived as invasion of privacy in some settings [29]. Users must be able to deactivate sensors and data transmission whenever they want [42].

Transparency of Information Flow between Formal and Informal CGs (T)
The computerization of patient health records and online access to readily available data enables the working ICGs to access information, receive psychosocial support from the professional counterpart and benefit from potential learning opportunities [45], but also brings into play the question of transparency. The main concerns of the formal CGs center on security and patient privacy and the ethics of processing and transmitting sensitive data outside the professional sphere; not only can this undermine the completeness, and hence usefulness, of the data transmitted to the ICG, but also can make the formal caregivers reluctant to share electronic data files with them [46]. Moreover, the ICG must be informed of the HIT tool's specific purpose and be allowed to gain a certain familiarity with it [36]. Adequate training and encouragement from others are essential in motivating family carers to use technology-based support services (ibidem). Those services (online support, helpdesk, etc.) can be valuable for older family carers in rural areas, for instance, where adopting new technology could help them and their care recipients regain social inclusion [47]. The language used must be familiar to the ICG, with interfaces that incorporate readily available lexicons written in simple, easy to follow terms [36]. Transparency of information flow is often impeded by fears over security and loss of privacy. A foundational concept of information sharing in healthcare is to avail patient health and biography information only to the authorized providers of care as concerns of confidentiality may be seen to outweigh benefits of quality of care [45]. Patient engagement and family caregiver access to portals may improve self-regulated disease management. The direct participation of the informal caregiver in the health IT ecosystem facilitates the continuity of care. This imposes a strong reliance on the level of use of the formal caregivers of the technology in order to record useable patient data.

4 Framing HIT Guiding Principles for CG Interaction

We applied a coding technique to the literature reviewed [48] that allowed for the emergence of open codes [49]. This generated key concepts that constitute the component of guiding principles of caregiver interaction for HIT (Table 2).

Table 2. Literature review coding results

DART	Key concepts/principles	Caregiver needs and interaction (Exemplary meaning units from the literature review)
Dialogue (D)	Reinforce dialog, communication and coordination	ICGs use HIT tools to reinforce their communication within the healthcare organization [26]. ICGs use HIT tools to foster communication among the circle of care [27]. ICGs place critical value on continuous communication with primary care providers [12], and coordination of care between formal and informal caregivers [30]
Access (A)	Provide access to data on the patient, understanding the course of disease and knowledge of treatment information	Online availability of data sources shortens the physical distance and improves CG effectiveness [31]. Access to consolidated health information about the patient, care professionals and care-related goals [28]. Knowledge of important treatment information improves interaction between caregivers [32]. Understanding of the course of the disease, the importance of caring for self, finding respite and remote care provision [35]
	Provide access to HIT tools that are easy to use	ICGs expect HIT tools to be easy to use whereby navigation ought to provide access to information quickly and easily [36]. Use of memory aids, visual aids, training programs [37], easy-to-read displays [38]
Risk-benefit (R)	Reduce the burden of ICT literacy on the caregiver to reduce chances of error	Well-designed assisted literacy resources require little training [34]. Easy access to disease information [20, 36, 57] and to online support, helpdesk, etc. [47] to lessen risk of errors [43]. Minimize pressure on ICG of checking equipment interpreting movements and having additional care

(continued)

Table 2. (*continued*)

DART	Key concepts/principles	Caregiver needs and interaction (Exemplary meaning units from the literature review)
		responsibilities [29]. Use of familiar language, and easy-to-understand interfaces [36]. Reduce risk to care recipient (e.g. prescription errors) by providing tools for prescription aid [43]
	Provide features to reduce CGs anxiety	Provide features to reduce ICG anxiety [40] and improve ICG's coping ability [39, 41], i.e. ICGs must understand the intent of the software (or device) and gain familiarity with it [36]
	Provide appropriate security and privacy	Technological data monitoring may be perceived as invasion of privacy [29]. Users must be able to deactivate their own data transmission whenever they want [42]. Overcoming significant challenges of security and privacy of sensitive data [46]
Transparency (T)	Provide features for transparent information flow between formal and informal CGs	Provide secure messaging for communication among the patient care team [28]. Enable working ICGs to access information, receive support from professional CG and network transparently [45]. Adequate training and encouragement to motivate family carers to use technology-based support services [36]. Workflow clarity is a vital component for role definition and understanding for transparency in information flow between formal and informal caregivers [29]
	Provide features for patient engagement to reduce challenges of security and privacy	Putting the patient in control of their health data privacy [28]. Patient engagement practices, enabled by secured portals, bring the control of who can have access to patient data [56]

We can therefore propose the following set of guiding principles for HIT developers that – once implemented - would potentially optimize informal caregiver interaction:

1. **Reinforce dialog, communication and coordination.** ICGs place critical value on continuous communication with primary care providers and depend on HIT tools to reinforce their communication within the healthcare organization. HIT tools should enable ICGs to receive recommendations, guidance and endorsement to sources of caregiving information in the early stages.

2. **Provide access to data on the patient, understanding the course of disease and knowledge of treatment information.** Online availability of data sources have shortened the physical distance between the source of information and the ICG, improving CG effectiveness. This enabled continued and easy access to consolidated health information about the patient. their care professionals contact information and care-related goals. Knowledge of important treatment information improves interaction between caregivers and the understanding of the course of the disease.

3. **Provide access to HIT tools that are easy to use.** ICGs expect HIT tools to be easy to use whereby navigation ought to provide access to information quickly and easily. Easy-to-use HIT tools provide fast, easy access to information, incorporating the use of memory aids, visual aids and easy-to-read displays.

4. **Reduce the burden of ICT literacy on the caregiver to reduce chances of error.** Well-designed assisted literacy resources require little training, easy access to disease information and to online support, helpdesk, etc. help lessen risk of errors. Designers of HIT ought to minimize the pressure on ICGs related to using, checking and interpreting equipment features. The use of familiar language and simple, easy-to-understand interfaces reduce risk to care recipient (e.g. lessening prescription errors by providing tools for prescription aid).

5. **Provide features to reduce CGs anxiety for caregiver effectiveness.** Provide features to reduce ICG anxiety and improve ICG's coping ability in order to improve the effectiveness of CGs, i.e. ICGs must understand the specific intent of the software (or device) and gain a certain familiarity with it in a short time.

6. **Provide appropriate security and privacy.** Technological data monitoring maybe perceived as invasion of privacy. Users must be able to deactivate their own data transmission whenever they want.

7. **Provide features for transparent information flow between formal and informal CGs.** Provide secure messaging for communication among the care team. This will enable working ICGs to access information, receive support from professional CG and network transparently. Adequate training and encouragement to motivate family carers to use technology-based support services would need to be bundled with the solution provided. In order to minimize tension among the care team, workflow clarity is a vital component for role definition and understanding for transparency in information flow between formal and informal caregivers.

8. **Provide features for patient engagement to reduce challenges of security and privacy.** Putting the patient in control of their health data privacy through engagement practices, enabled by secured portals, bring the control of who can have access to patient data.

5 Concluding Remarks

This research was driven by the belief that the ICG is an integral part of the healthcare ecosystem, i.e., that the ICGs are not 'additional' ICT consumers but "subjects bound into the systems as necessary functional components" [54: 169].

The study turns its attention to the needs of informal carers [18, 52], whose centrality emerges especially in those areas where - due to patient's health conditions - the service recipients cannot themselves act as co-producers, and supports the user-centered design principles (UCD) developed by earlier IT studies on the learnability of use of a tool, system or device [53]. Given that premise, what are the basic guidelines for the design of applications more tailored to the needs and experience of ICGs?

Our paper answers this question through the recommendation of the following basic key design principles. These principles concern:

1. Features to reinforce dialog, communication and coordination between the patient and formal, informal caregivers;
2. Access to data on the patient, for the understanding the course of disease and access to knowledge on treatment information;
3. Access to HIT tools that are easy to use;
4. Features for reducing the burden of ICT literacy on the caregiver;
5. Features to reduce CGs anxiety for caregiver effectiveness;
6. Features for appropriate security and privacy;
7. Features for transparent information flow between formal and informal CGs;
8. Features for patient engagement to reduce challenges of security and privacy.

Contribution

Our work brings a valuable contribution to the current literature that focuses on value co-creation among professionals in healthcare service [60]. While healthcare is one of the most important areas that can greatly benefit from the implementation and use of IT systems [50: 143], the extant research focuses far more on the information needs of the healthcare professionals than those of the actual patients, their informal caregivers and the wider public. The study's findings highlight the opportunities created by HIT to engage and enable ICGs. It is evident in the literature that the effective outcomes of such engagement are impossible to predict because these vary according to the different contingencies and interactions with other preexisting tools, norms and practices [18].

The use of the DART lens has provided an effective means for plotting the informal caregivers' interactions with the healthcare ecosystem. Drawing on concepts generated by the literature review has enabled the authors to propose a set of principles that categorizes the ICGs' needs, and hopefully, would be useful to guide both research and practice in the development of better-aligned HIT solutions.

In practice, "systems design is typically guided by the providers' perception of patients' information needs, rather than by actual needs assessment" [51: 476]. The development of HIT that promotes effective interaction ought to be founded on principles that enhance dialog between formal and informal caregivers, facilitate access to data by informal caregivers, provide features for risk mitigation and capitalizes on transparency of information flows between formal and informal CGs.

Effective interactions facilitate all participants in the service ecosystem to co-create inimitable values and experiences [23] where formal care providers, ICGs and patients, participate in the value realization (quality of care).

Limitations

Clearly, the preliminary principles outlined here need to be further refined and corroborated. For example, future research could explore the highly contextualized features [55] of HIT ecosystems. What works in a more or less centrally organized and fully public health system (e.g., Europe) might not work for a decentralized, market-oriented health system (e.g., United States). Economic principles and political decisions determine if a technology gets onto the market, if it is affordable and accessible also for informal caregivers.

The proposed principles are by no means exhaustive. Nevertheless, three aspects promise to be particularly instructive for the goals of the paper. First, at a macro level, the set of principles can serve to organize and promote synthesis across research findings, studies, and settings using consistent language and terminology to further stimulate conceptual development.

Second, the software developers can use the eight guiding principles to better align their solutions to the ICG practice, thus enhancing the quality of care in both non-clinical settings and the healthcare ecosystem.

Third and last, the set of principles supports the exploration of essential HIT evaluation issues to better grasp what works, where and how for the informal caregivers.

In conclusion, the paper extends prior work on user-centered design principles on usability conditions of HIT applications [58, 59], and adds to the emerging stream of research on informal caregiving to strengthen our understanding of the role of HIT as a key resource for the development of a true patient-centered health care system.

Compliance with Ethical Standards

The authors declared no potential conflict of interest with respect to the publication of this article, and did not receive any specific grant from funding agencies in the public, commercial, or not-for-profit sectors.

References

1. Payton, F.C., Pare, G., Le Rouge, C.M., Reddy, M.: Health care IT: process, people, patients and interdisciplinary considerations. J. Assoc. Inf. Syst. **12**(Special Issue), i–xiii (2011)
2. Burns, L.R., Pauly, M.V.: Accountable care organizations may have difficulty avoiding the failures of integrated delivery networks of the 1990s. Health Aff. **31**(11), 2407–2416 (2012)
3. Lehoux, P.: Patients' perspectives on high-tech home care: a qualitative inquiry into the user-friendliness of four technologies. BMC Health Serv. Res. **4**(1), 1–9 (2004)
4. Sheikh, A., Sood, H.S., Bates, D.W.: Leveraging health information technology to achieve the "triple aim" of healthcare reform. J. Am. Med. Inform. Assoc. **22**(4), 849–856 (2015)
5. O'Toole, M.S., Zachariae, R., Renna, M.E., Mennin, D.S., Applebaum, A.: Cognitive behavioral therapies for informal caregivers of patients with cancer and cancer survivors: a systematic review and meta-analysis. Psycho-oncology **26**(4), 428–437 (2017)

6. Mitchell, M.M., Robinson, A.C., Wolff, J.L., Knowlton, A.R.: Perceived mental health status of drug users with HIV: concordance between caregivers and care recipient reports and associations with caregiving burden and reciprocity. AIDS Behav. **18**(6), 1103–1113 (2014)
7. Batsch, N.L., Mittelman, M.S.: World Alzheimer Report 2012, Overcoming the stigma of dementia. Alzheimer's Disease International (2012). http://www.alz.org/documents_custom/world_report_2012_final.pdf
8. Woodford, J., Farrand, P., Watkins, E.R., Richards, D.A., Llewellyn, D.J.: Supported cognitive-behavioural self-help versus treatment-as-usual for depressed informal carers of stroke survivors. Trials **15**(1), 1–10 (2014)
9. Giordano, A., et al.: Low quality of life and psychological wellbeing contrast with moderate perceived burden in carers of people with severe multiple sclerosis. J. Neurol. Sci. **366**, 139–145 (2016)
10. Paraponaris, A., Davin, B., Verger, P.: Formal and informal care for disabled elderly living in the community. Eur. J. Health Econ. **13**(3), 327–336 (2012)
11. Milligan, C., Roberts, C., Mort, M.: Telecare and older people: who cares where? Soc. Sci. Med. **72**(3), 347–354 (2011)
12. Peterson, K., Hahn, H., Lee, A.J., Madison, C.A., Atri, A.: In the Information Age, do dementia caregivers get the information they need? BMC Geriatr. **16**(164), 179–187 (2016)
13. Sorrentino, M., De Marco, M., Rossignoli, C.: Health care co-production: co-creation of value in flexible boundary spheres. In: Borangiu, T., Drăgoicea, M., Nóvoa, H. (eds.) IESS 2016. LNBIP, vol. 247, pp. 649–659. Springer, Cham (2016). https://doi.org/10.1007/978-3-319-32689-4_49
14. Lundberg, S.: The results from a two-year case study of an information and communication technology support system for family caregivers. Disabil. Rehabilit. Assist. Technol. **9**(4), 353–358 (2014)
15. Adler-Milstein, J., Embi, P.J., Middleton, B., Sarkar, I.N., Smith, J.: Crossing the health IT chasm: considerations and policy recommendations to overcome current challenges and enable value-based care. J. Am. Med. Inform. Assoc. **24**(5), 1–8 (2017)
16. De Angelis, G., Davies, B., King, J., McEwan, J., Cavallo, S., Loew, L., Wells, G.A., Brosseau, L.: Information and communication technologies for the dissemination of clinical practice guidelines to health professionals: a systematic review. JMIR Med. Educ. **2**(2), 1–19 (2016)
17. Hillestad, R., Bigelow, J.H.: Health information technology: can HIT lower costs and improve quality. RB-9136 (2005)
18. Alzougool, B., Chang, S., Gray, K.: Modeling the information needs of informal carers. In: Proceedings of ACIS 2007, 5–7 December 2007
19. Barrett, J., Davidson, E., Prabhu, J., Vargo, S.L.: Service innovation in the digital age: key contribution and future directions. MIS Q. **39**(1), 135–154 (2015)
20. Vargo, S.L., Lusch, R.F.: It's all B2B... and beyond. Ind. Mark. Manag. **40**(2), 181–187 (2011)
21. Sorrentino, M., Badr, N.G., De Marco, M.: Healthcare and the co-creation of value: qualifying the service roles of informal caregivers. In: Za, S., Drăgoicea, M., Cavallari, M. (eds.) IESS 2017. LNBIP, vol. 279, pp. 76–86. Springer, Cham (2017). https://doi.org/10.1007/978-3-319-56925-3_7
22. Lee, T.W.: Using Qualitative Methods in Organizational Research. Sage, Beverly Hills (1999)
23. Prahalad, C.K., Ramaswamy, V.: Co-creation experiences: the next practice in value creation. J. Interact. Mark. **18**(3), 5–14 (2004)

24. Honka, A., Kaipainen, K., Hietala, H., Saranummi, N.: Rethinking health: ICT-enabled services to empower people to manage their health. IEEE Rev. Biomed. Eng. **4**, 119–139 (2011)

25. Sowa, J.F.: Top-level ontological categories. Int. J. Hum. Comput. Stud. **43**(5–6), 669–685 (1995)

26. Palm, E.: Who cares? Moral obligations in formal and informal care provision in the light of ICT-based home care. Health Care Anal. **21**(2), 171–188 (2013)

27. Matthew-Maich, N., et al.: Designing, implementing, and evaluating mobile health technologies for managing chronic conditions in older adults. JMIR mHealth uHealth **4** (2), 1–18 (2016)

28. Robben, S.H., et al.: Implementation of an innovative web-based conference table for community-dwelling frail older people, their informal caregivers and professionals. BMC Health Serv. Res. **12**(1), 251–263 (2012)

29. Eccles, A.: The complexities of technology-based care. Telecare as perceived by care practitioners. Issues Soc. Sci. **1**(1), 1–20 (2013)

30. Weinberg, D.B., Lusenhop, W., Gittell, J.H., Kautz, C.M.: Coordination between formal providers and informal caregivers. Health Care Manag. Rev. **32**(2), 140–149 (2007)

31. Greenhalgh, T., Stones, R.: Theorising big IT programmes in healthcare: strong structuration theory meets actor-network theory. Soc. Sci. Med. **70**, 1285–1294 (2010)

32. Gentles, S.J., Lokker, C., McKibbon, K.A.: Health information technology to facilitate communication involving health care providers, caregivers, and pediatric patients: a scoping review. J. Med. Internet Res. **12**(2), 1–17 (2010)

33. Torp, S., Hanson, E., Hauge, S., Ulstein, I., Magnusson, L.: A pilot study of how information and communication technology may contribute to health promotion among elderly spousal carers in Norway. Health Soc. Care Community **16**(1), 75–85 (2008)

34. Garcia, C.H., Espinoza, S.E., Lichtenstein, M., Hazuda, H.P.: Health literacy associations between Hispanic elderly patients and their caregivers. J. Health Commun. **18**(suppl), 256–272 (2013)

35. Lum, J.: Informal caregiving. In: Focus Backgrounder, pp. 1–11 (2011)

36. Duarte, J., Guerra, A.: User-centered healthcare design. Proc. Comput. Sci. **14**, 189–197 (2012)

37. Egan, M., Bérubé, D., Racine, G., Leonard, C., Rochon, E.: Methods to enhance verbal communication between individuals with Alzheimer's disease and their formal and informal caregivers. Int. J. Alzheimers Dis. (2010)

38. Cristancho-Lacroix, V., et al.: A web-based program for informal caregivers of persons with Alzheimer's disease: an iterative user-centered design. JMIR Res. Protoc. **3**(3), e46 (2014)

39. Madara Marasinghe, K.: Assistive technologies in reducing caregiver burden among informal caregivers of older adults: a systematic review. Disabil. Rehabilit. Assist. Technol. **11**(5), 353–360 (2016)

40. Gilliard, J., Hagen, I.: Enabling Technologies for People with Dementia: Cross-National Analysis Report' (Enable, 2004)

41. Lopez-Hartmann, M., Wens, J., Verhoeven, V., Remmen, R.: The effect of caregiver support interventions for informal caregivers of community-dwelling frail elderly: a systematic review. Int. J. Integr. Care **12**(5), 133 (2012)

42. Stowe, S., Harding, S.: Telecare, telehealth and telemedicine. Eur. Geriatr. Med. **1**(3), 193–197 (2010)

43. Lu, Y.-C., Xiao, Y., Sears, A., Jacko, J.A.: A review and a framework of handheld computer adoption in healthcare. Int. J. Med. Inform. **74**(5), 409–422 (2005)

44. Hamine, S., Gerth-Guyette, E., Faulx, D., Green, B.B., Ginsburg, A.S.: Impact of mHealth chronic disease management on treatment adherence and patient outcomes: a systematic review. J. Med. Internet Res. **17**(2), 1–15 (2015)
45. Andersson, S., Erlingsson, C., Magnusson, L., Hanson, E.: The experiences of working carers of older people regarding access to a web-based family care support network offered by a municipality. Scand. J. Caring Sci. **31**, 487–496 (2016)
46. Van Durme, T., et al.: Stakeholders' perception on the organization of chronic care: a SWOT analysis to draft avenues for health care reforms. BMC Health Serv. Res. **14**(1), 179 (2014)
47. Blusi, M., Asplund, K., Jong, M.: Older family carers in rural areas: experiences from using caregiver support services based on information and communication technology (ICT). Eur. J. Ageing **10**(3), 191–199 (2013)
48. Kaplan, B., Maxwell, J.A.: Qualitative research methods for evaluating computer information systems. In: Anderson, J.G., Aydin, C.E. (eds.) Evaluating the Organizational Impact of Healthcare Information Systems. HI, pp. 30–55. Springer, Berlin (2005). https://doi.org/10.1007/0-387-30329-4_2
49. Bowen, G.A.: Naturalistic inquiry and the saturation concept: a research note. Qual. Res. **8**(1), 137–152 (2008)
50. Venkatraman, S., Bala, H., Venkatesh, V., Bates, J.: Six strategies for electronic medical records systems. Commun. ACM **51**(11), 140–144 (2008)
51. Keselman, A., Logan, R., Smith, C.A., Leroy, G., Zeng-Treitler, Q.: Developing informatics tools and strategies for consumer-centered health communication. J. Am. Med. Inform. Assoc. **15**(4), 473–483 (2008)
52. Nguyen, L., Evans, S., Wilde, W., Shanks, G.: Information needs in community aged care. In: Proceedings of PACIS (2011)
53. Grudin, J.: Utility and usability: research issues and development contexts. Interact. Comput. **4**(2), 209–217 (1992)
54. Ekbia, H., Nardi, B.A.: Inverse instrumentality: how technologies objectify patients and players. In: Leonardi, P.M., Nardi, B.A., Kallinikos, J. (eds.) Materiality and Organizing, pp. 157–176. Oxford Univiersity Press, Oxford (2012)
55. Mettler, T.: Contextualizing a professional social network for health care: experiences from an action design research study. Inf. Syst. J. **28**(4), 684–707 (2018)
56. Walker, J., Darer, J.D., Elmore, J.G., Delbanco, T.: The road toward fully transparent medical records. N. Engl. J. Med. **370**(1), 6–8 (2014)
57. Lum, J., Hawkins, L., Ying, A.: Informal Caregiving and LGBT Communities. Canadian Research Network for Care in the Community, Ontario (2011)
58. Wright, P., McCarthy, J.: Experience-centered design: designers, users, and communities in dialogue. Synth. Lect. Hum. Cent. Inform. **3**(1), 1–123 (2010)
59. Nielsen, J.: Enhancing the explanatory power of usability heuristics. In: Proceedings of the SIGCHI Conference on Human Factors in Computing Systems. ACM (1994)
60. Zhang, L., Tong, H., Demirel, H.O., Duffy, V.G., Yih, Y., Bidassie, B.: A practical model of value co-creation in healthcare service. Proc. Manuf. **3**, 200–207 (2015)

Service Science Research and Service Standards Development

Reinhard Weissinger[1] and Stephen K. Kwan[2(✉)]

[1] Senior Expert, Research and Education,
International Organization for Standardization (ISO), Geneva, Switzerland
weissinger@iso.org
[2] Lucas College and Graduate School of Business, San José State University,
San José, CA, USA
stephen.kwan@sjsu.edu

Abstract. Recent increases in interest in the development of service standards among standards organizations follow the trend of growth in the service sector. This research in progress reviews the relationship between service science research and service standardization to determine whether there are areas of convergence and mutual influence and opportunities to increase exchanges between these two sides for mutual benefit. Service standards published by ISO and current service standards projects were categorized into Types (1) back stage, and (2) front stage of service activities. The definitions of "service" were also extracted from ISO standards to determine their commonality with service science concepts. It was found that ISO service standards were mostly related to back stage of service activities but some increase in projects with front stage orientation was seen. There was scant evidence that the definition of service used in standards had some commonality with service science concepts. Limitations to the research together with recommendations for further work that would foster mutual benefits for both service science research and standards development were discussed.

Keywords: Service science · Service science research · Service standards
Service standards development

1 Introduction

Many professional standards organizations, such as the American National Standards Institute (ANSI), a national standards organization, the European Committee for Standardization (CEN), a regional standards developer, and the International Organization for Standardization (ISO), an international standards organization, have, over the past few years, shown increased interest in developing service standards for different areas [1, 4, 5, 10]. In a recent survey by a working group of ISO's Committee on

---

The opinions expressed in this article are the authors' own and do not necessarily reflect the views of their affiliated organizations.

© Springer Nature Switzerland AG 2018
G. Satzger et al. (Eds.): IESS 2018, LNBIP 331, pp. 330–343, 2018.
https://doi.org/10.1007/978-3-030-00713-3_25

Consumer Policy, it was found that over half of the responding country members' national standards body have a services strategy [11]. This is not surprising since the world economy has been experiencing a dramatic shift to the service sector. According to the World Bank, about 70% of the world's combined GDP can be attributed to the service sector [26]. In the US, the service sector accounts for almost 80% of its GDP [op. cit.]. In many countries it has been recognized that the service sector has the potential to become the most important engine of growth, which resulted in significant investments in domestic services and service trade. Increasingly complex, online, cross-border, and connected services are being offered in competitive markets where consistency and dependency in service delivery and quality are being demanded by enterprises, public, and private customers. To meet this demand, many service standards were developed at industry, national, regional, and international levels and adopted by vendors. Often the demonstration of conformity with these standards through certification became a minimum requirement for market entry and to sustain competitiveness.

This article presents our research in progress where we review the relationship between service science research, a transdisciplinary academic approach to the study of macro and micro levels of service and service standardization in the industrial sector to determine whether there are areas of commonality, influence and opportunities to increase exchanges between these two sides for mutual benefit.

1.1   Questions Addressed in This Article

In this article, we address the following questions:

Question 1: Are there functions and aspects that dominate in the collection of existing service standards? If this is the case, how do they compare with core tenets of service science? Is there evidence of an influence, if not some commonality, of service science and service standardization?

Question 2: If we analyze core concepts used in service standards and compare these with fundamental concepts of service science, can we recognize a commonality in the concepts applied between these two sides? If this is the case, is there evidence for an uptick of influence of service science concepts in service standardization?

2   Service Science and Service-Dominant Logic

Service science is a transdisciplinary approach to understanding service systems with fundamental concepts including entities, interactions, outcomes, value propositions, governance mechanisms, resources, access rights, stakeholder roles, measures and ecology [13] as shown in Fig. 1. These entities interact to co-create value for the stakeholders. The views of service from different disciplines of marketing, economics, operations, industrial and systems engineering, operation research, computer science, information systems, social sciences and behavioral sciences are discussed in [19]. An annotated bibliography of service science related literature organized into areas of evolution, measures, resources and strategy can be found in [17].

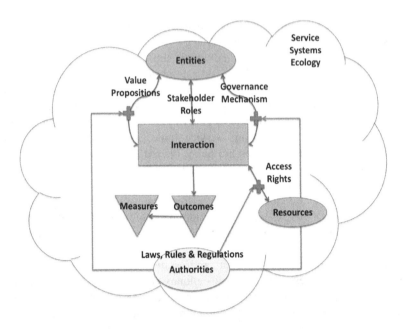

Fig. 1. Service science fundamental concepts (from [13])

Service-Dominant Logic (SD-Logic) is another approach to service research that offers a set of Fundamental Premises embodying both macro and micro views of service [15, 22]. The service science and Service-Dominant Logic approaches to service research are essentially intertwined and foundational [24]. Service science takes a broad perspective of service systems [16] with an entity/relationship view while Service-Dominant Logic takes a value/resources view of service. [25] provides some clarifications between the Fundamental Premises of SD-Logic and service science concepts. The most important aspect of both approaches is that stakeholder value, especially for the customer, is the economic driver of service provision and consumption. What is of value to a customer and service provider (and other stakeholders) is determined by shared factors communicated through value propositions [14].

3 Standards and Standardization

The International Organization for Standardization (ISO)[1] defines a standard as a "document, established by consensus and approved by a recognized body, that provides, for common and repeated use, rules, guidelines or characteristics for activities or their results, aimed at the achievement of the optimum degree of order in a given

---

[1] ISO is an independent non-government international organization that publishes and promotes international standards in almost every industry to facilitate international trade. Its members are over 160 national standards bodies and its over twenty thousand published standards are the sources of specifications for products, services, and systems for quality, safety, and efficiency around the world.

context" ([8] cl. 3.2). Typically, the use of standards is voluntary, although they can be required by market forces or made mandatory through regulations.

3.1 Standards and Innovation

Standards can be used as instruments for innovation, supporting technological change, process improvement and by assisting the transfer and dissemination of knowledge [7]. Blind shows that research results that are contained in patents and scientific publications are often transformed into standards and – through standards – become accessible to large numbers of societal players and trigger innovation ([2], p. 12).

3.2 Economic Importance of Standards

A recent publication about the macroeconomic impacts of standards and standardization summarizes the results from studies that have been undertaken over the last two decades as shown in Table 1 [3]. According to these studies, the contribution of standards to annual GDP-growth over the indicated periods of analysis range between 0.3% (Canada, 1981–2004 and the UK, 1948–2002) and 0.9% (Germany, 1961–1990).

Table 1. Contributions of standards to annual GDP-growth. Source: ([3], p. 39)

Country	France	Canada	Germany	Germany	UK	UK
Organization and Publication year	AFNOR (2009)	Standards Council of Canada (2007)	DIN (2000)	DIN (2011)	DTI (2005)	Cebr (2015)
Period of analysis	1950–2007	1981–2004	1961–1990	2002–2006	1948–2002	1921–2013
Estimated function	GDP-output	Labor productivity	GDP-output	GDP-output	Labor productivity	Labor productivity
Elasticity of stock of standards	0.12	0.36	0.07	0.18	0.05	0.11
Share of labor productivity growth, %	27.1	17	30.1	–	13	37.4
Growth rate of GDP, % p.a.	3.4	2.7	3.3	–	2.5	2.4
Share of GDP growth, %	23.5	9.2	27.4	–	11.0	28.4
Contribution of standards to GDP growth, % points	0.8	0.3	0.9	0.7	0.3	0.7

3.3   Standards for Services

With the growing importance of services around the world, government and industries have started to look towards services as a new area for growth and noted the potential for standards to contribute to efficient and consensus-based solutions. Standards are essential for easing interoperability among stakeholders and reduce barriers in allowing organizations including Small and Medium-sized Enterprises (SME's) to participate in commerce and trade. Standards also facilitate consumers in comparing and choosing among service value propositions that have incorporated standards.

Various standards organizations (e.g. ISO, CEN, ANSI and DIN, the German Institute for Standardization, to name just a few) have developed strategies, roadmaps or perspectives for service standardization [1, 4, 5, 10]. The European Commission has issued a mandate for the development of European service standards to support a European single services market through standards [6].

4   An Analysis of a Sample of ISO Service Standards

There are many definitions of the term "service" in standards from various standards development organizations. We based our analysis on a sample of standards published by ISO, the largest standards development organization in the world (see Footnote 1). The sample includes all standards and projects that were classified based on the International Classification for Standards (ICS) [9] into one of the following groups of service standards (in bold) and sub-groups as shown in Table 2.

Table 2. ISO service standards based on the international classification for standards. Source: www.iso.org retrieved on 2018/03/04

03.060 – Finances. Banking. Monetary systems. Insurance	03.200 – Leisure. Tourism
03.080 – Services	03.200.01 - Leisure and tourism in general
03.080.01 - Services in general	03.200.10 - Adventure tourism
03.080.10 - Maintenance services. Facilities management	03.200.99 - Other standards relating to leisure and tourism
03.080.20 - Services for companies	**03.220 – Transport**
03.080.30 - Services for consumers	03.220.01 - Transport in general
03.080.99 - Other services	03.220.20 - Road transport
03.180 – Education	03.220.40 - Transport by water
	03.240 – Postal services

The selection resulted in 361 published standards and 91 standardization projects (i.e. standards currently under development) classified into ICS-subgroups. These numbers amount to 1.6% of all currently published ISO standards and 1.9% of standards projects. The other standards and projects are classified into groups and subgroups that are typically more technical in nature.

4.1   Analysis of ISO Service Standards

We further classify each service standard into two stages based on its key functions as shown in Table 3. The differentiation between front and back stage is suggested by Teboul [21] who defined front stage as where service is performed as "... direct interaction with employees, equipment, décor and other customers" [op. cit. Fig. 1.5] and back stage as where "... operations to prepare products and components and process information" take place [op. cit. Fig. 1.7]. He also indicated that the relative size of front versus back stage depends on the type of service being performed.

Table 3. Service stages and associated standards

Service stage 1	Service stage 2
Service enablers: back stage	Service delivery/experience: front stage
Type 1 standards	Type 2 standards

The concept of service front and back stage is illustrated in Fig. 2. The Figure also shows the interaction between the service provider and the customer as the touchpoint where the customer experiences the service being delivered [12].

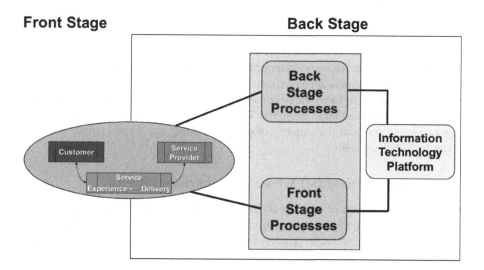

Fig. 2. Illustration of service front stage and back stage

Stage 1: Standards defining service enablers at the back stage - Service delivery relies typically on enablers, such as access to the Internet and the support of an IT platform. Enablers can also be rules and methods that are applied in the processes for the provision of services. This stage is often referred to as the back stage of service provision, as it is typically not visible to the customer (cf. below the line of visibility in a service blue print). Some of these back stage processes include reservation, payment settlement, marketing, CRM, scheduling, service assessment, etc. [op. cit.]

Stage 2: Standards addressing service delivery and experience at front stage - This stage includes the value proposition offered by the service provider to prospective customers, the decision by customers whether they wish to engage with the service provider, and the delivery of the service act. This is typically referred to as the front stage of service during which value is co-created between service providers and customers when the service act is experienced and consumed. There could be more than one instance of the exchange (involving the same or different service acts) during a customer's service journey with the provider. It is possible that parts or even the whole service act is automated so that the customer interacts with a machine and not a human being. Some of these front stage processes include service-scape maintenance, user interfaces, payment collection, feedback mechanisms (including after-service activities), etc. [op. cit.]

In line with the numbering of the service stages as 1 and 2, we refer to the standards that fall into these stages as Type 1 and Type 2 standards. This service model allows us to classify the focal areas of our sample of ISO service standards. In particular, we are interested to know how many of these standards fall into the service stage 2 and constitute Type 2 standards. Our classification is based on a review of the titles of the standards as well as the contents and scope of the standards. The results of the analysis (standards and projects) are shown in Tables 4 and 5.

Most of ISO service standards in our sample fall into stage 1 (around 74%). Only about 20% of the standards address service delivery (stage 2). In some exceptional cases, certain standards contain service elements without being mainly service standards. This is the case for some management system standards, which provide blueprints for an organizational management infrastructure, but also contain a number of service elements (e.g., a focus on customer orientation and customer satisfaction). For this reason, we identified these as Mixed Type 1 + 2, which constitutes a combination of the two types, but with the majority of elements falling into Type 1: these standards amount to 5.26% overall.

4.2   Analysis of Ongoing ISO Service Standardization Projects

The result of the analysis of ongoing standardization projects is shown in Table 6. The result shows that the percentage of projects categorized as Type 2 standards is higher among ongoing standardization projects than in the collection of published standards.

4.3   Results of the Analysis

In response to Question 1 from Sect. 1.1, we can conclude that most of the service standards consist of Type 1 standards and projects. Standards that fall into service stage 2, make up only around 20% of the sample. If we consider projects, this portion increases to around 26%. Whether this increase indicates a trend towards the development of more service standards of Type 2 is not clear and will have to await further analysis of the standards and projects based on time-ordered data points.

4.4 Comparing Definitions in ISO Standards with Fundamental Concepts of Service Science

In this section we try to respond to Question 2 by determining whether there is commonality between definitions of service in service science and ISO standards. For this purpose, we analyze a sample of definitions of the concept "service" in ISO standards and compare them with the definition of service and fundamental concepts of service science.

Table 4. Classification of service standards by types

ICS groups and sub-groups	No. of Standards	Type 1	Type 2	Mixed Type 1+2
03.060 – Finances. Banking. Monetary systems. Insurance	37	33	4	0
03.080 – Services				
03.080.01 - Services in general	3	0	3	0
03.080.10 - Maintenance services. Facilities management	3	2	0	1
03.080.20 - Services for companies	2	0	0	2
03.080.30 - Services for consumers	40	2	38	0
03.080.99 - Other services	14	2	2	10
03.180 – Education	5	0	3	2
03.200 – Leisure. Tourism				
03.200.01 - Leisure and tourism in general	8	0	8	0
03.200.10 - Adventure tourism	3	0	3	
03.200.99 - Other standards relating to leisure and tourism	14	0	14	
03.220 – Transport				
03.220.01 - Transport in general	135	132	0	3
03.220.20 - Road transport	97	96	0	1
03.220.40 - Transport by water	0	0	0	0
03.240 – Postal services	0	0	0	0
Total:	361	267	75	19
Percentage	100.00%	73.96%	20.78%	5.26%

The definition of service adopted in this study is: "Service is the application of resources (including competences, skills, and knowledge) to make changes that have value for another (system)" [20]. This aligns with SD-Logic where service is defined as the application of competences (knowledge and skills) by actors for the benefit of another party [23]. In both cases the emphasis is on value and the concept of value co-creation had become the requisite in defining services. The fundamental concepts of service science (including entities, interaction, resources, measures, outcomes, and authorities, see Fig. 1 [13, 18]) will also be used in the comparison.

The expected result from mapping ISO service concepts to the definition and fundamental concepts of service science is to gain an insight to which extent key ideas of service science have found their way into core concepts of ISO service standards. As our analysis is limited to the mapping of characteristics of the single concept "service", our results can only allow us to arrive at an initial and preliminary judgement as to whether there is some degree of commonality between concepts of service science and those applied in service standardization.

Table 5. Classification of ongoing service standards projects by types

ICS groups and sub-groups	No. of Projects	Type 1	Type 2	Mixed Type 1+2
03.060 – Finances. Banking. Monetary systems. Insurance	4	3	1	0
03.080 – Services				
03.080.01 - Services in general	1	0	1	0
03.080.10 - Maintenance services. Facilities management	1	0	1	0
03.080.20 - Services for companies	1	0	0	1
03.080.30 - Services for consumers	14	1	13	0
03.080.99 - Other services	6	2	1	3
03.180 – Education	2	0	1	1
03.200 – Leisure. Tourism				
03.200.01 - Leisure and tourism in general	2	0	2	0
03.200.10 - Adventure tourism	1	0	1	0
03.200.99 - Other standards relating to leisure and tourism	3	0	3	0
03.220 – Transport				
03.220.01 - Transport in general	28	26	0	2
03.220.20 - Road transport	28	28	0	0
03.220.40 - Transport by water	0	0	0	0
03.240 – Postal services	0	0	0	0
Total:	91	60	24	7
Percentage	100.00%	65.93%	26.37%	7.69%

4.5   Analysis Based on a Sample of ISO Definitions of "Service"

We extracted definitions of "service" from the complete collection of ISO standards by using ISO's public access Online Browsing Platform (www.iso.org/obp). This resulted in 68 different definitions. We limited our analysis to 15 definitions that either have high frequency in ISO standards or are in particular relevant to service stage 2. The results of our analysis are shown in Table 6 (see the "Appendix" for the explanation of sequence numbers that are used in the first column of this table).

To illustrate, in Table 6: ISO definition #6 for service is: "result of activities between a supplier and a customer, and the internal activities carried out by the supplier to meet the requirements of the customer". If we map the various concepts contained in this definition to the fundamental concepts of service science, we have (the fundamental concepts are expressed in square brackets and italics): Service is the "result [Outcome] of activities between [Interaction] a supplier [Provider] and a customer [Customer], and the internal activities carried out by the supplier [Provider] to meet the requirements of the customer [as met by the Value Proposition offered by the Provider]". Definition #6 maps into 5 of the 10 fundamental concepts (identified with an "X" in Table 6), so the degree of commonality is 50%.

Table 6. Mapping ISO service definitions to fundamental concepts of service science

SeqNo of definitions*	Service system ecology: Entities / Resources / Stakeholders				Interaction	Outcome	Value proposition	Access right	Measure	Value	Degree of coverage
	Provider (1:n)	Customer (1:n)	Information	Technology	(1:n)	(1:n)	(1:n)	(1:n)	(1:n)	co-creation	
1	X			X		X					3 (30%)
2				X		X					2 (20%)
3	X			X		X					3 (30%)
4	X	X			X	X					4 (40%)
5	(X)**					X					2 (20%)
6	X	X			X	X	X				5 (50%)
7		X			X	X	X				4 (40%)
8	X	X			X	X					4 (40%)
9	X	X			X	X					4 (40%)
10	X				X	X					3 (30%)
11		X			X	X	X				4 (40%)
12		X	X		X	X	X				5 (50%)
13		X	X		X	X	X	X			6 (60%)
14		X	X		X	X	X	X			6 (60%)
15	X	X			X	X					4 (40%)

* please refer to the full list of definitions in the Appendix of this paper with the sequence number

** provider is logically implied in this definition without being mentioned explicitly

4.6   Results of the Analysis

From the mapping in Table 6 we can see that the degree of commonality between the selected ISO service definitions in our sample and the fundamental concepts of service science ranges between 20 and 60%. However, only 4 out of 15 definitions (26%) show a substantial commonality of 50 to 60%.

This result leads us to conclude that, while there is certainly commonality between service science concepts and those applied in service standardization, such similarities are limited and not systematic. This means there is not yet direct or in any form systematic application of fundamental concepts of service science to service standardization.

5   Findings and Recommendations

From the analysis of ISO standards, we found that the vast majority of service standards fall into the category of service enablers (Type 1 standards). We further determined at this time, there is only scant evidence that service science research had an influence on the development of service standards in ISO. This is not surprising since standards have historically been developed for the technical specification and interoperability of products and this orientation favors the tendency to develop more standards in the familiar territory of the enabling side of services.

The analysis of the concept "service" in ISO standards has also shown that one of the fundamental concepts in service science of "service as value co-creation" has not entered the world of service standardization in ISO. In this world, service is often understood as the unidirectional delivery of something intangible to a customer usually as part of a product. This process is very similar to the delivery of a physical good and, in the lexicon of Service-Dominant Logic, reflects the prevalence of goods-dominant logic in the concepts applied in service standardization (see e.g., the definitions of service #4, #5, #7 and #9 in the Appendix).

Induce real-world
perspectives of
service industry

Service Standards   Service
& Standardization   Science

Induce more customer-
driven orientation and
alignment with customer
value expectations
and needs

Fig. 3. Exchanges and closer cooperation

We believe that more exchanges and closer cooperation between service science researchers (academic) and service standardization professionals (industry) will prove to be beneficial for both sides (see Fig. 3). For standardization, perhaps the most important benefit from a closer cooperation with service science could be the availability of a relatively consistent common theoretical framework that could prove particularly relevant to deal with the high diversity of different types of services. Such a framework could provide a bridge and common language between different sectors and areas of service standardization and could lead to more coordinated approaches in standards development. Consequently, a higher degree of consistency and, possibly, modularity, among different service standards and between service standards and other standards could be achieved which would facilitate the uptake and ease of implementation of these standards by end users. Furthermore, the adoption of service science concepts and models could provide a framework for more customer-driven orientation and alignment with customer value expectations and needs fulfillment. This would potentially lead to a gradual, timely and needed move from goods- dominant logic to Service-Dominant Logic in service standards development.

For service science, closer collaboration with standardization would provide increased scrutiny and a "real world test" into the applicability of some of its concepts through their use in standards for the service industry. Furthermore, standards as an instrument of knowledge dissemination and innovation, could contribute to increase the spread and influence of ideas, concepts and methods that have originated in service science.

We have indicated that some of the limitations of our research could be alleviated by more time-ordered analysis of service standards and projects as well as identifying concepts in service standards beyond the single concept of "service". Future research should extend the analysis beyond our sample of ISO standards and include other standard systems (e.g., ANSI) to determine whether they show similar characteristics. It should be investigated whether the shift towards an increased development of core service standards (Type 2 standards) can be confirmed both for ISO standards and other standards systems as well.

Appendix: List of Definitions of "Service" in ISO Standards

Source: ISO Online Browsing Platform (OBP), www.iso.org/obp. Data extracted on 2018/03/03

Sequence no.	Definitions of "Service" in ISO standards	# of ISO standards with this definition
1	Distinct part of the functionality that is provided by an entity through interfaces	14
1	Result of a process	6
3	Number of processes involving an organization in the provision of specific objectives	4
4	Output of an organization with at least one activity necessarily performed between the organization and the customer	4
5	Performance of activities, work or duties	4
6	Result of activities between a supplier and a customer, and the internal activities carried out by the supplier to meet the requirements of the customer	4
7	Means of delivering value for the customer by facilitating results the customer wants to achieve	3

(continued)

(continued)

Sequence no.	Definitions of "Service" in ISO standards	# of ISO standards with this definition
8	Operation performed on an entity by a user on behalf of other users	3
9	Result of at least one activity necessarily performed at the interface between the supplier and the customer, which is generally intangible	3
10	Action of an organization id meet a demand or need	
11	Independent, value-adding operation, which brings values to users, or applications providing benefits responding to user's needs	1
12	Means of delivering value for the user by facilitating results the user wants to achieve	1
13	Means of delivering value to users by facilitating results users want to achieve without the ownership of specific physical or logical resources and the risks related to ownership	
14	Means of delivering value to users by facilitating results users want to achieve without the ownership of specific resources and risks	1
15	Output of a service provider with at least one activity necessarily performed between the service provider and the customer	1

References

1. ANSI: American National Standards Institute - Focus on services standards. http://www.ansi.org/standards_activities/Focus-on-Services-Standards?menuid=3. Accessed 22 Mar 2018
2. Blind, K.: The impact of standards and standardization on innovation. Nesta Working Paper 13/15. Manchester Institute for Innovation Research (2013). www.nesta.org.uk/wp13-15
3. CEBR: Centre for Economics and Business Research Ltd. The economic contribution of standards to the UK economy. British Standards Institution, London (2015)
4. CEN: Strategic Plan on Services Standardization to implement the Ambitions 2020. CEN, Brussels (2016)
5. DIN: The German Standardization Roadmap – Services. DIN, Berlin (2015)
6. European Commission: European Commission Mandate M/517. Mandate addressed to CEN, CENELEC and ETSI for the programming and development of horizontal service standards. European Commission, Brussels (2013)
7. Gerundino, D.: Foreword to Standardization and Innovation. In: ISO-CERN Conference Proceedings, 13–14 November 2014. ISO, Geneva (2014)
8. ISO/IEC: ISO/IEC Guide 2:2004, Standardization and related activities – General vocabulary. ISO and IEC, Geneva (2014)
9. ISO: ISO International Classification of Standards. ISO, Geneva (2015)

10. ISO: ISO Strategy for Service Standardization. ISO, Geneva (2016)
11. ISO: Confirmed Minutes of the COPOLCO Working Group 18 Meeting, Bali, Indonesia, 7 May (2018)
12. Kwan, S.K., Min, J.H.: An evolutionary framework of service systems. J. Harbin Inst. Technol. **15**(Sup. 1), 1–6 (2008)
13. Kwan, S.K., Spohrer, J.C.: Fundamental concepts and premises of service science. In: Presented at the First International conference on Servicology (ICServ 2013), 16–18 October, Tokyo, Japan (2013)
14. Kwan, S.K., Hottum, P.: Maintaining consistent customer experience in service system networks. Serv. Sci. **6**(2), 136–147 (2014)
15. Lusch, R.F., Vargo, S.L.: Service-Dominant Logic—Premises, Perspectives, Possibilities. Cambridge University Press, Cambridge (2014)
16. Maglio, P.P., Vargo, S.L., Caswell, N., Spohrer, J.C.: The service system is the basic abstraction of service science. Inf. Syst. e-Business Manag. **7**(4), 395–406 (2009)
17. Spohrer, J.C., Kwan, S.K.: Service science, management, engineering, and design (SSMED): an emerging discipline - outline and references. Int. J. Inf. Syst. Serv. Sector **1**(3), 1–31 (2009)
18. Spohrer, J.C., Maglio, P.P.: Service science: toward a smarter planet. In: Salvendy, G., Karwowski, W. (eds.) Introduction to Service Engineering. Wiley, New York (2009)
19. Spohrer, J.C., Maglio, P.P.: Toward a Science of Service Systems: Value and Symbols. In: [17], pp. 157–194 (2010). https://doi.org/10.1007/978-1-4419-1628-0_9
20. Spohrer, J.C., Vargo, S.L., Caswell, N., Maglio, P.P.: The service system is the basic abstraction of service system. In: Proceedings of the 41st Hawaii International Conference on System Sciences (2008)
21. Teboul, J.: Service is front stage. We are all in services … more or less! INSEAD, Fontainebleau (2005)
22. Vargo, S.L., Lusch, R.F.: Evolving to a new dominant logic for marketing. J. Mark. **68**(1), 1–17 (2004)
23. Vargo, S.L., Lusch, R.F.: Service-dominant logic: what it is, what it is not, what it might be. In: Vargo, S.L., Lusch, R.F. (eds.) The Service-Dominant Logic of Marketing: Dialog, Debate, and Directions, pp. 43–56. ME Sharpe, New York (2006)
24. Vargo, S.L., Akaka, M.A.: Service-dominant logic as a foundation for service science: clarifications. Serv. Sci. **1**(1), 32–41 (2009)
25. Vargo, S.L., Lusch, R.F., Akaka, M.A.: Advancing service science with service-dominant logic: clarifications and conceptual development. In: [17], pp. 133–156 (2010)
26. World Bank Homepage. http://data.worldbank.org. Assessed 28 July 2018

From Data to Service Intelligence: Exploring Public Safety as a Service

Monica Drăgoicea$^{1(\boxtimes)}$ ⓘ, Nabil Georges Badr^2 ⓘ, João Falcão e Cunha^3 ⓘ,
and Virginia Ecaterina Oltean^1 ⓘ

^{1} Faculty of Automatic Control and Computers, University Politehnica of Bucharest,
Splaiul Independenței 313, 060042 Bucharest, Romania
{monica.dragoicea,ecaterina.oltean}@upb.ro
^{2} Superior Institute for Public Health, USJ Lebanon, Beirut, Lebanon
nabil@itvaluepartner.com
^{3} Faculty of Engineering, University of Porto,
Rua Dr. Roberto Frias, 4200-465 Porto, Portugal
jfcunha@fe.up.pt

Abstract. This paper describes an exploration process aligned with the core domain of Service Science inside a critical sector of Society, aiming at developing City in a sustainable, responsible, inclusive way. The paper focuses on defining the *Public Safety as a Service* concept in an inclusive and responsible value co-creation urban design vision for liveable cities. It explains how service intelligence can act on immaterial artefacts to transform data into information to generate value co-creation processes whose outcomes are applied to the evolution of knowledge in public safety services. Public safety is approached within a service ecosystem perspective, following the global targets of the Sendai Framework for Disaster Risk Reduction as an application perspective. Managerial implication are approached from two perspectives: establishment of governance principles with the help of Elinor Ostrom's works, and a Viable Systems Approach on the response to disasters operating rules.

Keywords: Service ecosystems · Service intelligence · Liveable cities
Public safety · Sustainable institutions · Viable Systems Approach

1 Introduction

As we are advancing at fast speed towards an urban-dominant form of habitation [1], several design perspectives for future cities have been imagined, advanced, and partially implemented over the last decades. Highly supported by technology-based companies [2–5], the Smart Cities concept was mainly introduced as an approach to develop smart IT-oriented solutions for specific problems facing cities' operations, with deeply networked digital information technologies. Today it is recognized that this solution development concept fails to address a larger vision on robust city design as it lacks an *inclusive urban design* perspective [6,7] that should be based on a clear *engineering* understanding of

© Springer Nature Switzerland AG 2018
G. Satzger et al. (Eds.): IESS 2018, LNBIP 331, pp. 344–357, 2018.
https://doi.org/10.1007/978-3-030-00713-3_26

transforming cities to become sustainable, safe, resilient, resource efficient and secure spaces of living [8,9].

In the perspective of the Smart Cities' way of solution development, other two concepts evolved - Internet of Things (IoT) and Internet of Services (IoS) - and lead to the creation of *service ecosystems* that use high volumes of data [10]. IoT refers to -Things- as *"any active participants in business, information and social process where they are enabled to interact and communicate among themselves and with the environment by exchanging data and information sensed about the environment"* [11]. *"IoTs are able to react autonomously to the real/physical world events and influencing it by running processes that trigger actions and create services with or without direct human intervention"* [12]. Services are offered as part of an interconnected ecosystem and fuelled by data from a large number of interconnected smart devices and transformed into smart services within an IoS [13]. IoS provide data and aggregate it to be used for decision support systems so to provide effective sensor to decision chain dataflow in crisis management [14]. Advanced IoS leverage the power of a large amount of connected devices as IoT [15] and stress the need to make sense of the pervasively collected and stored data [16]. Solutions based on IoT and IoS technologies aim to improve the efficiency of public safety agencies and establishing new business operating models that push digital transformation through massively connected devices [17,18].

In this perspective, *Public Safety as a Service* (PSaaS) term was introduced as referring to a *"highly innovative software business and delivery model that reinvents the way agencies acquire, use and integrate technology"* [19], possibly using the cloud services for cross jurisdictional sharing of applications and services among public service stakeholders [20]. However, transitioning to more *liveable cities* [21] requires new visions that associate the rapid grow of big data with diverse opportunities for service advances [22]. Taking advantage of these opportunities means successfully fostering processes and knowledge in a service oriented world, where companies are powered by data, while people drive the value co-creation processes in service exchange. Therefore, the value of the new service business operating models and the scientific prospects of service phenomena [23] may be rooted in data availability, creating a *service experience advantage* through data generation, data processing, and knowledge extraction [24] supporting *service intelligence*.

Value co-creation in public services, a process actively involving public service users as main stakeholders in designing, managing, delivering, and evaluating public services [25], was approached from different perspectives. Some have focused on evaluating the role of companies as co-creators of value in B2B interactions at city operation level [26], others explained the role of public organizations in public service co-creation [27], or represented the integration of citizen engagement as a major component in service design landscape [28], to name but a few examples.

This paper presents a value co-creation based approach to designing public services supporting the vision of *liveable cities*. We approach *public safety* within

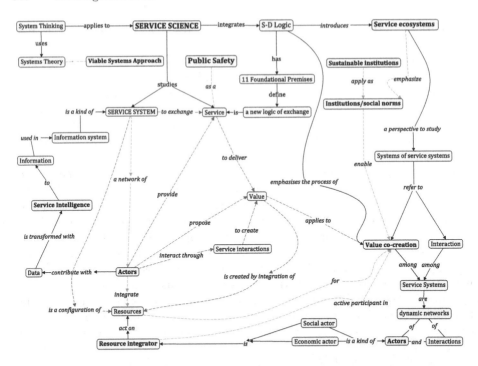

Fig. 1. Exploring service intelligence: positioning the *Public Safety as a Service* concept inside the Service Science and Service Dominant (S-D) Logic body of knowledge

a service ecosystem perspective, further elaborating towards the conception of the *public safety as a service*. Fig. 1 summarizes the set of concepts used in this exploration paper and describes how its core findings are positioned inside the Service Science domain. All findings are aligned to recent developments of Service Dominant (S-D) Logic.

From a methodological perspective, drawing research guidelines, we follow the recommendations of United Nations Office for Disaster Risk Reduction fostering multi-stakeholder coordination of the regional platforms for Disaster Risk Reduction (DRR). It supports standardization of disaster risk reduction related terminology, it recommends updated indicators to monitor the global targets of the *Sendai Framework for Disaster Risk Reduction* [29], and it follows their operationalization. As well, it provides annual results-based monitoring reports on framework implementation at national level [30].

The Sendai Framework for DRR 2015–2030 was adopted in 2015, as a successor to the Hyogo Framework for Action (HFA) 2005–2015 [31]. The framework provides guidelines and priorities for action aimed at *"the substantial reduction of disaster risk and losses in lives, livelihoods and health and in the economic, physical, social, cultural and environmental assets of persons, businesses, communities and countries"* [32]. The framework has been used to support studies in community resilience [33] that strengthen preparedness [34], safeguard life during disasters [35], and assist long-term recovery [14,36]. The seven global targets of the Sendai Framework, aiming to increase *"the availability of and access to multi-hazard early warning systems and disaster risk information and assessments to the people by 2030"* [29], support *automating* related *activities* in Early Warning Systems (EWS) based on various Information and Communication Technologies (ICT) [37]. Specifically, *Key Priority 1* of the Sendai Framework, *Understanding disaster risk*, advises for the *"collection, analysis, management and use of relevant data and practical information and ensure its dissemination, taking into account the needs of different categories of users, as appropriate"* [29].

Henceforth, the paper is organized as follows. Section 2 introduces a discussion on public safety and how it relates to the four priorities of action described by the Sendai Framework for Disaster Risk Reduction. Key data intensive activities from these recommendations are extracted. Section 3 discusses the importance of user data in services and how service performance may be tuned based on data collected from customer experiences. It elaborates on a case study related to public transport services, as an example, and introduces the definition of the *Public Safety as a Service* vision. Section 4 briefly describes two managerial implications in service ecosystems, using two knowledge bases: institutions and viability in service systems, with a discussion on a Viable Systems Approach (vSa) perspective on the response to disasters operating rules. Section 5 concludes the paper, followed by Sect. 6 that draws further research directions.

2   Related Work

While there is no universal definition, *public safety* is a term most often associated with law enforcement, typically with fire and emergency medical services [38]. *Public Safety as a Service* is not a new topic of discussion. As far as literature reveals, it originates from two lines of thought. First, the term Public Safety as a Service defines a model of acquiring and integrating public safety and security software technology, aiming towards a special category of stakeholders, i.e. public safety and security agencies [19]. Massive integration of wearable and integrated sensing technologies with cloud computing support the development of large-scale monitoring applications to provide public safety as a service [39]. Further, mission critical requirements in the public safety domain requisite enhanced data services from utility service providers [40,41]. Second, the term *safety as a service* relates to some education management tools in *transportation* [42]. It refers mainly to a cloud-based Driver Management System useful to *transportation companies* for on-line driver training, driver education, and driver safety on roads.

Nevertheless, public safety has been a central objective in the larger adoption of Sendai Framework action plan for greater resilience to natural and human-induced hazards [43–45]. Within the scope of this work, we rely on the role of Sendai Framework for DRR in drawing guidelines for advancing public safety as a major dimension of action in *resilience*-prone *safety critical business processes* of *liveable cities*. The framework introduces four priorities for action:

Table 1. Sendai Framework for DRR: four priorities for action

Priority	Objective	Data-centric activity
1 (*Assess*)	- Identify means of assessment of potential risks, dissemination of information, loss accounting and understanding of exposure; - Build awareness/capacity (consider different audience);	- Collect risk related data (historical, demographic, geospatial, etc. for past disaster mapping (GIS) to inform research and decision-making; - Provide real time access to data;
2 (*Develop*)	- Develop initiatives required for disaster risk governance; - Oversight on progress;	- Gauge the level of preparedness; - Establish mitigation measures; - Assess risk response capacity;
3 (*Promote*)	- Promote policies, plans and investments (budget allocation) to reduce risk;	- Provide risk evaluation data for disaster risk planning and insurance with financial protection mechanisms;
4 (*Monitor*)	- Implement disaster preparedness and contingency policies, plans and programs (including disaster preparedness, response and recovery exercises, for instance);	- Collect data for disaster preparedness assessment and apply data analytics for predictive assessment of disaster related risks (potential use of Early Warning Systems);

Priority 1. Understanding disaster risk.

Priority 2. Strengthening disaster risk governance to manage disaster risk.

Priority 3. Investing in disaster risk reduction for resilience.

Priority 4. Enhancing disaster preparedness for effective response and to "Build Back Better" in recovery, rehabilitation and reconstruction.

Each priority stresses the importance of *data intensive key activities* that provide the raw material needed for prioritization, decision making and preparedness (Table 1). To understand disaster risk (**Priority 1**), the framework stipulates the requirement of multidimensional data collection (such as demographic, risk related, or geospatial data) for the assessment of potential risks, dissemination, loss accounting, and understanding of exposure. This includes the collection of historical and climate data for past disaster mapping using Geographic Information Systems (GIS), for example, to inform research and decision making stakeholders.

Recognizing the importance to strengthen disaster risk governance and better manage disaster risk (**Priority 2**), the framework suggests the provisioning of real time access to - non-sensitive - data (i.e. data that do not reveal personal, financial information) by all entities (citizens, planners, government and research) and build awareness among stakeholders to strengthen the technical capabilities and scientific capacities required to minimize potential adverse the effect of an eventual disaster. This necessitates a higher capacity of risk response assessment through information gathering on the level of preparedness and mitigation measures, which would be then correlated to inform policy makers, for example. Additionally, focused at driving resilience into DRR, **Priority 3** underscores the need for policies, plans and investments to reduce risk including the necessary budget allocation at each level local, regional and national. Lastly, **Priority 4** emphasizes a continual effort of data collection for disaster preparedness assessment with data analytics for predictive assessment of disaster related risks for the use as Early Warning Systems. This vision is further developed in Sect. 3 from a service ecosystems perspective.

3   *Public Safety as a Service* and Service Ecosystems

Ecosystems vision plays a special role in advancing knowledge on service, providing an *inclusive* framework for studying systems of service systems. A service ecosystem is defined as a *"relatively self-contained, self-adjusting system of resource-integrating actors connected by shared institutional arrangements and mutual value creation through service exchange"* [46]. Additionally, the Viable Systems Approach (vSa) draws a new line of thinking, according to which a service ecosystem may be considered as being a *viable system of service systems* connected internally and externally by mutual *value creation interactions* realized through service exchange and resources integration [47–49].

Service value creation can be improved with data [50] in a special type of service - *information-intensive service* (IIS) - in which value is created primarily via *information interactions* [51], while *technology* - as an *operant resource* [52] - fosters information sharing across service ecosystems, therefore enabling value co-creation as a service strategy [53].

In this article, we briefly elaborate on this line of thought, borrowing from the principles of service design and incorporating data-driven service creation [54], that suggests the use of feedback loops to tune service performance (Fig. 2). We consider data collected from the customer journey inside the service as input to the service system for corrective and predictive action. Feedback loops reflect a brief return to alter or refine an earlier decision, inform a continuation from the point in the service development process, and provide a data source for predictive action [55]. Through iterative feedback from the output of the service system, the performance of the system receives input from the customer journey for corrective/predictive action following a cyclic process [56].

Following the case study on public transport services in times of emergency approached in [57], some data-based corrective and predictive actions may be formulated as an example at hand. For instance, data collected from previous events (risks) may be useful in predicting a certain trajectory for a potential recurrence of such event. For illustration, we could review a case where a train is proceeding over a bridge, listening to and recording vibrations caused by the passage. The train could report these data captured to a central analytical

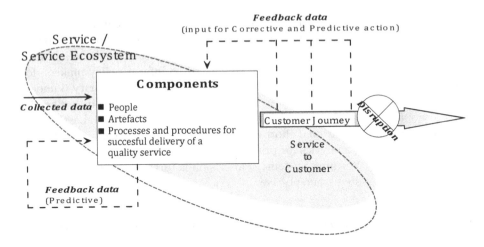

Fig. 2. Tuning service performance based on data collected from the customer journey

processor using an IoT device for instance. The next train on this route would perform the same data capture. A third train, follows suite, etc. Data is now available for the multiple passages for comparison. The processor may detect a dangerous variation in the captured information and proceed with acting. Action could be either to communicate the potential risk to the following trains with a notice to reduce speed, or flag this bridge as "maintenance required" for the next budget cycle. For instance, using data gathered, *a priori*, from risk response capacity assessment, may dispatch service crews to the area earlier than normal.

Hence, the components of a service system (people, artefacts and processes) are supplemented by a data analytics, potentially transforming data with *service intelligence* that helps *predict* emergencies, *react* in time of emergencies or even *anticipate* potential emergencies. Furthermore, the ability of the system to analyse the growing database entering the system and to transform it into *information* and *knowledge* [58,59] using *service intelligence* would be useful to inform collaborations between authorities, operators and potentially researchers to transform otherwise concealed information into effective strategies to strengthen the resilience of the transport system [60].

Returning to the general idea of public safety services, the definition of PSaaS (as a software business and delivery model) used in [19] - without using the value co-creation dimension - highlights the role of technology (software technology) in providing *integration services* to support public safety and security. Consequently, stressing the role of *technology as an operant resource* [52] helps in advancing service using data [22] by actively changing the nature of service provision [61], and requires new data-based value creation processes [50] where technology itself is the set of practices and processes through which value may appear in new forms [52,62] in service ecosystems.

Drawing further ideas based on above discussion, we introduce in this paper the following definition of *Public Safety as a Service* in service ecosystems.

Definition 1. *Public Safety as a Service is defined as an inclusive and responsible value co-creation urban design vision for liveable cities, fostering the expansion of service knowledge to multiple public service contributors, potentially transforming data into information using service intelligence to create sustainable, resilient, and trustworthy service ecosystems.*

This definition supports a working perspective on developing a data-intensive, information-based vision fostering public safety service development to support Disaster Risk Reduction related activities, according to the Sendai Framework. At the same time, it helps at developing an institutional based vision for collecting, processing and mining public safety related data to foster resilience for liveable cities, to align information intensive processes to the current digital transformation.

4   A Governance Perspective on the Response to Disasters Operating Rules

Advancing service based on data raises concerns not only on new theoretical formulations, but also on managerial guidelines [22]. In this section we briefly describe two managerial implications in service ecosystems, using two knowledge bases: institutions and viability in service systems.

First, our previous work introduced in [57] proposes a new perspective for the application of the line of thought introduced by Elinor Ostrom on institutions, as seen as *"formal and informal rules that are understood and used by a community"* [63] for the design of sustainable public transport services in times of emergency. This is a special case of application for the four priorities of action of Sendai Framework and the extraction of key data intensive activities. From a managerial perspective, in order to support the operational activities of the transport service company, the sustainable institutions principles are evaluated for this case study. Hence, categorizing the response to disasters based on a set of rules and operating conditions which define *who*, *where*, *when*, and *how* can appropriate the resources (physical and conceptual) in order to have a quick emergency response.

Second, the Viable Systems Approach (vSa) systemic theory [64–66] applied to the study of service systems, if used as a governance approach, may lead to the identification of two distinct logical areas in the study of viable systems: decision making and operations [67]. In this perspective, vSa supports the development of Service-oriented Enterprise Engineering (SoEE) modelling discipline for strategic decision-making, describing an enterprise as a *"a viable system (service system). A viable system (service systems) is an organization founded on interconnections and interdependence among its internal components (sub-systems) and the components of other systems (supra-systems) in order to evolve, develop and improve over time its conditions of survival"* [47].

Viability is attained, through *consonance* and *resonance*, by a system capable of developing harmonious relationships (through corrective and predictive feedback mechanisms) with its interconnected sub- and supra-systems. Consonance expresses the potential compatibility between the interrelated sub- and supra-systems and it is a structural performance of the relational network. Dynamically, resonance indicates the harmonic effective interaction between system's elements, operating distinctly towards a common goal [66]. Subsequently, business processes, as event-driven dynamic systems, evolve in complex and partially controllable environments [68], so there is no guarantee that resonance is attained, or that the network of system's elements remains consonant. This is one of the possible situations that can occur in case of disruptive events requiring for DRR key data intensive activities, for example, for transportation systems in time of emergency.

A solution for building and maintaining short-term viability can be created based on the smart character of a service system [66,69]. In this regard, the ICT infrastructure plays a major role, by augmenting people with cognition mediators (*consonance*) and ensuring high-speed adequate reactions to events occurrences (*resonance*). This includes the adequate response of smart urban public services, equipped with IoS functionality, to indicate occurrence of disruptive events. A long-term approach for building and maintaining viability concerns the extension to wise service systems [69,70], targeting the improvement of interaction opportunities not only for the present urgent demands, but also for estimated future needs. In this view, a viable service system evolves as a triple loop learning system, aimed to improve (i) efficiency (plans), (ii) effectiveness (goals) and sustainability (relationships and resources) [71] as feedback for corrective and predictive action.

5   Conclusions

In a viable service system, the decision-making component of the governance entity has to identify sustainable goals, creating and enhancing the wise character of the service system. At the *operational level*, the management component of the governance entity draws efficient plans and interaction networks implementing the decided goals, thus configuring the smart character of the service system. In the *decision-making area*, resource identification and allocation is the fundamental decision to be iterated. In the operational area, already existing resources are continuously monitored and managed. This focus on resources in the decisional and operational interconnected processes build the bridge to Ostrom's principles, aimed to provide a successful management of common finite resources that a network of viable systems must share.

Considering the above mentioned example, the case of the public transport system facing disruptive (emergency) events, the problem is not only to manage the already prepared resources, foreseen in the emergency plans, but also to develop the capability and agility to decide to *re-locate resources* to be shared with other stakeholders. This is a response to unpredicted situations occurring in the rescue process. It must be fulfilled in such a way that the *recovery* after the emergency event would not be compromised. This capability is part of the sustainability of the transport service.

These concepts reinforce the suggestion that Ostrom's five rules of sustainable governance principles described in [57] can be regarded as *operational vectors* and *decisions support* for *consonance* achievement in time of emergency. All the rules adjust *the governance of the relationships* developed by the service organization (i.e. transport) with its sub- and supra-systems:

Rule 1. Boundaries of the common resources must be specified and foreseen by contracting agreements (i.e.*Assess*).

Rule 2. Resource allocation decisions and the resulting interactions must be adapted to the environmental conditions (i.e. *Develop*).

Rule 3. Participations of individuals in the elaboration of operational rules implies relations building consequences for preparing successful interactions (i.e. *Promote*).

Rule 4. Monitoring the evolution of resources defines the observation function that is crucial to proper decisions (i.e. *Monitor/Evolution*).

Rule 5. The design of a system of sanctions and penalties, by the authorities, is part of relations building (i.e. *Monitor/Governance*).

The goal of these rules is the achievement of resonance in time of emergency, with minimization of the risk of recovery failure. Just as indicated by the Sendai Framework that dispatches its first priority to identify means of assessment of potential risks, dissemination of information, loss accounting and understanding of exposure. The successive priorities guide the development of the initiatives required for disaster risk governance, promoting policies, plans and investments (budget allocation) to reduce risk and implement disaster preparedness and contingency policies, plans and programs with conscious oversight on progress. In other words, establishing the proper set of operating rules for a response to disasters is the kernel of structural building of harmonic relations of a viable service organization prepared to face disruptive (emergency) events and a support for decisions in resource allocation.

6   Further Work

In this exploration paper we conceptualize on a common knowledge base to describe a service ecosystem comprising networked public safety services and service systems, exploring various areas of application and research (see also Fig. 1). The requirements guidelines are drawn following the Sendai Framework for DRR recommendations. We conceptualize on *Public Safety as a Service* in service ecosystems as an inclusive vision for liveable cities. We ground related developments to driving service exchange based on value co-creation via information interactions. As further research directions, it is envisioned that this inclusive vision will drive further the development of platform-based, data centric, and information intensive business process models needed to suitably allow information extraction exactly from the point of interest. Several questions may be formulated along with this approach and they must be further explored to advance Society in its core values and processes. From an application point of view, *how* can we really transform data to support intelligence in services? How can we build a pyramid on "intelligence" in services helping to advance services inside Society? Can we create a roadmap to advance service intelligence with data, i.e. to find opportunities successfully fostering processes and knowledge in a service oriented world, where companies are powered by data, while people drive the value co-creation processes in service exchange?

References

1. UN: World Urbanization Prospects: The 2014 Revision. United Nations, Department of Economic and Social Affairs, Population Division, ST/ESA/SER.A/366 (2015)
2. IBM: Making Sense of a Sensored World. IBM Smarter Planet. https://www.ibm.com/
3. Cisco: Smart+connected Communities. Changing a City, a Country, the World. Cisco White paper. https://www.cisco.com/
4. Siemens: Siemens City Cockpit White paper. City Cockpit in Singapore: Collective Intelligence - Real Time Government. Siemens. https://www.siemens.com/
5. Microsoft: Microsoft Citynext White paper. Accelerating the Digital Transformation of Smart Cities and Smart Communities. Microsoft. https://enterprise.microsoft.com/
6. Greenfield, A.: Against the Smart City (the City is Here for You to Use, Part i). Amazon Digital Services LLC, Kindle edn, Do Projects, 1.3 edn (2013)
7. UrbanScale: Design for Networked Cities and Citizens. UrbanScale. http://urbanscale.org/
8. LiveableCities: Transforming the Engineering of Cities for Global and Societal Wellbeing. Liveable Cities. http://liveablecities.org.uk/
9. UN: Sustainable Development Knowledge Platform. Goal 11: Make Cities Inclusive, Safe, Resilient and Sustainable. United Nations (2017). http://www.un.org/
10. Ng, I.C., Wakenshaw, S.Y.: The internet-of-things: review and research directions. Int. J. Res. Mark. **34**(1), 3–21 (2017)
11. Gubbi, J., Buyya, R., Marusic, S., Palaniswami, M.: Internet of things (IoT): a vision, architectural elements, and future directions. Future Gener. Comput. Syst. **29**(7), 1645–1660 (2013)
12. Sundmaeker, H., Guillemin, P., Friess, P., Woelfflé, S.: Vision and challenges for realising the Internet of Things. Clust. Eur. Res. Proj. Internet Things Eur. Comm. **3**(3), 34–36 (2010)
13. Soriano, J., et al.: Internet of services. In: Bertin, E., Crespi, N., Magedanz, T. (eds.) Evolution of Telecommunication Services. LNCS, vol. 7768, pp. 283–325. Springer, Heidelberg (2013). https://doi.org/10.1007/978-3-642-41569-2_14
14. Moßgraber, J., et al.: The sensor to decision chain in crisis management. In: Boersma, K., Tomaszewski, B. (eds.) CoRe Paper - Universal Design of ICT in Emergency Management, Proceedings of the 15th ISCRAM Conference - Rochester, NY, USA, May 2018 (2018)
15. Mims, C.: The Internet of Things isn't things, it's services. Wall Str. J. (2016). https://www.wsj.com/
16. Estevez, P.A.: CIS in the next decade. IEEE Comput. Intell. Soc. (2016). http://cis.ieee.org/

17. Leather, A.: Internet of Things in public safety. Frost & Sullivan Webcasts (2015). https://ww2.frost.com/
18. Sims, K.: Top public safety and security tech trends for 2017. Hexagon Safety & Infrastructure (2017). www.hexagonsafetyinfrastructure.com/
19. Tiburon: PSaaS: The Next Evolution of Public Safety and Security is Here. Tiburon. http://www.gradicom.com/
20. Concepts To Operations: Public Safety as a Service (PSaaS). Concepts to Operations. http://www.concepts2ops.com/psaas/
21. The Strategic Research and Innovation Agenda - Transition Towards Sustainable and Liveable Urban Futures. JPI Urban Europe (2015). http://jpi-urbaneurope.eu/
22. Lim, C., Kim, M.J., Kim, K.H., Kim, K.J., Maglio, P.P.: Using data to advance service: managerial issues and theoretical implications from action research. J. Serv. Theory Pract. (2017). https://doi.org/10.1108/JSTP-08-2016-0141
23. Spohrer, J.C., Kwan, S.K., Demirkan, H.: Service science: on reflection. In: Cinquini, L., Minin, A.D., Varaldo, R. (eds.) New Business Models and Value Creation: A Service Science Perspective, pp. 7–24. Springer, Milano (2013). https://doi.org/10.1007/978-88-470-2838-8_2
24. Macbeth, S., Pitt, J.V.: Self-organising management of user-generated data and knowledge. Knowl. Eng. Rev. **30**(3), 237–264 (2015)
25. Osborne, S.P., Radnor, Z., Strokosch, K.: Co-production and the co-creation of value in public services: a suitable case for treatment? Publ. Manag. Rev. **18**(5), 639–653 (2016)
26. Tokoro, N.: The Smart City and the Co-creation of Value: A Source of New Competitiveness in a Low-Carbon Society. Springer, Tokyo (2015). https://doi.org/10.1007/978-4-431-55846-0
27. Magno, F., Cassia, F.: Public administrators' engagement in services co-creation: factors that foster and hinder organisational learning about citizens. Total Qual. Manag. Bus. Excell. **26**(11–12), 1161–1172 (2015)
28. Rytilahti, P., Miettinen, S., Vuontisjärvi, H.-R.: The theoretical landscape of service design. In: Marcus, A. (ed.) DUXU 2015. LNCS, vol. 9186, pp. 86–97. Springer, Cham (2015). https://doi.org/10.1007/978-3-319-20886-2_9
29. UN: Sendai Framework for Disaster Risk Reduction 2015–2030. In: United Nations, Third UN World Conference on Disaster Risk Reduction, Sendai, Japan, 18–20 March 2015
30. UNISDR: UNISDR annual report 2016. United Nations Office for Disaster Risk Reduction (UNISDR) (2017). https://www.unisdr.org/
31. UN: Hyogo Framework for Action (HFA) 2005–2015. United Nations Office for Disaster Risk Reduction (2005). https://www.unisdr.org/we/coordinate/hfa
32. UN: Chart of the Sendai Framework for Disaster Risk Reduction. United Nations (2015). https://www.unisdr.org/
33. Norris, A., Gonzalez, J., Martinez, S., Parry, D.: Disaster e-Health framework for community resilience. In: Hawaii International Conference on System Sciences 2018, Hawaii International Conference on System Sciences (HICSS), pp. 35–44 (2018)
34. Grasso, N., Lingua, A.M., Musci, M.A., Noardo, F., Piras, M.: An INSPIRE-compliant open-source GIS for fire-fighting management. Nat. Hazards **90**(2), 623–637 (2018)
35. Navas, S.R., Miyazaki, K.: Exploring "Big Data" applications for disaster management: a scientific keyword word co-occurrence network analysis. JAIST Repository,

Japan Society for Research Policy and Innovation Management (2016). https://dspace.jaist.ac.jp/

36. Heikkurinen, M., Schiffers, M., Kranzlmüller, D.: Environmental computing 1.0: the dawn of a concept. In: International Symposium on Grids and Clouds (ISGC) 2015, PoS(ISGC2015)030. Academia Sinica, Taipei (2015)

37. Grasso, V., Singh, A., Pathak, J.: Early warning systems: state-of-art analysis and future directions. In: United Nations Environment Programme, Division of Early Warning and Assessment (2012). http://www.unep.org

38. Roles of local government - Public Safety. Michigan Townships Association. https://www.michigantownships.org/publicsafety.asp

39. Berrahal, S., Boudriga, N., Bagula, A.: Cooperative sensor-clouds for public safety services in infrastructure-less areas. In: 2016 22nd Asia-Pacific Conference on Communications (APCC), pp. 222–229. IEEE (2016)

40. Raza, A.: LTE network strategy for smart city public safety. In: IEEE International Conference on Emerging Technologies and Innovative Business Practices for the Transformation of Societies (EmergiTech), pp. 34–37. IEEE (2016)

41. Manzalini, A., Crespi, N.: An edge operating system enabling anything-as-a-service. IEEE Commun. Mag. **54**(3), 62–67 (2016)

42. Safety as a Service. http://www.safetyasaservice.com

43. UN: Americas to Agree on Sendai Framework Action Plan. United Nations (2017). https://www.unisdr.org/archive/52207

44. Public Safety Canada. https://www.publicsafety.gc.ca/

45. UN: National framework in order to reduce earthquakes by multi-stakeholder participation in Turkey: National earthquake strategy and action plan of Turkey (UDSEP-2023). United Nations (2013). https://www.unisdr.org/we/inform/publications/50178

46. Vargo, S.L., Lusch, R.F.: Institutions and axioms: an extension and update of service-dominant logic. J. Acad. Mark. Sci. **44**(1), 5–23 (2016)

47. Rafati, L.: Capability-actor-resource-service: a conceptual modelling approach for value-driven strategic sourcing. Ph.D. thesis, Ghent University, Belgium (2018)

48. Pels, J., Barile, S., Saviano, M., Polese, F., Carrubbo, L.: The contribution of VSA and SDL perspectives to strategic thinking in emerging economies. Manag. Serv. Qual. **24**(6), 565–591 (2014)

49. Vargo, S.L., Akaka, M.A.: Value cocreation and service systems (re)formation: a service ecosystems view. Serv. Sci. **4**(3), 207–217 (2012)

50. Lim, C., Kim, K.H., Kim, M.J., Heo, J.Y., Kim, K.J., Maglio, P.P.: From data to value: a nine-factor framework for data-based value creation in information-intensive services. Int. J. Inf. Manag. **39**, 121–135 (2018)

51. Lim, C.H., Kim, K.J.: Information service blueprint: a service blueprinting framework for information-intensive services. Serv. Sci. **6**(4), 296–312 (2014)

52. Akaka, M.A., Vargo, S.L.: Technology as an operant resource in service (eco)systems. Inf. Syst. e-Bus. Manag. **12**(3), 367–384 (2014)

53. Vargo, S.L., Akaka, M.A., Vaughan, C.M.: Conceptualizing value: a service-ecosystem view. J. Creat. Value **3**(2), 117–124 (2017)

54. Toots, M., et al.: A framework for data-driven public service co-production. In: Janssen, M., et al. (eds.) EGOV 2017. LNCS, vol. 10428, pp. 264–275. Springer, Cham (2017). https://doi.org/10.1007/978-3-319-64677-0_22

55. Johnson, S.P., Menor, L.J., Roth, A.V., Chase, R.B.: A critical evaluation of the new service development process: integrating innovation and service design. In: Fitzsimmons, J.A., Fitzsimmons, M.J. (eds.) New Service Development: Creating Memorable Experiences, pp. 1–32. Sage Publication, Thousand Oaks (2000)

56. Clayton, R.J., Backhouse, C.J., Dani, S.: Evaluating existing approaches to product-service system design: a comparison with industrial practice. J. Manuf. Technol. Manag. **23**(3), 272–298 (2012)
57. Drăgoicea, M., Salehpour, S., Nóvoa, H., Oltean, V.E.: Towards a proposal for the sustainability through institutions in public transport services in times of emergency. In: Za, S., Drăgoicea, M., Cavallari, M. (eds.) IESS 2017. LNBIP, vol. 279, pp. 355–369. Springer, Cham (2017). https://doi.org/10.1007/978-3-319-56925-3_28
58. Liew, A.: DIKIW: data, information, knowledge, intelligence, wisdom and their interrelationships. Bus. Manag. Dyn. **2**(10), 49–62 (2013)
59. Baškarada, S., Koronios, A.: Data, information, knowledge, wisdom (DIKW): a semiotic theoretical and empirical exploration of the hierarchy and its quality dimension. Aust. J. Inf. Syst., [S.l.] **18**(1) (2013). https://doi.org/10.3127/ajis.v18i1.748. ISSN 1449-8618
60. Mattsson, L.G., Jenelius, E.: Vulnerability and resilience of transport systems-a discussion of recent research. Transp. Res. Part A: Policy Pract. **81**, 16–34 (2015)
61. Bitner, M.J., Zeithaml, V.A., Gremler, D.D.: Technology's impact on the gaps model of service quality. In: Maglio, P., Kieliszewski, C., Spohrer, J. (eds.) Handbook of service science, pp. 197–218. Springer, Boston (2010). https://doi.org/10.1007/978-1-4419-1628-0_10
62. Arthur, W.B.: The Nature of Technology: What it is and How it Evolves. Simon and Schuster, New York City (2009)
63. Ostrom, E.: Design principles and threats to sustainable organizations that manage commons. In: Paper for Electronic Conference on Small Farmer's Economic Organizations, Organized by Julio A. Berdegue. Santiago, Chile, March 1999
64. Barile, S., Polese, F.: Linking the viable system and many-to-many network approaches to service-dominant logic and service science. Int. J. Qual. Serv. Sci. **2**(1), 23–42 (2010)
65. Spohrer, J., Golinelli, G.M., Piciocchi, P., Bassano, C.: An integrated SS-VSA analysis of changing job roles. Serv. Sci. **2**(1–2), 1–20 (2010)
66. Barile, S., Polese, F.: Smart service systems and viable service systems: applying systems theory to service science. Serv. Sci. **2**(1–2), 21–40 (2010)
67. Polese, F., Tommasetti, A., Vesci, M., Carrubbo, L., Troisi, O.: Decision-making in smart service systems: a viable systems approach contribution to service science advances. In: Borangiu, T., Drăgoicea, M., Nóvoa, H. (eds.) IESS 2016. LNBIP, vol. 247, pp. 3–14. Springer, Cham (2016). https://doi.org/10.1007/978-3-319-32689-4_1
68. Badinelli, R., Barile, S., Ng, I., Polese, F., Saviano, M., Di Nauta, P.: Viable service systems and decision making in service management. J. Serv. Manag. **23**(4), 498–526 (2012)
69. Spohrer, J., Bassano, C., Piciocchi, P., Siddike, M.A.K.: What makes a System Smart? Wise? In: Ahram, T., Karwowski, W. (eds.) Advances in The Human Side of Service Engineering. AISC, pp. 23–34. Springer, Cham (2017). https://doi.org/10.1007/978-3-319-41947-3_3
70. Siddike, M.A.K., Iwano, K., Hidaka, K., Kohda, Y., Spohrer, J.: Wisdom service systems: harmonious interactions between people and machine. In: Freund, L.E., Cellary, W. (eds.) AHFE 2017. AISC, vol. 601, pp. 115–127. Springer, Cham (2018). https://doi.org/10.1007/978-3-319-60486-2_11
71. Spohrer, J., Maglio, P.P., Bailey, J., Gruhl, D.: Steps toward a science of service systems. Computer **40**(1), 71–77 (2007)

Managing Patient Observation Sheets in Hospitals Using Cloud Services

Florin Anton[(⊠)], Theodor Borangiu, Silviu Raileanu, Iulia Iacob, and Silvia Anton

Department of Automation and Applied Informatics,
University Politehnica Bucharest, Bucharest, Romania
{florin.anton, theodor.borangiu, silviu.raileanu,
iulia.iacob, silvia.anton}@cimr.pub.ro

Abstract. In many hospitals all over the world there is an acute lack of physicians; in addition, doctors who are working in hospitals are overwhelmed by the number of patients and other administrative duties which they must do. Due to the specific of the work many operations/procedures/activities must be done manually and there are no automated systems which could improve the quality of the medical service and the efficient usage of the physician's time. In this paper we propose a service system designed to help physicians to automate the work of registering patient clinical observations into the patient clinical observation sheet. The procedure of registering observations can be time consuming in some situations due to the numerous parameters which must be registered. The proposed system uses voice to text conversion engine to register the observations; thus, doctors spend much less time to review the clinical observations and eventually make corrections if necessary.

Keywords: Hospital services · Hospital information system
Patient electronic health record · Cloud services · Mobile computing

1 Introduction

The paper describes the research carried out to design and implement an electronic service system managing patient observation activities made by the medical staff working in healthcare organizations: surgeons, radiologists, medical doctors, and therapy specialists in hospitals, laboratories for medical analyses and recovery clinics.

The service system is conceived to assist physicians in directly observing the health status and evolution of patients, interpreting and documenting important medical acts like clinical examinations, operations, analyses, therapy procedures and effects of medical treatments. For all these medical activities, the physician has to formulate interpretations, propose investigations, treatments, surgical acts and post-operatory procedures based on current medical rules, modus operandi and best practices; these interpretations and medical decisions may differ from one physician to another because they also depend on the professional knowledge, experience and predictive capacity derived from medical case history and past learning.

© Springer Nature Switzerland AG 2018
G. Satzger et al. (Eds.): IESS 2018, LNBIP 331, pp. 358–370, 2018.
https://doi.org/10.1007/978-3-030-00713-3_27

In most healthcare organisations, these medical observations, interpretations and decisions resulting from the direct interactions between the physician and the patient are manually saved in written form: first in a synthetic, abbreviated form during the medical act (daily medical visit) and then in extended detailed form at the end of the daily shift. In the first stage the doctor verbalizes the information for a medical assistant who writes down a draft document; this stage may be skipped for the second one when the physician himself writes the detailed report of what he has observed/found out during the medical act (visit, examination, operation, etc.)

These written medical observations are included in forms which are standardized at various levels: hospital (lowest), medical specialty (e.g., ORL), national – set up by the national healthcare system, insurance system, etc. (highest). There are many types of medical forms: general clinical observation sheet [medical specialty] in hospitals with more than one medical specialty; clinical assessment; surgical hospitalization form for patients who will be operated; surgical form for patients who were operated; post-surgery recovery procedure form, a.o. Important observations in certain sections of these forms describing: hospitalisations (entry and exit dates, diagnostic, treatment, evolution, a.o.), surgery (operation plan, type of anaesthesia, monitored patient's parameters, medical team, etc.), periodical family physician examinations, etc. should be translated in time in the patient's Electronic Medical Record (EMR) which synthetises the health status and care history of a person in time [1–3]. These informations from EMR if described in detail can serve in resident student training based on medical cases.

The patient's general clinical observation sheet for a medical specialty has an identification number, context and activity-related data fields, to which the medical observations are added as text sections. Examples of such data fields are: patient identification (name, date of birth, sex, residence address, citizenship, height, weight, blood group, allergies, identity document, employer, education level, diagnostic at entry/exit in/from hospital); patient social security (type of insurance, no. of national card, type of entry in hospital); surgical actions (surgical team, data and hour of start/finishing, transfer between hospital's sections); therapy procedures (code, type, date, functional explorations), general clinic examination (general state, specialty exams, laboratory analyses, radiology examination, echography examination).

The medical observations added daily to the clinical observation sheet represent a vital information for the follow up of the patient's post-operation healing, of the effect of prescribed medication on the patient's parameters affected by a disease and of the evolution of the patient's status [4]. Also, disposing of an electronic observation sheet would allow quickly informing other specialists who, together with the physician who has created the medical information, can take better treatment decisions.

The research is devoted to an electronic system using cloud services to add medical observations about the patient's health state evolution; the information is translated automatically from the physician's voicing to the text in electronic format using a SpeechToText SaaS service on an external cloud. For security reasons, both audio and text records are kept for time periods imposed by the medical system. Also, the cloud SpeechToText service must provide protection against data tempering and prevent patient information disclosure; authenticating the mobile devices used to register the doctor's observation as audio record, storing and accessing medical information in various formats must be secured with high priority imposed by this sensitive domain [5].

Smart composite services for patient observation, interpreting and documenting medical diagnostics, decisions and activities with final information integration in the patient's EMR strongly rely on knowledge intensive, information and data-driven services (KIS) that use advanced ICT, tools and software: cognitive computing; agent-based web and cloud services; big data analytics; micro service composition [6].

One such KIS is the SpeechToText (Sp2Tx) service used in this research to convert audio files generated by physicians while speaking out medical observations during patient visits into text that is then integrated in the patient's electronic observation sheet. The Sp2Tx service is based on speech recognition – a complex task relying on many sources of knowledge (acoustics, phonetics, linguistics, semantics) and complex techniques (signal processing, acoustic-phonetic models, neural networks, statistical language models). Speech recognition accuracy highly depends on the type of data to be processed and is typically measured in terms of word error rate which can be as low as few percentages for some tasks and as high as 40% on very challenging tasks.

Sp2Tx services use Speech APIs to convert audio to text by applying neural network models. All Sp2Tx service providers offer APIs with the functions: language identification, audio and speaker segmentation, speech-to-text conversion, and speech-text alignment in three submission modes: file, streaming, and real-time. In addition to the speech transcription Web service, some providers offer services for batch processing of large quantities of data and for the development of customized models [7]. The result of the conversion is a fully annotated XML document including speech and non-speech segments, speaker labels, words with time codes, high quality confidence scores and punctuations. This XML file can be directly indexed by a search engine, or alternatively can be converted into plain text. Some models can detect multiple speakers; this may slow down the service's performance.

Speech recognition can be very sensitive to input audio quality. It is necessary to ensure that the input audio quality is as good as possible; for best accuracy, close, speech-oriented microphones are recommended. In addition to audio quality, speech recognition systems are sensitive to nuances of human speech, e.g., regional accents and variance in pronunciation and may not always successfully transcribe audio input.

A speech recognition model indicates the language in which the audio is spoken and the rate at which it is sampled. For most languages, the Sp2Tx service supports two models: (a) a *broadband model* for audio that is sampled over 16 kHz; the broadband model is recommended for responsive, real-time applications such as live-speech applications; (b) a *narrowband model* for audio that is sampled at 8 kHz; this rate is used typically for telephonic audio. Custom models are offered by some providers like IBM; they include: (a) custom language models which expand the service's base vocabulary with terminology from specific domains; (b) custom acoustic models which adapt the service's base acoustic model for the acoustic characteristics of environment and speakers. Several Sp2Tx service programming interfaces are available:

- The WebSocket interface offers an efficient, low latency and high throughput implementation over a full-duplex connection.

- The HTTP REST interface allows transcribing audio with/without establishing a session with the service; audio data can be transmitted to the service in one of two ways: (i) one-shot delivery; (ii) streaming. It can be used with command-line HTTP clients (cURL) or with HTTP client libraries for C/C++, PHP, Java or JavaScript [7].

Speech APIs support any device that can send a REST or gRPC request including smart phones, PCs, tablets and IoT devices (e.g., cars, TVs, speakers).

Sp2Tx service providers offer various usage plans: pay as you go, daily plan, batch plan, a.o. For large quantities the price is on the order of 0.01 euro per minute [7, 8] or 0.024 \$US per minute for up to 106 min/month [9]. Pricing is based on speech duration, i.e. silences are not counted and there is no minimum cost per submission. The cost of using a speech-to-text system is proportional to the system's error rate.

Audio files can be uploaded in the request or integrated with the provider's cloud storage, e.g. by accessing Watson services on the IBM Cloud or Sp2Tx on Google Cloud Storage. The integration of the Sp2Tx transcript service in the smart composite service for patient observation management can be realized by storing mp3 audio files containing the physicians' observations in the hospital's server. The server will be connected to the external cloud using SSL Encryption to generate Sp2Tx service requests for the mp3 files; the SaaS conversion produces xml files which are sent back to the server and placed in the same database as the mp3 audio files.

The remainder of the paper is organized as follows: Sect. 2 describes the system infrastructure; Sect. 3 presents the service flow; the components and the services offered are discussed in Sect. 4; Sects. 5 and 6 present the private cloud services and the database; and Sect. 7 describes experimental results and offers conclusions.

2   The System Infrastructure

The system is based on an infrastructure containing the main components (Fig. 1):

- A set of smart phones, running iOS or Android, distributed to the physicians and used to register and transmit the observations during patient evaluation visits.
- A private cloud which offers services for the system: authentication using Kerberos protocol, a database for storing and retrieving medical information including the doctor's observations, an agent for the connection to online Speech to Text SaaS service for voice conversion into text, and a web interface for users (most of them physicians) allowing them to view and modify the information recorded (this is done in most case to correct the transcript provided by the Speech to Text service).
- A SaaS service accessed remotely to convert voice in transcript (Sp2Tx service).
- An infrastructure and a set of protocols used to secure the communications inside and outside the hospital.

The mobile phones are distributed to doctors and used to register the observations into the system; the phones can be owned by the physician or be the property of the hospital and used only inside it. The application used on the phone for observation recording uses only the photo camera, the microphone and the Wi-Fi connection, so the phone doesn't need a SIM card. The device could be also a tablet, but the mobile phone has been selected because it can be handled and used much easier.

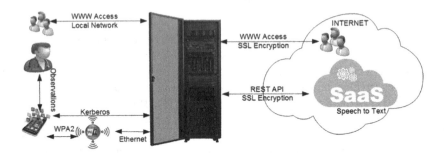

Fig. 1. The system infrastructure

The private cloud system offers a set of services which can be accessed and used inside the hospital, being provided by virtual machines; the implementation can be also made using a single server, but in this case the system will lack high availability, fault tolerance and scalability which a cloud system offers.

The SaaS Speech to Text service is accessed using an agent which runs on the private cloud and monitors the database. When a new observation is added to the database the agent sends the mp3 file containing the observation to the Sp2Tx service and receives the transcript which is also added to the patient observation sheet in the database. Vocapia and Google Sp2Tx systems were tested in SaaS models [7, 9].

The communication infrastructure allows the mobile phones used for this service to connect to the private hospital Wi-Fi access points, and hence to access the hospital network in a secure way; this network is connected to the private cloud system where the internal composite services can be accessed.

3   The Service Flow

When the physician starts a visit to hospitalized patients to check their health state and evolution, he connects his smart phone to the hospital's Wi-Fi network (a password is requested for network access). The Wi-Fi network is a private one using WPA2 protocol which encrypts the data transferred between the Wi-Fi access points and the mobile devices connected wireless to the network.

Once the connection to the network performed, the physician starts an application on the phone which allows recording observations. When the application is starting it asks a username and password which the user should provide. After authenticating (based on the Kerberos protocol) the physician can start his visit. When consulting a patient, he will first read the information already printed in the clinical observation sheet attached next to the patient bed, then he will consult the patient, and when he is ready to record his current observations he will use the application on the mobile phone to scan a QR code which is printed on the patient observation sheet. This QR code contains two identification data: the name of the patient which is displayed on the mobile phone screen to be validated by the physician and an ID number which is used by the application to identify the patient in the database.

The physician verifies that the name displayed on the screen is the name of the patient printed on the observation sheet he had just examined, and then pushes the record button on the application interface (Fig. 2) starting to register his observations keeping the phone at the ear as in a normal phone conversation.

Fig. 2. The mobile application

If, during the registration of the observation the phone rings or another application is started on the phone in the foreground, the application pauses the recording; this action can be done also voluntarily by pressing the pause button. A recording placed in pause can be resumed at any time by pressing the "resume" button on the interface. If necessary, the physician can reset the recording and restart it. When the doctor has finished recording, he can send it to the database for the patient clinical observation sheet by pushing the "send" button; the recording stops and an mp3 file containing the record is created; the mobile application contacts the database and sends the data:

- The username of the physician (the username which has been used to authenticate to the application)
- The patient ID (read from the QR code which was scanned from the patient observation sheet)
- The mp3 file which stores the physician observation

When the information is received by the database, it is stored into the patient's current observation sheet. The database is scanned by a SaaS agent which checks if for each mp3 file there exists a transcript placed into the database; if there are mp3 files which do not have related transcripts, then the SaaS agent retrieves the mp3 file from the database and sends the file to the Speech to Text cloud service which returns the transcript of the file in the form of XML file or in plain text. The transcript is placed into the database by the SaaS agent.

The private cloud offers another service – a web application that allows physicians or other users to access the information inside the database. After finishing the visit, the physician can review the observations and eventually correct the transcript if this is necessary; other users like resident students can access the application to learn from the clinical cases stored in the database. If needed, the application can be accessed remotely by other physicians if more qualified opinions or expertise are necessary from outside the hospital.

4 The Mobile Application

The application running on the smart phone was created in C++ using the RAD Studio development environment from Embarcadero [10]. RAD Studio was selected because it can be used for rapid deployment; the application development is based on using components which are added to the application using drag and drop, then they can be customized and integrated with each other. Another reason for selecting RAD Studio is because a project for creating a mobile application can be compiled to build the application for Android and iOS without changing the code; this eliminates the effort to migrate the application on another platform from the start.

The mobile application offers the following set of functions:

- Allows the users to remotely authenticate by help of the Kerberos protocol version 5 which uses Advanced Encryption Standard (AES) encryption [11–13].
- Allows the physician to insert the patient data (ID and name) using a QR code scanning module rather that introducing the data by using the phone keyboard (the user has the possibility to introduce these informations by hand using the keyboard, but it is much easier to scan the QR code).
- Allows the physician to record his observations in mp3 format and to store the information into a database offered as a service into a private cloud connected to the hospital network, the mp3 file being stored into the database along with the patient ID and the user ID of the physician.

From the point of view of security, the application does not store the credentials for authentication. Also, after the mp3 file is sent to the database the identification information about the patient and the mp3 file are erased from the phone.

The risk of intercepting patient information from the phone is not entirely removed because if the phone is owned by the physician being also used for other personal scopes, then there is no control over the applications which are installed, or the websites visited; in this case the phone may contain malware which can intercept data. On the other hand, if the phone is owned by the hospital being used only for recording medical observations, restrictions can be applied to increase the security.

5 Private Cloud Services

The private cloud offers a set of four services inside the hospital network, from which one can be accessed also from internet:

1. The *authentication service* for the mobile application. It benefits of the Kerberos v5 authentication protocol which is implemented on a Linux virtual machine.
2. The *database* is used to store patient information, which can be inserted by using the web application and/or the mobile application (only for inserting observations into an already created patient observation sheet). The information stored into the database can be changed, but all changes are stored separately, and the information is not overwritten. This facility was implemented through a versioning application at the level of the database. By consulting the history, all changes can be identified, and the user is registered for each change he made. The system was designed in this way to prevent losing data and to help investigating malpractice cases.
3. The private cloud also offers a *SaaS agent* which is written in C++, the agent monitors the database and retrieves the mp3 files which do not have a transcript associated. The files are transmitted to a Sp2Tx service in cloud using REST API or Code::Blocks [14] and AutoHotKey [15]. The connection between the private cloud and the Speech to Text service is secured by help of Secure Sockets Layer protocol (SSL) which uses encryption for data transfer. The transcript in form of XML file or clear text is then received by the SaaS agent and stored back into the database.
4. *The web access to the database* is the last service offered by the private cloud. The access can be granted to users which are connecting to the system from the local network, or remotely from the internet. The connections from internet are allowed only to view data and not to modify it; these connections are granted only to outside experts who can access the patient clinical observation sheet in some cases to offer additional expertise regarding the diagnostic and the treatment which should be applied. If the web application is accessed from the internal network, the users have the possibility to insert or modify data depending on the permissions associated to each user. There are multiple roles defined in the web application and privileges which can be associated with each user; for example, the "administrator" can access any information (he can read or write information into the database), can add users and associate permissions or roles to the users. The role "office" can create new observation sheets and register patients, and the role "physician" can insert data and modify data in the patient's observation sheet. Another role is "student" allowing resident students to access a selected set of patient observation sheets from which they can learn based on case studies how some cases have been documented, followed, and treated in the hospital. Some fields can be hidden from users depending on associated permissions; e.g. the role "student" is not authorized to view the patient's name or other personal data which are irrelevant to the case study that is analysed.

The web application also offers an integrated mp3 player service that can be used to heed voice observations. The physician can use this player service to correct faster the transcript. If the transcript is in the form of XML file and contains timestamps, when the user clicks on a certain word, the player updates the cursor in time to the moments when the selected word was pronounced; the physician can thus listen and make fast corrections in the transcript when the Sp2Tx system doesn't recognize the word. Figure 3 shows the relationship between the components and services offered by the system.

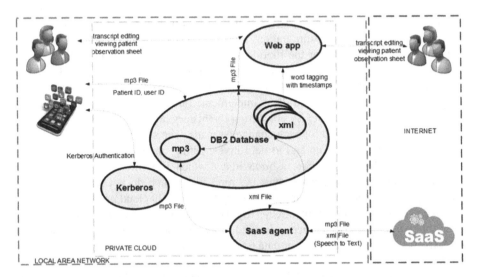

Fig. 3. The relationship between services

6   The Database

The designed system allows the physician to record the observations (in Romanian) in mp3 file, then extracts from this file and stores the transcript using a Speech to Text SaaS service, and later the physician can review and correct the transcript if needed.

For the storage of the information generated in this application a relational database compatible with the SQL language and specifications was used, having the features: relational database system, client/server architecture, SQL compatibility, Sub-SELECTs, views, stored procedures, triggers, all conceivable character sets, user interface, replication, transactions, platform independence, high operation speed. The database should be accessible through a large number of programming languages.

We have currently chosen MySQL due to its easy operation with the free and open source graphical user interface tool, but for the final implementation we intend to use IBM DB2 due to its increased performance and the possibility to self-tune memory management. For efficient data storage and processing the database structure shown in Fig. 4 was designed. This structure, defined by tables, fields and relations, aims at ensuring data integrity and minimizing redundant information.

The database is composed of 5 tables containing data about: (a) system users, with (b) access data/passwords, (c) patients, (d) observation sheet data, and (e) patient observation data:

(a) **Users' table (ped_users)**: contains information about the users of the system, who fall into three categories: *administrator* (can alter the database structure), *doctor with read/write access*, and *doctor with read access*. The composing fields are: unique identifier, full name, personal identification number (the unique ID from the person's identity document), and login.

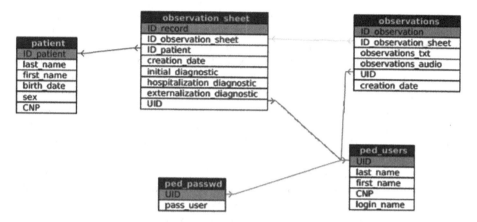

Fig. 4. Database structure and relations between composing tables

(b) **Password table (ped_passwd)**: it has only two columns (user identification and encrypted password) which make the association between a login name and its encrypted password.

(c) **Patient table (patient)**: contains information about the monitored patients - unique identifier, full name, personal identification, birthdate, and sex.

(d) **Observation sheet table (observation_sheet)**: makes the link between: the creator of the sheet; the patient; the initial diagnostic, the evolution during hospitalization and the diagnostic at externalization; the observation sheet. The above categories of data represent the fields of the table.

(e) **Observation table (observations)**: contains all the observations made by the system's users. Information here is incremental, meaning that everything is logged, and nothing is erased. The main fields of this table are: the unique identifier; the identifier of an observation sheet; the identification of the user that made the changes, observations in text mode, and observations in audio mode (stored as BLOB – Binary Large Object). The information from the observation sheet table points to the last observation made, but all the previous observations are available.

The relations between the database tables rise as a consequence of the fact that we want to eliminate redundant information and preserve data integrity. Thus, two types of relations appear:

- *One to many* among tables: (i) patient (primary key patient identifier) and observation sheet (foreign key patient identifier), (ii) users (primary key user identifier) and observations, password and observation sheet (foreign key user identifier).

- *Many to many* among tables: (i) observation sheet (primary key observation sheet identification) and observations (foreign key observation sheet identification).

7 Experimental Results and Conclusions

The paper presents a service system designed to support the process of registering medical observations in the patient electronic observation sheet in hospitals. The voice observations are recorded using an application which runs on smart phones and is then converted into text and placed into a database. The data can be accessed using a web application in the cloud which permits the physician to consult the information in the observation sheet and to make corrections. The web application also allows other specialists to consult the observation sheet and express their opinion on the medical case, resident students to learn from the cases stored in the database and the hospital staff to register patients, update and print the observation sheet. Also, the way that data is stored in the database can help in investigating malpractice cases.

In order to evaluate the efficiency of the system three sets of tests, each of them using 20 mp3 files containing detailed observation in Romanian made by physicians during their visits to patients (some of the physicians are also professors accompanied by resident students) have been executed using two Speech to Text services offered by Vocapia and Google; each file has a duration of 60 to 120 s:

1. The first test used files containing noise and voices in the background, which had an important impact on the results. The resulted transcripts contained 56.92% of Romanian words recognized correctly by Vocapia and 75.24% by Google.
2. The second test used filtered files; the filter diminished the background noise and voices; 73.45% of the Romanian words were recognized by Vocapia and 84.73% by Google.
3. The third test used files with observations made in an environment where the noise was reduced, and no other voices were recorded; the observations were made with constant voice tone without inflexions (this was possible because the noise was introduced mainly by the resident student attending to the visit; they were asked to be silent during the registration, and the physician was trained to use a uniform, steady tone in the voice in order to have a better registration). The recognized Romanian words in this case were 86.33% for Vocapia and 93.41% for Google.

Interpreting these tests we can conclude that the system is not 100% accurate regarding the word recognition; this is because a number of medical words are currently not included in the dictionary of the Speech to Text services, but this will be solved in future development by adding new Romanian words to the dictionary.

From the point of view of time efficiency, we estimate that the system offers about 75% reduction of the time spent by physicians to add new observations to the patient sheet, being thus relieved from writing down medical comments for about 35–40 patients during a visit, in a limited time interval; they must only connect themselves to the system, verify and correct the transcript. This new service allows doctors to spend more time in the medical process than filling data in documents. Usually, data about the patient's current health state, evolution, prescribed drugs or procedural recommendations is filled in much later in the day, with the risk of omitting important details. Hence, the quality of the medical process and service will be improved. The time efficiency is higher when the observations are larger and contain more than 30 s of recorded voice.

From the point of view of security, the system is focussed on protecting patient data. The connections inside the hospital use private Wi-Fi access points with WPA2 protocol that encrypts the data; the authentication uses Kerberos v5 protocol based on AES encryption algorithms with 256 bit Keys, and the connections to the SaaS Speech to Text and the web application are accessed using SSL encryption. The passwords are encrypted into the database and no data is stored on the mobile devices. The access to the data stored on the private cloud is based on privileges and roles and in most cases the users don't have access to personal data. The entire system is designed to respect the new EU GDPR regulation, which protects the patient personal data to disclosure, and allows deleting, at patient request, the information from the system.

Unlike other similar solutions used in healthcare, the project which implements the solution presented in the paper is integrating the components that allow tracking the patient in the hospital by creating the personal record, adding multiple observation files which are updated with voice and text observations in an integrated mode. Other solutions in the market come with a speech to text solution which is not integrated into the patient management system, and most of them are based on pc, which is not allowing the physician's mobility. This project is, to our knowledge, unique because the personal mobile phones of the physicians are used in the process, which follows the industry trend "Bring Your Own Device" (BYOD); also, only this system supports the Romanian language in this type of healthcare applications.

The system architecture, the mobile application, the web application, the database structure and the security system were developed from scratch, the other components like the cloud system, Speech to Text service, Kerberos authentication and the database management system were implemented using technologies and products already available.

The system is under development and testing, we are estimating that at the end of 2018 it will be ready to be deployed into production. Up to now, in the tests we carried out in hospitals, the practitioners proved to be favourable to the new service we propose.

Acknowledgments. This scientific work was supported by a grant of the Romanian National Authority for Scientific Research and Innovation, CNCS/CCDI - UEFISCDI, project number PN-III-P2-2.1-PED-2016-1336, within PNCDI III.

References

1. Wang, H., Wu, Q., Domingo-Ferrer, J.: FRR: fair remote retrieval of outsourced private medical records in electronic health networks. J. Biomed. Inform. **50**, 226–233 (2014)
2. Capterra: Electronic Medical Records (EMR) Software (2018). https://www.capterra.com/electronic-medical-records-software/
3. Reisenwitz, C.: EHR Software Industry User Report, Capterra Research (2015). https://www.capterra.com/electronic-medical-records-software/user-research

4. Shanley, L.A., et al.: Structure and function of observation units in children's hospitals: a mixed-methods study. Acad. Paediatr. **15**(5), 518–525 (2015)

5. AHIMA: Requirements for the disclosure of protected health information. J. AHIMA **84**(11) (2013). http://library.ahima.org/PB/Disclosure#.WprkVdhJmM8

6. Borangiu, T., Polese, F.: Introduction to the special issue on exploring service science for data-driven service design and innovation. Serv. Sci. INFORMS **9**(4), v–x (2017). https://doi.org/10.1287/serv.2017.0195

7. Vocapia Research: Speech to Text Services (2018). http://www.vocapia.com/speech-to-text-services.html

8. IBM Watson: Speech to Text (2018). https://www.ibm.com/watson/services/speech-to-text/

9. Google: Cloud Speech API. Speech to text conversion powered by machine learning (2018). https://cloud.google.com/speech/

10. RAD Studio 10.2.3: (2018). https://www.embarcadero.com/products/rad-studio

11. Todd, C., Johnson Jr., N.L.: Chapter 3 – Kerberos server authentication. In: Hack Proofing Windows 2000 Server, pp. 63–104 (2001)

12. De Clercq, J., Grillenmeier, G.: Chapter 6 – Kerberos. In: Microsoft Windows Security Fundamentals, pp. 303–408 (2007)

13. Russell, D.: High-level security architectures and the Kerberos system. Comput. Netw. ISDN Syst. **19**(3–5), 201–214 (1990)

14. Code::Blocks: http://www.codeblocks.org/license

15. AutoHotKey: https://autohotkey.com/docs/AutoHotkey.htm

Design Science Research in Services

Bringing Design Science Research to Service Design

Jorge Grenha Teixeira[1]([⊠]), Lia Patrício[1], and Tuure Tuunanen[2]

[1] INESC TEC, University of Porto, Porto, Portugal
jteixeira@fe.up.pt
[2] Faculty of IT, University of Jyväskylä, Jyväskylä, Finland

Abstract. Service design is a multidisciplinary field dedicated to create new and innovative services. To accomplish this goal, service design resorts to contributions from other disciplines such as service management, marketing, information systems and interaction design. However, service design lacks dedicated methods and models that integrate the contributions from these disciplines. Design science research (DSR) offers a solid methodology to develop such artifacts and is already starting to be used in service research. To show how DSR can support service design, this article presents two new service design methods that have been developed using DSR and examines the process followed for developing them. Building on these methods, the article discusses how DSR can leverage service design characteristics of multidisciplinarity, human-centeredness and creativity, to develop further knowledge contributions for service design. Finally, the challenges posed by using DSR in service design and service research are also discussed, as well as ways to address those challenges.

Keywords: Service design · Service research · Design science research

1 Introduction

Service design is a human-centered, multidisciplinary and creative approach for service innovation [1]. Service design involves understanding customers and service providers, their context and social practices, and translating this understanding into the development of evidence and service systems interaction [2, 3]. Service design integrates multiple contributions from service research fields such as service management, marketing and operations, as well as from technology-related fields, such as information systems and interaction design [4]. Still, service design lacks dedicated, cross-disciplinary models and methods and design science research (DSR) has been considered a valuable methodology to support the development of such artifacts [5].

DSR is well suited to support this challenge of service design. DSR supports the advancement of knowledge in a field, through the development of artifacts that have both practical and theoretical relevance [6]. March and Smith [7] identified four types of knowledge, or artifacts, produced by DSR: constructs, models, methods and instantiations. A complementary stream of research has also recognized design theories as knowledge contributions [8–10].

© Springer Nature Switzerland AG 2018
G. Satzger et al. (Eds.): IESS 2018, LNBIP 331, pp. 373–384, 2018.
https://doi.org/10.1007/978-3-030-00713-3_28

DSR is a well-established methodology in the information systems field [11, 12] and brings a solid foundation, that can support service design in the development of new and dedicated artifacts and design theories. As service design strives to develop these artifacts, it needs to ascertain that the artifacts effectively address both research and practical challenges, i.e. that the artifacts developed are both relevant and rigorously developed [13]. Following iterative cycles of relevance and rigor [13], DSR is a methodology for understanding organizational phenomena in context and advances research by creating and evaluating artifacts that solve organizational problems [6]. Whereas service design practice may generate new services that solve specific problems, DSR creates novel knowledge (artifacts and design theories) that advances the service design and the service research fields through an iterative process of conceptualization and validation.

DSR is now starting to be acknowledged in service research and service design [5, 14, 15]. Following such efforts and building on two applications of DSR in a service design context where new methods (artifacts) were developed, this article examines the potential contributions of DSR to service design. These two research projects were selected because of their novelty and thoroughness in following the DSR methodology, as defined by Peffers et al. [16]. In fact, these two research projects are, to the best of our knowledge, among the first service design artifacts developed resorting to DSR that are published in scientific journals [15, 17]. As such, these two studies offer a unique opportunity to study how DSR contribute to service design. These new artifacts (methods) are entitled MINDS (Management and Interaction Design for Service) and SD4VN (Service Design for Value Networks). This article briefly presents each of these methods, emphasizing the DSR process followed to develop them. Building upon these two applications of DSR in a service design context, this article discusses how DSR can be applied to advance service design research. Finally, the challenges posed by bringing DSR to service design are discussed, as well as possible ways to address such challenges.

2   Management and Interaction Design for Service

MINDS is an interdisciplinary method to support the design of innovative technology-enabled services that integrates and leverages contributions from two service design perspective: the managerial one, coming from service marketing, service management and operations management and a technology-oriented perspective, coming from information systems and interaction design [15]. The management perspective is focused on creating new value propositions and orchestrating multiple interfaces and backstage support processes to enhance the customer experience. The technology perspective is supported on interaction design contributions that are visually richer and loosely structured, enabling the depiction of customer interactions with technology.

Although service design resorts to models and methods from these two perspectives, they were used separately, without considering their complementarities. Following a DSR methodology, MINDS offered a new method and new integrated models, building on existing models from each perspective, such as multilevel service design models [18], affinity diagrams [19], storyboards [20] and interaction sketches [15].

The MINDS method and its models were then applied in two different service design projects, where their ability to address real-world problems was assessed. As such, using a DSR methodology that is described in the next section, MINDS contributed to service design with new dedicated models, a method and two instantiations that leverage different multidisciplinary perspectives and contribute towards the advancement of this field.

2.1 Using DSR to Develop MINDS

DSR involves the iteration between building an artifact and evaluating how well the artifact addresses real-world problems and advances the knowledge base [6]. The development of MINDS followed Peffers et al. [16] DSR process model to design MINDS, create its instantiations, and to guide its evaluation by stakeholders. This process is depicted in Fig. 1. The development of MINDS, was supported qualitative research methods as described below.

Starting by the identification of the problem and motivation, a literature review identified the gap in the knowledge base, namely the importance of connecting technology and service design for service research, the lack of dedicated service design models and the need to pursue interdisciplinary research [5, 21]. With these gaps identified, the objectives of a solution were defined: the integration of two service design perspectives (management and interaction design) to leverage the role of technology, fuel service innovation, and enhance the customer experience. To achieve these objectives a design and development phase ensued, with the conceptualization of a set of models and an integrated method, the MINDS method.

The ability of this method to address the identified challenges and objectives was then tested by applying it in two instantiations in distinct service industries (media and healthcare). These instantiations showed how MINDS was able to support the design of complex technology-enabled services and provide new contributions over dispersed models. In both cases, qualitative data was collected through semi-structured interviews with customers and other value network actors to understand the context of application of MINDS and, later, through focus-groups, to evaluate the method. In the first instantiation, MINDS supported the development of a new service to watch football. Data collection involved 17 in-depth interviews with residential customers from a media company to understand their customer experience and a focus group with the service design team to evaluate MINDS application. In the second instantiation, MINDS supported the development of a service for skin cancer prevention and treatment. Data collection involved 20 in-depth interviews with dermatologists and patients, and a focus group to evaluate MINDS. As these two applications emphasize the relevance of this research, the adherence to both DSR methodology and qualitative research canons, ensured the rigor of the contributions.

The evaluation of MINDS included both an ex ante evaluation focused on the artifact development process, and an ex post evaluation focused on the outcome of the MINDS. Regarding the outcome of MINDS, it achieved its purpose regarding its utility to solve real-world problems. MINDS was successfully applied in two research projects, helping to bridge a managerial-oriented and an interaction-oriented mindset and supporting the design of two services in very different settings. These two instantiations demonstrated the method's applicability and usefulness in real-world settings. Regarding the process, MINDS was evaluated through multiple workshops and meetings with stakeholders relevant to both instantiation. In the first instantiation, evaluation of the process was performed by a multidisciplinary design team and by the partner company. A team from the company provided feedback at five different moments during the design process, resulting in multiple changes on the models used and throughout the method. Later, the service design team evaluated MINDS during a workshop that resulted in several changes in the structure of the method. In the second instantiation, the service design team used MINDS and discussed its application monthly during the 18 months of the project. A final workshop was also performed to evaluate the method. Participants found it integrative, easy to understand and use, and a suitable communication tool for involving numerous stakeholders.

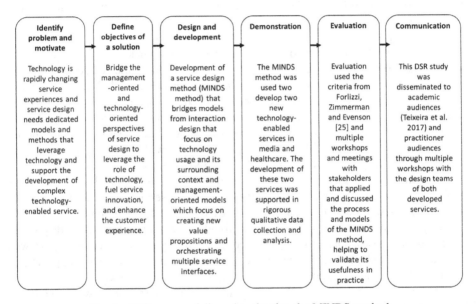

Fig. 1. DSR process followed to develop the MINDS method

MINDS was also evaluated through Forlizzi et al. [22] design research evaluation criteria of process, invention, relevance and extensibility. Regarding process, academic publishing describes the design process in detail so that it can be replicated by other researchers and improved upon. Invention concerns the novelty of the contribution and relevance concerns its importance to research and practice. MINDS and its instantiations show it can leverage on the two service design perspectives to design innovative technology-enabled services. Extensibility concerns the ability of the method to be applied in other settings. The two instantiations in different service industries suggests MINDS can be applied to other contexts.

Finally, as highlighted by Venable et al. [23] evaluation also concerns contributions to the knowledge base. This relates to the last of the steps suggested by Peffers et al.

[16], involve the dissemination of the artifact so that it can integrate the knowledge base of field. This was successfully accomplished through academic publication in a research journal [15] and, to practitioner audiences, through multiple workshops with stakeholders.

3 Service Design for Value Networks

SD4VN, is a method for creating services from a value network perspective. While other service design methods adopt an organizational perspective where the service provider develops offerings for its customers, SD4VN was developed to support the interactions between multiple parties in complex networks, such as healthcare, and their systemic context. Following a DSR methodology, SD4VN contributes with a method and a set of models to enable understanding the complex web of network interactions and designing a service to support them. Similar to MINDS, to assess the applicability and suitability of SD4VN to address real-world problems, SD4VN was used to design a service that supports a complex value network, i.e. a nation-wide electronic health record.

To achieve it purpose, SD4VN adapts and improves current service design models, such as the Actor Network Mapping [24], to take into consideration a network perspective. As such, SD4VN is focused on identifying and prioritizing relevant actors and mapping the relationships they establish within a network. It then maps the different actor's goals and expectations to support the design of services that offer a balanced response to actor's needs and avoid local optimizations that could hamper the overall well-being of the network. SD4VN thus departs from other service design methods that are focused on the organizational level and supports the design of new services on the network level through adapted and specifically created models. As described in the next section, following a DSR methodology ensured that SD4VN was rigorously developed and evaluated and thus offered a relevant contribution for service design research and practice.

3.1 Using DSR to Develop SD4VN

The development of SD4VN followed Peffers et al. [16] process to conduct DSR combined with a qualitative approach [25] to support the understanding of the real-world problems surrounding SD4VN instantiation (the development of the electronic health record) and its evaluation by stakeholders. The DSR process followed to develop SD4VN is illustrated in Fig. 2.

Regarding the problem identification and definition of research objectives, literature review on value co-creation in health care value networks, and service design provided the research challenge and theoretical support for the development of the SD4VN method. To address these challenges, the objective of this research was developing a new method to design services from a network perspective.

The design and development stage of SD4VN was supported by theoretical foundations and research methodologies to ensure that its contributions are relevant and rigorously pursued [8]. As such, the development of SD4VN evolved existing service design approaches from a dyadic perspective towards a network perspective. Finally, the demonstration phase used the artifact to design a service for a complex value network, a national healthcare system. The SD4VN method was used to develop the Portuguese electronic health record (EHR), a service to support a complex value network of interdependent actors (e.g. citizens, doctors, nurses, pharmacists). A qualitative approach, encompassing focus groups, in-depth interviews and participatory design sessions with over 170 participants at different service design stages supported the application of the SD4VN method.

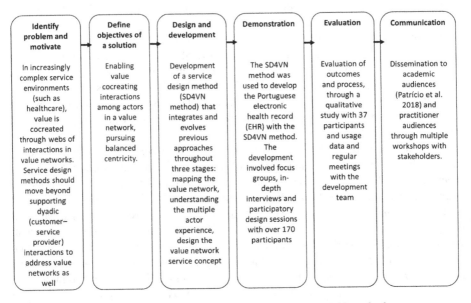

Fig. 2. DSR process followed to develop the SD4VN method

The evaluation of the SD4VN method, similarly to the MINDS method, also included an ex ante evaluation focused on the artifact development process, an ex post evaluation focused on the outcome and the evaluation of its design research contribution [22]. Regarding the outcomes, a qualitative study involving 37 healthcare professionals was performed to understand the usage of the EHR by healthcare actors and its perceived usefulness. This study concluded the Electronic Health Record had an overwhelmingly positive effect on physicians and nurses practice. Usage data also shows that, two years after its launch, one million citizens, 600 institutions and 41,000 health care professionals were registered, with an average of 18,000 health care practitioners and 6,000 citizens daily logged in to access health care information. Healthcare professionals and citizens are not obliged to register or use this EHR, making these numbers a good indicator of its success.

Second, SD4VN was evaluated regarding its process, through the regular meetings with the stakeholders. The feedback received showed that the human-centered and network approach of SD4VN provided the development team with a holistic view of the healthcare system and helped them to make design decisions without losing a systemic perspective. Finally, SD4VN was evaluated according to established criteria for design research contributions [22], namely process, invention, relevance, and extensibility. Overall, this threefold evaluation highlights that SD4VN offers valid and relevant research contributions and has been successfully applied in a real-world setting, where it supported the design of a service for a complex value network.

Finally, regarding communication, SD4VN has been disseminated to practitioner audiences through multiple workshops with stakeholders. Regarding academic communication, the SD4VN method was also published in a research journal [17]. Building on these two cases and in DSR literature, the next chapter discusses the potential contributions of DSR to advance service design. The final section concludes with the challenges of developing DSR in service research more general and in service design research more specifically and proposes ways to overcome these challenges, especially those concerning academic publication.

4   DSR Knowledge Contributions for Service Design

Being a developing multidisciplinary field, service design often resorts to "borrowed" or "enhanced" tools, that are not entirely adapted to the service setting. MINDS and SD4VN show how DSR can contribute to the development of new service design knowledge, namely new and dedicated service design methods that are relevant and rigorously developed. However, the potential contributions of DSR to service design can be more comprehensive than the ones demonstrated on MINDS and SD4VN. With the support of DSR, service design can develop its own artifacts and even increasingly complex and abstract design theories that are able to anchor the field. Also, DSR can leverage service design characteristic of multidisciplinarity, human-centeredness and creativity, to develop all four types of DSR contributions: invention, improvement, exaptation, and routine [8]. Following a heuristic theorizing process [26], the knowledge gained through the developing the service design artifacts (MINDS and SD4VN) was compared with the before mentioned four DSR contribution types in order to theorize on how DSR can contribute to service design research. We argue that distinct contributions emerge, which are (1) contextualized to service design, (2) supported by the DSR approach (3) and which offer potential knowledge contributions to service design research. These are summarized in Table 1.

Regarding invention, by following an iterative cycle of exploration, ideation and implementation service design can define new problems and develop new solutions. DSR offers a structured approach to develop new service artifacts and validate them, ensuring that new and valid research knowledge is created. What's more, new innovative services, instantiations, drive prescriptive knowledge formation of constructs, models, methods, or design theories that can further service design knowledge.

Second, for improvement we see that service design has a wealth of artifacts at its disposal. However, coming from different origins of research and business as well, these artifacts may not have been adapted to service design. DSR can here support the improvement and adaption of these service artifacts to known service design problems, thus offering new research contributions. Thus, the context situated service instantiations can offer improvements compared to existing artifacts in the original field of research or business. Constructs, methods, models, and design principles, in turn, may lead to new descriptive knowledge.

Third, service design is a multidisciplinary field that draws knowledge from many other areas, such as management, marketing, interaction design, information system, software engineering and others that can be used for exaptation. DSR can help in adapting existing artifacts that are known in one field to new service design problems, thus offering nontrivial and interesting extensions to current service design practice, research opportunities and new descriptive and prescriptive knowledge contributions.

Table 1. DSR knowledge contribution framework for service design research.

Knowledge type	Service design context	DSR's offering	Knowledge contribution(s)
Invention: Invent new solutions for new problems	Service design follows iterative cycle of exploration, ideation and implementation that supports the definition of new problems and the development of new solutions for these problems	DSR offers a structured approach to develop new artifacts and validate them, ensuring that new and valid research knowledge is created	New innovative services (service instantiations) drive prescriptive knowledge formation of constructs, models, methods, or design theories
Improvement: Develop new solutions to know problems	Service design has a wealth of artifacts at its disposal. However, coming from different origins, these artifacts are not adapted to service design	DSR can then support the improvement and adaption of these artifacts to known service design problems, thus offering research contributions	Context situated service instantiations offer improvements vs. existing artifacts. Constructs, methods, models, and design principles are proposed as research improvements that may lead to descriptive knowledge
Exaptation: Extending known solutions to new problems	Service design is a multidisciplinary field that draws knowledge from many other areas, such as management, marketing, interaction design, information system, software engineering and others	DSR can help in adapting existing artifacts that are known in one field to new problems, thus offering research opportunities and knowledge contributions through exaptation	The extension of known design knowledge into a service design that is nontrivial and interesting. Both descriptive and prescriptive knowledge contributions to service design are offered

(continued)

Table 1. (*continued*)

Knowledge type	Service design context	DSR's offering	Knowledge contribution(s)
Routine design: Applying known solutions to known problems	Service design as a professional activity is growing and it applies existing knowledge to organizational problems	DSR offers a way to link service design practice with research opportunity and contribution to knowledge	Application of existing knowledge to service design problems using "best practice" artifacts (constructs, models, methods, and instantiations)

Finally, with routine design we look at applying known solutions to known service design problems. DSR offers a way to link service design practice with research opportunity and contribution to knowledge. This adds to knowledge by application of existing solutions to service design problems using "best practice" artifacts (constructs, models, methods, and instantiations).

The development of each of these contributions should follow a process, such as the one described by Peffers et al. [16]. This process can also be adapted to a service design context, as follows:

1. Identify problem and motivate: being a young field and following a holistic approach, service design can tackle a great number of problems. It is also a human-centered field, striving for improving service provision to a broad group of stake-holders. As such, the problems addressed by service design are well-suited for DSR, that aims to produce scientific knowledge with a real-world contribution
2. Define objectives of a solution: service design deals with the so-called "wicked problems" [27], i.e. ill-defined problems that do not have an optimal solution, but rather a satisficing one. As such, the definition of objectives of a DSR solution in a service design setting can seldom be quantitatively assessed. A qualitative description can offer a suitable alternative to define the objectives of a service design solution developed using DSR.
3. Design and development: the two cases described in this article show new service design models and methods developed with DSR. This builds on a strong tradition of service design visualization tools that are used in research in practice. However, DSR can also support the development of more fundamental building blocks, such as constructs, or more comprehensive design theories.
4. Demonstration: service design is intrinsically oriented towards applying new designs in concrete cases. This application might range from prototyping new solutions to more complete case research.
5. Evaluation: following the definition of the solution, the evaluation of service design artifacts can be supported by qualitative methods.
6. Communication: being an emerging field, service design is still consolidating its research outlets for scholarly communication. However, following MINDS and SD4VN examples, suitable alternatives for scientific publication can be found in service science and service research outlets.

By offering a structured and rigorous process that supports the development of artifacts and can be adapted to the service design context, it is argued that DSR can provide a needed approach to leverage service design throughout the four types of research contributions. However, the application of DSR in a service design still faces numerous challenges. In the next section, these challenges are discussed, along with ways to resolve them.

5 Challenges Facing Us and Ways to Resolve Them

The application of DSR in a service design setting holds great promise, and MINDS and SD4VN were successfully developed using DSR methodology and had the acceptance from both practitioner and academic audiences. Still, using DSR in a service design context faces several challenges. First, DSR is still largely unknown in both service research and service design and, as such, special care should be given to introduce and explain it. To address this challenge we found that the DSR publication schema and knowledge contribution framework proposed by Gregor and Hevner [8] are particularly useful. Also, the DSR process systematized by Peffers et al. [16] offers a clear path to present the development of the artifacts and their relevance for research and practice. Other DSR methodologies should be also considered such as design-oriented information systems research [12] or action design research [28].

Second, the nature of the contribution of a research work develop with DSR can be questioned. In fact, while service design is recognized as a multidisciplinary field [29] and engaging in interdisciplinary research is a service research priority [5], the audiences from different fields have different epistemological understandings on what a research contribution is. The design field, for example, debates the positivist and constructivist approaches [30], with contributions building on Simon [31] being considered an effort to position design as an orthodox and positivist research field [32], and those building on Schon [33] being the constructivist answer. From such a perspective, DSR, in its effort to bring rigor and structure to the design process, is aligned with a positivist view in the design field. However, in other fields with a strong influence from approaches coming from the natural sciences, DSR, with its focus on building artifacts, is aligned with a more constructivist view. To solve potential clashes between these perspectives, DSR can build upon the relevance of its contributions. In fact, both in management [34] and IS [9], research argued that an overreliance on approaches building on positivist, natural sciences leads to a decrease on the relevance and innovation of the research contributions. Also, combining DSR with a qualitative approach, such as the one employed to develop MINDS and SD4VN, can support abductive research that is rigorously grounded, while addressing relevant and "wicked" problems [27].

Finally, evaluation of DSR contributions is often a challenge and as Peffers et al. [35] concisely describe "research papers are unlikely to be published in influential outlets, unless authors can make persuasive arguments that artifacts were appropriately evaluated". Also, Gregor and Jones [9] present a related concern that "both students and more experienced researchers struggle with the problem of expressing design knowledge in an acceptable form in theses and journal articles". Indeed, clearly showing how developed artifacts contributed to the claimed results was a challenge while publishing MINDS and SD4VN methods in service research outlets. Still, there were ways for successful evaluation of MINDS and SD4VN artifacts. First, research design was planned so that data on the methods usage by the stakeholders was collected along the process, both through workshops and focus groups. Second, qualitative research approaches were also employed to ensure that the feedback about the artifacts was rigorously collected and analyzed. Third, Forlizzi et al. [22] criteria for evaluating design research contributions was useful to thoroughly evaluate the artifacts through their process, relevance, invention and extensibility. Also, Hevner et al. [6] design evaluation methods and Peffers et al. [35] evaluation types can provide guidance to choose the most adequate way to evaluate one artifact. Despite these challenges, MINDS and SD4VN were successfully developed using DSR, hopefully paving the way for further application of this methodology in service research and service design, extending it along the lines of the knowledge contribution types discussed in this article.

References

1. Yu, E., Sangiorgi, D.: Service design as an approach to implement the value cocreation perspective in new service development. J. Serv. Res. **21**, 40–58 (2018)
2. Holmlid, S., Evenson, S.: Bringing service design to service sciences, management and engineering. In: Hefley, B., Murphy, W. (eds.) Service Science, Management and Engineering Education for the 21st Century. SSRI, pp. 341–345. Springer, Berlin (2008). https://doi.org/10.1007/978-0-387-76578-5_50
3. Yu, E., Sangiorgi, D.: Service design as an approach to new service development: reflections and futures studies. In: ServDes. 2014. Fourth Service Design and Innovation Conference "Service Futures", pp. 194–204. (2014)
4. Patrício, L., Gustafsson, A., Fisk, R.: Upframing service design and innovation for research impact. J. Serv. Res. **21**, 3–16 (2018)
5. Ostrom, A.L., Parasuraman, A., Bowen, D.E., Patrício, L., Voss, C.A.: Service research priorities in a rapidly changing context. J. Serv. Res. **18**, 127–159 (2015)
6. Hevner, A.R., March, S.T., Park, J., Ram, S.: Design science in information systems research. MIS Q. **28**, 75–105 (2004)
7. March, S.T., Smith, G.F.: Design and natural science research on information technology. Decis. Support Syst. **15**, 251–266 (1995)
8. Gregor, S., Hevner, A.R.: Positioning and presenting design science research for maximum impact. MIS Q. **37**, 337–355 (2013)
9. Gregor, S., Jones, D.: The anatomy of a design theory. J. Assoc. Inf. Syst. **8**, 312 (2007)
10. Walls, J.G., Widmeyer, G.R., El Sawy, O.A.: Building an information system design theory for vigilant EIS. Inf. Syst. Res. **3**, 36–59 (1992)
11. Winter, R.: Design science research in Europe. Eur. J. Inf. Syst. **17**, 470 (2008)
12. Kuechler, W., Vaishnavi, V.: The emergence of design research in information systems in North America. J. Des. Res. **7**, 1–16 (2008)
13. Hevner, A.: The three cycle view of design science research. Scand. J. Inf. Syst. **19**, 87 (2007)

14. Beloglazov, A., Banerjee, D., Hartman, A., Buyya, R.: Improving productivity in design and development of information technology (IT) service delivery simulation models. J. Serv. Res. (JSR) **18**, 75–89 (2015)

15. Teixeira, J.G., Patrício, L., Huang, K.-H., Fisk, R.P., Nóbrega, L., Constantine, L.: The MINDS method: integrating management and interaction design perspectives for service design. J. Serv. Res. **20**, 240–258 (2017)

16. Peffers, K., Tuunanen, T., Rothenberger, M.A., Chartterjee, S.: A design science research methodology for information systems research. J. Manag. Inf. Syst. **24**, 45–77 (2007)

17. Patrício, L., Pinho, N., Teixeira, J.G., Fisk, R.P.: Service design for value networks: enabling value cocreation interactions in health care. Serv. Sci. **10**, 76–97 (2018)

18. Patrício, L., Fisk, R.P., Cunha, J.F.E., Constantine, L.: Multilevel service design: from customer value constellation to service experience blueprint. J. Serv. Res. **14**, 180–200 (2011)

19. Beyer, H., Holtzblatt, K.: Contextual Design: Defining Customer-Centered Systems. Morgan Kaufmann Publishers, San Francisco (1998)

20. Truong, K.N., Hayes, G.R., Abowd, G.D.: Storyboarding: an empirical determination of best practices and effective guidelines. In: Proceedings of the 6th Conference on Designing Interactive Systems, pp. 12–21. ACM (2006)

21. Ostrom, A.L., et al.: Moving forward and making a difference: research priorities for the science of service. J. Serv. Res. **13**, 4–36 (2010)

22. Forlizzi, J., Zimmerman, J., Evenson, S.: Crafting a place for interaction design research in HCI. Des. Issues **24**, 19–29 (2008)

23. Venable, J., Pries-Heje, J., Baskerville, R.: A comprehensive framework for evaluation in design science research. In: Peffers, K., Rothenberger, M., Kuechler, B. (eds.) DESRIST 2012. LNCS, vol. 7286, pp. 423–438. Springer, Heidelberg (2012). https://doi.org/10.1007/978-3-642-29863-9_31

24. Morelli, N., Tollestrup, C.: New representation techniques for designing in a systemic perspective. In: Nordic Design Research Conference, Stockholm, Sweden (2007)

25. Corbin, J.M., Strauss, A.L.: Basics of Qualitative Research: Techniques and Procedures for Developing Grounded Theory. Sage Publications Inc., Thousand Oaks (2008)

26. Gregory, R.W., Muntermann, J.: Research note—heuristic theorizing: proactively generating design theories. Inf. Syst. Res. **25**, 639–653 (2014)

27. Buchanan, R.: Wicked problems in design thinking. Des. Issues **8**, 5–21 (1992)

28. Sein, M.K., Henfridsson, O., Purao, S., Rossi, M., Lindgren, R.: Action design research. MIS Q. **35**, 37–56 (2011)

29. Blomkvist, J., Holmlid, S., Segelström, F.: Service design research: yesterday, today and tomorrow. In: Stickdorn, M., Schneider, J. (eds.) This is Service Design Thinking, pp. 308–315. BIS Publishers, Amsterdam (2010)

30. Dorst, K., Dijkhuis, J.: Comparing paradigms for describing design activity. Des. Stud. **16**, 261–274 (1995)

31. Simon, H.A.: The Sciences of the Artificial. MIT Press, Cambridge (1969)

32. Cross, N.: Designerly ways of knowing: design discipline versus design science. Des. Issues **17**, 49–55 (2001)

33. Schon, D.A.: The Reflective Practitioner: How Professionals Think in Action. Basic Books, New York (1984)

34. Aken, J.E.: Management research based on the paradigm of the design sciences: the quest for field-tested and grounded technological rules. J. Manag. Stud. **41**, 219–246 (2004)

35. Peffers, K., Rothenberger, M., Tuunanen, T., Vaezi, R.: Design science research evaluation. In: Peffers, K., Rothenberger, M., Kuechler, B. (eds.) DESRIST 2012. LNCS, vol. 7286, pp. 398–410. Springer, Heidelberg (2012). https://doi.org/10.1007/978-3-642-29863-9_29

Scaling Consultative Selling with Virtual Reality: Design and Evaluation of Digitally Enhanced Services

Osmo Mattila[1]([⊠]), Tuure Tuunanen[2], Jani Holopainen[1],
and Petri Parvinen[1]

[1] Department of Forest Sciences, University of Helsinki,
P.O. Box 27, 00014 Helsinki, Finland
{osmo.mattila, jani.m.holopainen,
petri.parvinen}@helsinki.fi
[2] Department of Computer Science and Information Systems,
University of Jyväskylä, P.O. Box 35, 40014 Jyväskylä, Finland
tuure@tuunanen.fi

Abstract. Virtual, augmented, and mixed reality technologies allow creation of powerful customer experiences and illustrative demonstrations especially in use cases that benefit from spatial visualizations. Our study focuses on the natural resource management sector and digitalizing of consultative selling process. More specifically, we look at how to improve customer engagement with the use of virtual reality (VR) and thus digitally scale consultative selling. In this process, a VR application is used to demonstrate various management operations and their economic results. Design research methodology is applied to a pre-development phase and three application development iterations between 2016 and 2018. Data consists of user interviews and video observations (N = 129) during various development iterations and three application development plans. The results show that VR offers an emotionally engaging and illustrative tool in consultative selling. Further, it opens a novel way for interaction between the salesperson and customer and possibilities to scale consultative selling digitally, emphasizing the role of trust.

Keywords: Consultative selling · Design science research methodology
Framework for evaluation in design science · Virtual reality

1 Introduction

Tightening competition pushes companies to develop services that create memorable events to their customer, and the ability to create positive customer experiences helps companies in differentiating [1]. Because of their recent technological advancements, various visually immersive computer-mediated realities such as virtual reality (VR), augmented reality and augmented virtuality have attracted attention in research and media. Currently available VR-technologies in the consumer markets allow the creation of personal and strong emotional experiences [2]. This enables novel paths for companies to create customer experiences to interact with a firm. However, it is not known

© Springer Nature Switzerland AG 2018
G. Satzger et al. (Eds.): IESS 2018, LNBIP 331, pp. 385–398, 2018.
https://doi.org/10.1007/978-3-030-00713-3_29

how to systematically design these digitally supported services to be a natural part of a customer journey aiming to increase customer value and sales for firms.

Gaining an ability to create strong customer experiences has become a leading management objective in many firms [3]. Customer engagement constitutes touch points along the customer journey [3] and it is a psychological state that occurs by virtue of interactive, co-creative customer experiences [4]. For companies considering how to increase the scalability of their business models, customer engagement is a central mechanism as it concerns the value proposition the firm offers to its customers [5]. A decision about the level of customization is usually related to scalability and it is closely related to the degree to which a firm's operations or customer's experiences are made possible by digital technologies [5].

In this research, we focus on customer engagement in consultative selling utilizing a novel VR application in visualization and scalability of this service. Currently, VR content platforms are still on their way towards institutionalization [6] and they should currently be considered as tools for improving customer and employee interaction and business performance [7].

In this research, DSR methodology is used to describe and evaluate the development process [8] of a VR application currently under development – the described process consisting of a pre-development phase and three application development iterations implemented between 2016 and 2018. The digitally supported consultative selling of forest management services was selected as the use case. The application is aimed to increase scalability in a market environment in which the currently dominating consultative sales process is challenged by e.g. long physical distances and fragmented customer base [9]. The main challenge of the application in the use context was recognized to be related to the user acceptance, i.e., how to a fit new technological solution into the sales situation. In this research, customer engagement as a service scalability mechanism [5] is selected as the main guideline for application development evaluation.

2   Literature

2.1   Design Science Research Process and Evaluation

Design science in information technology is a research approach aiming to create and evaluate artifacts to solve identified organizational problems [10]. In this research, organizational problems arise from the need to digitalize personal selling practices to improve scalability and to offer engaging customer experiences. DSR – a popular framework for planning and evaluating service development especially in information systems research [8] – is applied. The framework provides a nominal model for doing design science research consisting of six steps: (1) problem identification and motivation, (2) definition of the objectives for a solution, (3) design and development, (4) demonstration, (5) evaluation, and (6) communication.

To ensure the usability of the application it is important that it is tested in the real use situations [11] and design can be integrated as a major component of research [12]. Evaluation of design artifacts and design theories have become a central part of DSR

[10, 13] and it may be tightly coupled with design itself [14]. Venable et al. [14] argue that because design artifacts and design theory evaluation are used to actually design, develop, or 'build' new artifacts, they are more relevant, important, and specific to DSR than other research paradigms.

Venable et al. [14] have developed a DSR evaluation framework, FEDS, to complement the existing evaluation frameworks and to offer a new evaluation design process for applying that framework. FEDS is designed to give an answer to the question of "What would be a good way to guide the design of an appropriate strategy for conducting the various evaluation activities needed throughout a DSR project and to bridge the gap between evaluation goal and evaluation strategies?" [14]. DSRM, in turn, describes purpose of the DSR evaluation as whether the purpose is to help (1) formatively to improve the outcomes of the process under evaluation or (2) summatively to judge the extent that the outcomes match expectations. What's more, Peffers et al. [8] argue that we should also see whether the DSR evaluation is (1) artificial and tests the research hypotheses nearly always in a positivist and reductionist way [15] not excluding the possibility to use interpretive techniques or (2) naturalistic and explores the performance of a solution in its real environment [14].

Firstly, FEDS helps in concretizing, why an artifact is evaluated, in other words, whether the reason is to support decision making formatively by concentrating expected consequences or to evaluate the meaning of an artifact to support the selection of the evaluand [16]. Secondly, FEDS describes the timing of when to evaluate, whether to predictively evaluate the impact of future situations, or to assess the value of the implemented system [15]. FEDS provides four steps for evaluation: (1) explicating the goals, (2) choosing strategies for the evaluation, (3) determining the properties to evaluate, and (4) designing the individual evaluation episodes [14].

2.2   Scaling Consultative Selling with Virtual Reality

Consultative sales behavior is a practice of a salesperson trying to help their customers to make purchase decisions that will satisfy customer needs [17]. Consultative selling is defined to be a "process of providing information in a professional fashion to help customer take intelligent action to achieve their business activities" [18]. More broadly, consultative selling is recognized as one of the value-related salesperson's behaviors that aim at understanding the customer's business model, crafting the value proposition and communicating customer value [19].

Zhang et al. [5] have proposed that three mechanisms are central for the scalability of digital business models: engaging both paying and non-paying customers, organizing customer engagement to allow self-customization, and orchestrating network value chains. Sources of scalability for these mechanisms are proposed to originate from dynamics of (1) learning by using, (2) network externalities, (3) economic scale in production and distribution, (4) informational increasing returns, (5) technological interrelatedness [20–22] and (6) distributed resourcing [5]. Requirements for customization and human interaction in consultative selling are high compared to many entirely digitally realized services. Therefore, the focus of our research at the current development level of the application is more on mechanisms engaging customers than orchestrating network value chains.

This research aims at digitalizing a part of consultative selling in a business case that has two perquisites for potentially benefitting from it: the case relies on consultative sales tradition and is challenged by digitalization that increases business performance requirements. For this purpose, we build on customer experience and engagement literature and VR technologies.

Customer experience is a multidimensional construct that can be defined as the internal and subjective response that customers have to any direct or indirect contact with the company [23, 24]. The experience is customer's personal and emotional reaction to an event, interaction with a brand or a firm [25]. By adopting the interpretation of representative heuristics [26], total customer experience is composed by a customer judging events not by the entirety of an experience, but by prototypical moments. Firms are broadening their thinking about marketing by designing and managing the entire processes the customers go through in more systematic ways [3] and tracking experiences at customer touch points helps in developing understanding how an experience can be enhanced for the customer [27]. The design, delivery, and management of the customer experiences can be divided into multiple perspective including for instance the firm's point of view, customer's point of view, and the co-creation perspective [3].

Customer engagement, in turn, focuses on the extent to which the customer reaches out and initiates contact with a firm [3]. Customers can be cognitively and affectively committed to an active relationship with the brand as personified by computer-mediated entities that are designed to communicate brand value [28]. Customer engagement attempts to distinguish customer attitudes and behavior beyond purchases and it can be classified into (1) cognitive, (2) emotional, and (3) behavioral responses to the firm's offerings on the part of the customer [4]. For companies, the value of customer engagement is measured in forms of purchasing behavior, referral behavior, influencer behavior and knowledge behavior [29].

Finally, mixed reality encompasses both augmented and VR technologies largely covering concepts that mix virtual and real-life experiences [30]. More specifically, VR covers computer technologies that use software to create realistic sensations that represent an immersive environment and simulate user's physical presence in this environment [31]. Scholz and Smith [32] also point out that these technologies can prompt interaction between various parties, even between users and bystanders. Thus, VR aims at creating sensorial stimulations while trying to avoid awareness of intrusion, in other words, the presences of experience in another world is accepted naturally [2]. Presence is related to the emotion of 'being there' [33] VR technologies allow illusion of immediately to be transported into the computer world beyond the head mounted displays [34] and the creation of personal and strong emotional experiences [2]. Interaction in VR is becoming an increasingly important research topic. People in VR can understand and empathize when they comprehend another person's subjective experience and environment allowing people to understand each other [35]. This justifies the use of VR in studying how to scale consultative selling.

3   Methodology and Data

3.1   Research Approach and Structure

Methodologically, this paper follows the DSRM [8] analyzing the performance of the application in a naturalistic way in its real environment [14]. FEDS is used to evaluate the development outcomes focusing on customer engagement as a service scalability mechanism [5]. The development process consists of a pre-development phase and three iterations that all include a software development cycle resulting in an 'artifact', which is tested by users and evaluated by the researchers. The user experience research phases were implemented empirically and they were artificial and summative in their nature. Results of the user tests were used to formulate development proposals for managerial purposes in each development iteration. They were naturalistic and for-mative in their nature with an aim to improve the outcomes of the service design process.

The main challenge of this VR application is related to the user acceptance, i.e., that the application will not fit well into the sales situation. Therefore, our goal for eval-uating the artifacts with a focus on customer engagement as a mechanism by which a business model attempts to gain scale. Scalability in terms of economic scales in production and distribution at this point of the development are marginal, even though not dispensed. Therefore, emotional customer experiences and perceived usability were recognized as the most central features to evaluate. These are related to emotional and cognitive customer engagement. Factors related to behavioral customer engagement such as social context and word-of-mouth became important when testing the appli-cation with real customers. Finally, behavioral customer engagement was recognized to be even more important making the role of the application as a part of the customer journey focal (Fig. 1).

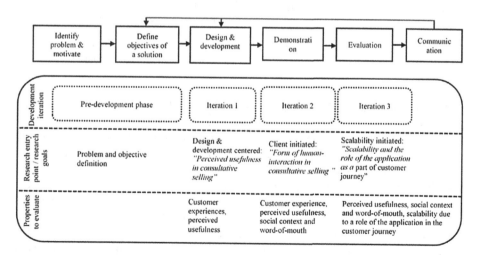

Fig. 1. Evaluating customer engagement and scalability of VR enhanced consultative selling.

Our DSR evaluation efforts – summarized in Table 1 – consists of user interviews, observations, and open-ended survey questions of altogether 129 users, and three research and development plan documentations. In all the phases, the interviews were recorded on audio and transcribed. Further, we applied Biocca's [36] approach to study people both in a virtual world observed simultaneously as avatars in a virtual world but also their person in the physical space. This allows the study of participant's behavior without interrupting them [37]. Also the recorded video data was transcribed. Further, iteration two included also multiple quantitative survey questions where the user was asked to rate perceived realness of the environment, usefulness, learning and behavioral intentions of e.g. sharing the experience by using a five-point Likert scale. Here, the effects of the form of presence of the service person was tested by using Mann-Whitney U-test [38]. In addition, an e-mail address of a friend or a relative was asked and it was coded as yes/no-answer in the analysis to indicate actual willingness to share the experience when comparing remote and present appearance of the service person.

Table 1. Summary of the data.

Development iteration and time	Form of interaction	Data	Customer engagement approach (E = emotional, C = cognitive, B = behavioral)	N
Iteration one, autumn 2016	Present	User interviews and video observations + research and development plan	Customer experiences (E), perceived usefulness (C)	50
Iteration two, autumn 2017	Present vs. remote	Customer interviews and video observations + research and development plan	Customer experience (E), perceived usefulness (C), social context and word-of-mouth (B)	64 (37 + 27)
Iteration three, spring 2018	Individual use, present in sales situation	Customer interviews and video observations + research and development plan	Perceived usefulness (C), social context and word-of-mouth (B), scalability due to a role of the application in the customer journey (B)	15 in pre-test 6 of them in-depth (each user participating 3 test rounds)

The interviews allowed the participants to reflect on the use of the application from his/her own perspective [39]. The user experience interviews and managerial research and development plans were analyzed qualitatively by using Atlas.ti—software. By following the interpretations by Kahn [40] and Hollebeek [41], the user experiences were categorized into emotional (E), cognitive (C) and behavioral (B) elements and depending on the case into customer experience, perceived usefulness, social context and word-of-mouth, and scalability due to a role in the customer journey. To improve the reliability of an interpretive analysis by ensuring that the observation represent the practices they claim to represent [42] the team had expertise in service and land owner research. To improve the validity of the findings, the researchers met multiple times to discuss the themes and the empirical evidence.

3.2   Pre-development Phase

Before starting the application development, three business meetings and one workshop were organized in the spring 2016. As there was no relevant content available in the natural resource management context, twelve professionals of that specific topic were familiarizing themselves with the VR technology and existing applications in the fields on architectural and industrial maintenance.

In the pre-development phase, it was recognized that emerging mixed reality technologies will have various use cases in the natural resource management context and more specifically, in consultative sales. The main potential value drivers were related to the good availability of the natural resource inventory data to support scaling the application, and traditions of the consultative selling simultaneously suffering from high travelling costs. As a result of these discussions, VR application to land owners' decision making process was seen as a potential and feasible use case to develop a VR application.

3.3   Iteration One – Design and Development Focus

The first version of the application was created in the autumn 2016 in a research project of two universities. The objective was to determine whether it is possible to develop an environment in virtual reality that could be used to visualize various management operations. The solution needed to fulfill sufficient visual and functional quality requirements of the users. To open paths for possible further development, it needed to demonstrate that an environment based on real natural resource inventory information can be modeled, if the first requirement was fulfilled.

A land site was captured by using a stationary terrestrial laser scanner and 360-degree photos to help the users to evaluate and compare the virtual experience with an experience in nature. This resulted in a colored 3D point cloud representing 25×25 m precisely scanned area. One 360-degree photo and a simplified point cloud were imported to a gaming engine and the environment was generated based on this data by using basic terrain textures and simple assets. This enabled an interactive environment where the user was able to gain money by removing trees and explore the area by either taking a few physical step or by teleporting. In addition, a bear was placed to wander around the area. The main gaming area was surrounded by a larger space allowing free movement but only plain terrain. Besides interaction with the asset-based environment, the user was able to watch the 360-degree photo and visit the point cloud where it was possible to move by walking physically in 2.5×2.5 m area and to see the scene before any trees were removed. Figure 2 presents three sample views of the application with an overview from a hill, laser-scanned point cloud visualization and a 360-degree photo. A VR-headset with two controllers was used as a user interface and there was a computer running the system.

Fig. 2. Screenshot of the application on the development phase 1.

During three demonstration days, 50 users consisting mostly of invited business managers tested the management application and were interviewed. All the use tests were recorded on video resulting to 13 h of video material and interviews after the experience were recorded and transcribed resulting in 6 h of audio material and 96 pages on transcribed interview material. The form of the interview was open and followed the customer engagement framework starting from asking the user to describe the experience which led to a description of emotional experiences and functionalities that were possible to conduct in the application. The interviewee then continued by asking the user to more cognitively reflect the usefulness and utilities the user would feel, the interviewer continued by asking how these kinds of experiences could be derived to values or goals related either to this application or in interviewee's own business. This interview process was continued until the saturation point of the interviewee having nothing more to say. The transcribed interview material was analyzed by classifying the comments whether they were covering emotional experiences or cognitive analysis covering utilities the application could offer either in user's own business or related to the tested application. Finally, this material was coded based on emotional customer experiences or cognitively perceived usefulness [41]. Further, a research plan and development proposal for the next application development phase was created.

3.4   Iteration Two – Demonstration Focus

The second version of the application was created in the autumn 2017 in collaboration with a technology development company, a university and an industrial company. During this iteration, a land area covering 10 ha was captured by using a portable terrestrial laser scanning and 360-degree photos. Based on the point cloud data, open access terrain data and existing natural resource inventory information, an interactive 3D model of the area was created by using a gaming engine. The user interface was the same as in the previous iteration, i.e. a virtual reality headset with two controllers and a computer running the system. The application allowed the users to examine detailed information about trees, gain money by cutting single trees and making large-scale management operations such as clear cutting, move by physically walking or teleporting by using the controllers (Fig. 3) and visit 360-images. The user was able to use these functionalities in three areas. Each area represented a different kind of nature and included an information sign telling about the area and providing a management proposal. The user was able to choose whether to test any of the management operation and to compare the revenues of these actions in monetary terms. Finally, the user was able to cancel all the operations already done.

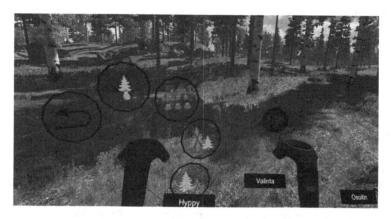

Fig. 3. Screenshot of the second version of the application

Seven days of demonstrations were organized and altogether 64 users tested the application; 19 females and 45 males. This number includes 10 pre-testers who were service personnel of the industrial company. The rest of the users were landowners. The age of the users varied between from 25 to 90. The user tests were recorded by using 360-degree cameras capturing both the user actions in the real world and in VR [36] resulting in 11 h of video material. In addition, this iteration included a comparison of customers using the application with a salesperson guiding 37 users in-person and 27 remotely. In remote contact, a web camera view and screen presenting the view in VR were shared via a voice over internet protocol application. The users were interviewed shortly before the use and more comprehensively after the experience using open ended survey questions. The interviews were recorded resulting to 8 h of audio interview recordings. A total of 140 pages of observation notes and interview recordings were transcribed and analyzed by focusing on customer engagement as follows: emotional customer experience, cognitively perceived usefulness, and behavioral social context and word-of-mouth. Based on the results, a research and development proposal was formulated for the next development phases.

3.5   Iteration Three – Evaluation Focus

Third version of the application was implemented in 2018. During this iteration, an ethnographic research approach will be applied to analyze the use of this application in persuading current customers to make wood trade in an acquisition of new customers. This research phase will be focused on deeply understanding users who represent the selected customer segments of the industrial company. Data collection and analysis for this phase will be finished by autumn 2018.

At the first stage, 15 users were recruited from a seminar targeted to land owners and the interested ones were briefly interviewed to identify willing test subjects for the next stage. Two users from each customer segments will be selected, one of them representing a current customer and the other a prospect customer recruited from the first stage. The application versions to be tested are (1) an improved version of the second research phase, and (2) a mobile version of the application. The mobile version is an updated mobile version of the previous one with modified visuals and features. For this iteration, a smartphone with a virtual reality headset and one controller will be used. From the customer engagement point of view the focus is on cognitively perceived usefulness, and behavioral social context and word-of-mouth and scalability by considering the role of the application in the customer journey.

4   Findings and Discussion

The results open various interesting development paths considering the future and scalability of VR tools for consultative selling. Even though there are other applications available for participative planning, VR was considered as a very useful and illustrative tool. Considering the version 2, in which the usefulness of the system was asked, 82% agreed or strongly agreed that the system would be useful with no statistical difference whether the service person was present or the instructions were given remotely. 36% of the users of version 2 commented that the way of visualizing different management operations was illustrative, concrete or helped them to understand the results. VR also makes it easy for the user to concentrate on the content.

By allowing customers to participate the process, the company can learn about their preferences. In line with Zhang et al. [5], as the customer is doing a part of the production work themselves, the costs of satisfying their specific need are minimized. Starting from the development version 1 it was found out that communication during the interaction was easy. By offering possibilities for a customer to participate in planning by trying out various scenarios, it becomes easy for a salesperson to observe the user and to discuss in order to – by following Terho et al. [19] – understand customer's value model, craft the value proposition and communicate customer value. This increases understanding of the products, its value, adaptation and scale [5]. During all the development iterations, users were active in participating idea generation of how to improve the application. Willingness to participate in development can be interpreted being behavioral engagement to interact with the firm [43]. What's more, during development iteration two, it quickly became clear that the close service contact made the opportunity for fruitful discussions as the users were actively telling about their feelings during the use. Majority of the users were talking or commenting something during the use and right after removing the headset.

When considering network externalities, this willingness to participate can be utilized when the adoption of the technology still requires human interaction VR [7].

By helping customers to participate in co-creation in their own social network opens ways for scaling the service. In iteration three, the use of the application in various parts of the customer path was explored by giving the mobile devices to users. This may help them to make decisions with the larger group of stakeholders which is the practical case in land owning where the estate is often owned by a group of owners only one of them able to meet a salesperson of the company. From the scalability point of view, the company could use key contributors – such as active landowners or entrepreneurs – to help diffusing the service [5]. In line with Scholz and Smith [32] on behavioral engagement, technology was prompting interaction between users and bystanders, as the users were active in persuading the next users to put the headset on to share the experience and hence to increase network externalities [5].

Even though the users expressed much positive emotion by e.g. laughing and talking during the use, based on the video observations, the users testing the interaction remotely were the most reserved which was expressed by not talking and joking as much as the users in the other group. Interestingly, only 19% of the users who were served via the remote connection gave the contact information of their friends or relatives compared to 51% in face-to-face service which may be interpreted as an indication or mistrust. However, we argue that building trust in multiple ways [44] is the key where the most obvious one could be e.g. offering a familiar service person to lower this effect. In line with Hirschman and Holbrook [45], the emotional aspects of decision-making and experience should be recognized more broadly when designing the customer experience. From the company perspective, it is also easy to remind customers from strong emotional customer experience later. Some users in case one took photos of others using the system in iteration 1. Considering technological interrelatedness, the expressed high willingness to share the experience can be scaled easily e.g. by videoing the use and sharing in social media.

Remote consultation was rather easy to implement and it works well especially when the focus of the interaction is in transmitting information. Therefore, scalability potential by e.g. establishing call centers is good. This is also related to economic scale in production and distribution as well as to technology interrelatedness [5]. The quick development in multiplayer features and possibilities to easily make recordings in VR also open paths for various network-marketing strategies. Further, integrating data queries from databases including up-to-date information about prices and volumes of timber help in scaling the service when considering of orchestrating network value chains.

5   Conclusions and Further Research

The application is targeted to create value in a expertise service sales in a case that is challenged by long physical distances. In line with Elbamby [46], scalability of VR-services are still limited by factors such as availability of the equipment and high broadband speed among consumers, and technical incompleteness of the equipment. Base on the results of iteration one, it was found out that in marketing use such as in trade shows, it is crucial to keep a continuous flow of users to test the system is crucial for success. When the headset is not in use and it is not possible to observe other using the system beforehand, the setup is easily considered bizarre creating a social barrier to be the first one to test the system. These findings are in line with the notions of Sharafi et al. [47] about modes of negative engagement that are related to the user's avoidance or hesitation because of the user's feeling of lack of skills. Currently, beyond professionals and enthusiastic users, guidance is still often needed to ensure a good user experience. The need for help when using the headset also makes physical contact natural. This can be used to increase trust and to make the experience more enjoyable. Despite the rapidly developing multiplayer features, there are still many technological challenges to be refined, such as haptic feedback and eye contact, before an interaction of avatars in VR feels natural.

We are currently working on how to test new levels of social use context and service scalability by providing the VR enabled smartphones to the land owners so that they can use and familiarize themselves with the application in other social context with their friends and family. With this research, the interplay of scalability, trust and engagement will be further investigated [48]. By giving better tools to a customer for decision-making may increase customer engagement and simultaneously help in scaling the service by empowering the customer(s) to participate to the co-defining the solution. With future research, we are planning to investigate how scaling consultative selling with VR can be adapted to retail sales setting and more specifically furniture sales for built-to-order high-end condominiums. With this study, we are also interested to evaluate the developed artifacts based on their effects on the actual sales performance of VR enabled consultative selling vs. traditional retail store based consultative selling. In other words, we aim to evaluate the artifacts by real sales figures for the client firm.

References

1. Pine, B., Gilmore, J.: The Experience Economy: Work is Theatre & Every Business is a Stage, 1st edn. Harvard Business School Press, Boston (1999)
2. LaValle, S.: Virtual Reality. Cambridge University Press. http://vr.cs.uiuc.edu/ (2017)
3. Lemon, K., Verhoef, P.: Understanding customer experience throughout the customer journey. J. Mark. 80(6), 69–96 (2016)
4. Brodie, R., Hollebeek, L., Juric, B., Ilic, A.: Customer engagement: conceptual domain, fundamental propositions, and implications for research. J. Serv. Res. 14(3), 252–271 (2011)
5. Zhang, J., Lichtenstein, Y., Gander, J.: Designing scalable digital business models. In: Baden-Fuller, C., Mangematin, V. (eds.) Advances in Strategic Management. Emerald Press, Bingley (2015)
6. Heitner, D.: Content Remains the Most Important Aspect of Virtual Reality to Solve. Inc., Technology (2018). https://www.inc.com/darren-heitner/content-remains-most-important-aspect-of-virtual-reality-to-solve.html. Accessed 25 June 2018
7. Forni, A.: Transform Business Outcomes with Immersive Technology. Gartner, 5 May 2017. https://www.gartner.com/smarterwithgartner/transform-business-outcomes-with-immersive-technology. Accessed 25 June 2018
8. Peffers, K., Tuunanen, T., Rothenberger, M., Chatterjee, S.: A design science research methodology for information systems research. J. Manag. Inf. Syst. 24(3), 45–78 (2007)

9. Häyrinen, L., Mattila, O., Berghäll, S., Toppinen, A.: Forest owners' socio-demographic characteristics as predictors of customer value: evidence from Finland. Small Scale For. **14**, 19–37 (2015)
10. Hevner, A., March, S., Park, J., Ram, S.: Design science in information systems research. MIS Q. **28**(1), 75–105 (2004)
11. Nunamaker, J., Briggs, R.: Toward a broader vision for information systems. ACM TMIS **2** (4), 20 (2011)
12. Nunamaker, J., Chen, M., Purdin, T.: Systems development in information systems research. J. Manag. Inf. Syst. **7**(3), 89–106 (1990)
13. March, S., Smith, G.: Design and natural science research on information technology. Decis. Support Syst. **15**(4), 251–266 (1995)
14. Venable, J., Pries-Heje, J., Baskerville, R.: FEDS: a framework for evaluation in design science research. Eur. J. Inf. Syst. **25**(1), 77–89 (2016)
15. Walls, J., Widmeyer, G., El Sawy, O.: Building an information system design theory for vigilant EIS. Inf. Syst. Res. **3**(1), 36–59 (1992)
16. William, D., Black, P.: Meanings and consequences: a basis for distinguishing formative and summative functions of assessment? Br. Educ. Res. J. **22**(5), 537–548 (1996)
17. Saxe, R., Weitz, B.: The SOCO scale: a measure of the customer orientation of salespeople. J. Mark. Res. **19**(3), 343–351 (1982)
18. Liu, A., Leach, M.: Developing loyal customers with a value-adding sales force: examining customer satisfaction and the perceived credibility of consultative salespeople. J. Pers. Sell. Sales Manag. **21**(2), 147–156 (2001)
19. Terho, H., Haas, A., Eggert, A., Ulaga, W.: 'It's almost like taking the sales out of selling'— towards a conceptualization of value-based selling in business markets. Ind. Mark. Manag. **41**, 174–185 (2012)
20. Arthur, W.: Competing technologies: an overview. In: Dosi, G., et al. (eds.) Technical Change and Economic Theory, pp. 590–607. Columbia University Press, New York (1988)
21. Baldwin, C., Clark, K.: Design Rules: The Power of Modularity, vol. 1. MIT Press, Cambridge (2000)
22. Gawer, A.: Bridging differing perspectives on technological platforms: toward an integrative framework. Res. Policy **43**, 1239–1249 (2014)
23. Meyer, C., Schwager, A.: Understanding customer experience. Harv. Bus. Rev. **85**(2), 116–126 (2007)
24. Verhoef, P., Parasuraman, A., Roggeveen, A., Tsiros, M., Schlesinger, L.: Customer experience creation: determinants, dynamics, and management strategies. J. Retail. **85**(1), 31–41 (2009)
25. Otto, J., Ritchie, B.: The service experience in tourism. Tour. Manag. **17**(3), 165–174 (1996)
26. Fredrickson, B., Kahneman, D.: Duration neglect in retrospective evaluations of affective episodes. J. Pers. Soc. Psychol. **65**(1), 45–55 (1993)
27. Schmitt, B.: Customer Experience Management: A Revolutionary Approach to Connecting with Your Customers. The Free Press, New York (2003)
28. Mollen, A., Wilson, H.: Engagement, telepresence and interactivity in online consumer experience: reconciling scholastic and managerial perspectives. J. Bus. Res. **63**(9), 919–925 (2010)
29. Kumar, V., Aksoy, L., Donkers, B., Venkatesan, R., Wiesel, T., Tillmans, S.: Undervalued or overvalued customers: capturing total customer engagement value. J. Serv. Res. **13**(3), 297–310 (2010)
30. Milgram, P., Kishino, F.: A taxonomy of mixed reality visual displays. IEICE Trans. Inf. Syst. **77**(12), 1321–1329 (1994)

31. Steuer, J.: Defining virtual reality: dimensions determining telepresence. J. Commun. **42**(4), 73–93 (1992)
32. Scholz, J., Smith, A.: Augmented reality: designing immersive experiences that maximize consumer engagement. Bus. Horiz. **59**(2), 149–161 (2016)
33. Biocca, F., Levy, M.: Communication in the Age of Virtual Reality. Lawrence Erlbaum Associates, Hillsdale (1995)
34. Smyth, M., Benyon, D., McCall, R., O'Neill, S., Carroll, F.: Patterns of place: an integrated approach for the design and evaluation of real and virtual environments. In: Lombard, M., Biocca, F., Freeman, J., Ijsselsteijn, W., Schaevitz, R. (eds.) Immersed in Media, pp. 237–260. Springer, Cham (2015). https://doi.org/10.1007/978-3-319-10190-3_10
35. Shin, D.: Empathy and embodied experience in virtual environment: to what extent can virtual reality stimulate empathy and embodied experience? Comput. Hum. Behav. **78**, 64–73 (2018)
36. Biocca, F.: The cyborg's dilemma: progressive embodiment in virtual environments. J. Comput. Mediat. Commun. **3**(2), 324 (1997)
37. Mackellar, J.: Participant observation at events: theory, practice and potential. Int. J. Event Festiv. Manag. **4**(1), 56–65 (2013)
38. Singh, K.: Quantitative Social Research Methods. SAGE Publications, Delhi (2007)
39. Babbie, E.: The Practice of Social Research. Wadsworth Publishing Company, Belmont (1989)
40. Kahn, W.: Psychological conditions of personal engagement and disengagement at work. Acad. Manag. J. **33**(4), 692–724 (1990)
41. Hollebeek, L.: Demystifying customer brand engagement: exploring the loyalty nexus. J. Mark. Manag. **27**(7–8), 785–807 (2011)
42. Arminen, I.: Institutional Interaction: Studies of Talk at Work. Routledge, New York (2017)
43. Vivek, S., Beatty, S., Dalela, V., Morgan, R.: A generalized multidimensional scale for measuring customer engagement. J. Mark. Theory Pract. **22**(4), 401–420 (2014)
44. McKnight, D.H., Choudhury, V., Kacmar, C.: Developing and validating trust measures for e-commerce: an integrative typology. Inf. Syst. Res. **13**(3), 334–359 (2002)
45. Hirschman, E., Holbrook, M.: Hedonic consumption: emerging concepts, methods, and propositions. J. Market. **46**, 92–101 (1982)
46. Elbamby, M., Perfecto, C., Bennis, M., Doppler, K.: Toward low-latency and ultra-reliable virtual reality. IEEE Netw. (2018). https://arxiv.org/abs/1801.07587. Accessed 25 June 2018
47. Sharafi, P., Hedman, L., Montgomery, H.: Using information technology: engagement modes, flow experience, and personality orientations. Comput. Hum. Behav. **22**(5), 899–916 (2006)
48. Serban, C., Chen, Y., Zhang, W., Minsky, N.: The concept of decentralized and secure electronic marketplace. Electron. Commer. Res. **8**(1–2), 79–101 (2008)

Designing Value Co-creation with the Value Management Platform

Geert Poels[1]([⊠]) (iD), Ben Roelens[2] (iD), Henk de Man[3],
and Theodoor van Donge[3]

[1] Faculty of Economics and Business Administration, Ghent University,
Tweekerkenstraat 2, 9000 Ghent, Belgium
geert.poels@ugent.be
[2] Faculty of Management, Science and Technology, Open University,
Postbus 9260, 6401DL Heerlen, The Netherlands
ben.roelens@ou.nl
[3] VDMbee, Schietboom 2, 3905TD Veenendaal, The Netherlands
{hdman,tvdonge}@vdmbee.com

Abstract. The Value Delivery Modeling Language (VDML) is an Object
Management Group (OMG) standard for the analysis and design of value creation and value capture in enterprise operations. Although the VDML 1.0
specification was published in October 2015, little is known about the application of value modeling with VDML. We report in this paper (an earlier,
unpublished version of this paper was presented at the 12th International
Workshop on Value Modeling and Business Ontologies, which was held in
Amsterdam on 26–27 February 2018) on the practice of applying VDML for
value co-creation design using the Value Management Platform (VMP) tool of
the Dutch company VDMbee. Neither the VMP user guide nor the VDML
specification prescribe how to perform value modeling. Therefore, we analyze
value co-creation design with the VMP in a case-study of a low-cost carrier. By
identifying, extracting, and making explicit the applied method of value co-
creating design, we contribute to a better understanding of the practice of value
modeling with VDML.

Keywords: Value modeling · Value co-creation design
Value Delivery Modeling Language · Value Management Platform

1 Introduction

A value model shows how value is co-created and delivered in a network of actors
(e.g., a supply chain, a consumer market, a smart grid, a healthcare system). As such,
value models are used for strategic analysis and design of value networks. Since the
early 2000's, different value modeling approaches have been proposed, each with a
specific emphasis (e.g., value exchange, value impact and value creation analyses [1],
e-service design [2]). The Value Delivery Modeling Language (VDML) [3], which
integrates business model concepts into value modeling, has been adopted as a standard
by the Object Management Group (OMG) [4]. Tool support for VDML is offered by

© Springer Nature Switzerland AG 2018
G. Satzger et al. (Eds.): IESS 2018, LNBIP 331, pp. 399–413, 2018.
https://doi.org/10.1007/978-3-030-00713-3_30

the Dutch company VDMbee [5]. VDMbee's Value Management Platform (VMP) provides a user-friendly tool for working with VDML using different kinds of canvas/map templates and storytelling/mapping techniques for model building. The transparent creation of VDML meta-model instantiations via visual interfaces targets value management professionals with a business-oriented profile.

A complete modeling approach requires a modeling procedure which guides the creation and analysis of models [6]. Currently, there is no research on how to apply VDML as related research focuses on ontological analysis of the conceptualization of value [7–10], VDML extensions and analysis techniques for applications like business model analysis [11, 12], compliance engineering [13] and reputation systems design [14], and integration of VDML into enterprise architecture modeling approaches (e.g., ArchiMate [15, 16]). Regarding the method underlying the use of the VMP, an overview can be found in [17], which is a VDMbee blog post that is deliberately high-level and introductory. Apart from largely anecdotic evidence of (showcase) VDML applications (e.g., case-studies reports [5]), knowledge of the practice of value modeling with VDML is largely tacit. This knowledge is embedded in the VMP code and in its documentation, video tutorials and other training materials, not forgetting the tacit knowledge 'embodied' by the tool developers themselves.

The goal of this paper is to make this tacit knowledge explicit. We intend to contribute procedural knowledge (i.e., how to?) of value co-creation design by making the method underlying the use of the VMP explicit. By describing such method, it can be investigated whether the use of the VMP results in a more complete value modeling approach for VDML. Further, by demonstrating the VMP-supported approach as a transparent practice of VDML, we hope to help increasing the maturity of value modeling practice and boosting the adoption of VDML.

To identify and extract this method, we follow the Design Science Research Method for Information Systems Research [18]. In particular, we conducted client/context-initiated design-based research. According to [18, p. 56] "a client/context-initiated solution may be based on observing a practical solution that worked; it starts with activity 4 [*Demonstration: find suitable context & use artifact to solve problem*], resulting in a D[*esign*]S[*cience*] solution if researchers work backward to apply rigor to the process retroactively". In our case, the client is VDMbee that asked us for help to make tacit knowledge about their method explicit. The context is the application of the VMP in a low-cost carrier (LCC) case-study. This case-study was developed by the VDMbee Academy to be used for training value management professionals in the use of the VMP. VDMbee provided us access to the case-study documentation and data (i.e., VMP files created) and the two VDMbee consultants who were involved in the development of the case-study, joined the research team. Their participation greatly helped us to reconstruct the development process, while ensuring the correct interpretation of the case-study materials.

The evaluation of the extracted method is outside the scope of the paper. This is also only a first iteration of retroactively applying rigor to the method design process as we focus first on describing the method underlying the use of the VMP as observed from a particular instantiation (i.e., the LCC case-study). Consequently, the maturity level of the knowledge contribution of our research is still low, being of the type 'situated implementation of artifact' [19].

The paper is structured as follows. The second section of the paper introduces the LCC case-study. The third section analyses how the VMP was used in the case-study to identify and extract the underlying method. The fourth section discusses the most distinctive features of the method, the limitations of the research, and the next steps in research. The final section concludes by outlining our contribution and its implications for practice and research.

2   Case-Study

The LCC studied differentiates itself from a full-service carrier through the adoption of the 'no-frills' business model pattern [20]. Passengers can buy cheap tickets, but basically any other service must be bought as add-ons at a premium price. The ancillary revenues generated by the add-ons combined with a cost-cutting focus allows keeping ticket prices low. Apart from low costs, other key values of the LCC under study are environmental sustainability and passenger satisfaction, which is determined by ticket prices, promptness and the variety of destinations within the European Union that are served.

Despite the success of the current business model, the LCC's management team realizes that action is required to cope with emerging challenges and threats, including a worsening reputation (in terms of lack of customer service quality and bad treatment of personnel), projected market (share) growth (necessitating investments in hundreds of new aircrafts), increased competition by so-called ultra-low-cost carriers, and the Brexit (potentially resulting in a reversing of deregulation and liberalization for flights to and from the UK). Several ideas for business model innovation are actively being pursued, including operating long-haul flights (e.g., inter-Atlantic flights), setting up short-haul operations in the Middle-East, and offering connecting flights (i.e., feeder lines) for long-haul airlines. Also, several alternatives of the current business model are being explored, to find the best basis for future growth (e.g. lowering of fares, leasing of planes).

To guide this strategic thinking and foster new value co-creation design initiatives, the VMP was used. Using the VMP, business model innovation is moved from ideation to prototyping, allowing managers to take informed decisions on the adoption and implementation of new/modified business models and on the phasing of the business transformation. The VMP informs managerial decision-making by means of prognoses and analyses of scenarios regarding the value impact of the continuation of current business models and/or adoption of new business models.

3   Value Co-creation Design Method

A first subsection presents the conceptual framework of strategic planning on which the development of the VMP was based. Using this framework, the value co-creation design process performed for the LCC under study was analyzed. The second subsection looks into what activities were performed, for which purpose, in what order, and how they were organized as to who is involved and where they took place. The third subsection then looks specifically at the different techniques of the VMP tool that support value co-creation design activities.

3.1   Conceptual Framework for Value Co-creation Design

VDML integrates business model concepts into value modeling. When using the VMP, the business model is considered as the elementary unit of strategic planning. Given the vision and goals of an enterprise, a strategic plan is devised that guides the efforts of the enterprise towards the achievement of objectives that quantify the enterprise goals. The objectives that VDML focuses on are those pertaining to value delivery, where value is defined as *a measurable factor of benefit, of interest to a recipient* [4].

Objectives can be defined per phase in the strategic plan. For each phase, alternatives can be defined to explore different approaches for achieving the phase's objectives. Strategies are incorporated into the plan as business models for phases/alternatives that define how the enterprise operates in each phase (i.e., business model evolution) and alternative (i.e., business model variation). The strategic plan is thus decomposed in a set of interacting business models, together covering all enterprise operations, both customer-facing and internal.

It is important to note that when using the VMP, a business model is not just considered as the formulation of a product-market strategy (e.g., product differentiation, cost leadership). A business model is seen as the blueprint of a value proposition and activity system used to deliver value to customers [21]. The activity system is formulated in terms of customer and partner relationships, business activities and required competencies. As such, it is the design of a value co-creation system. Hence, strategic planning using the VMP entails the design of value co-creation networks.

In the case-study, the VMP was used to define a strategic plan for the LCC, distinguishing three phases: current situation (year 0), growth in year 1, and growth in year 2. The growth phases take into account both market growth (in terms of prognoses of tickets sold) and the projected growth of LCC's business. For the growth in year 1 phase, three alternatives were defined: unchanged policy, lower fares, and lower fares & partial lease. The first alternative concerns a slight increase in market size and market share, without change of operating model. The second alternative explores a significant increase in market share as a result of dramatic lowering of ticket prices, but also necessitating a significant increase in fleet size. The third alternative builds on the second one, but reflects also a change in LCC's policy of fleet ownership by considering leasing planes instead of buying them.

3.2   Activities

When looking into the activities performed, Fig. 1 shows that in the LCC case-study, the VMP was used in three different stages. These stages (i.e., Discover, Prototype and Adopt) provide a high-level structuring of the value co-creation design method.

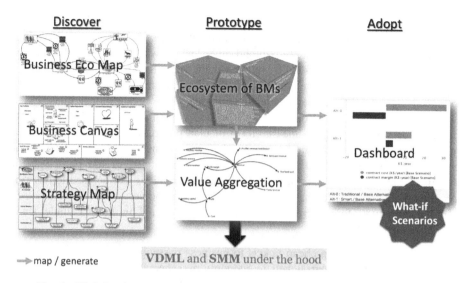

Fig. 1. High-level overview of strategic planning with the VMP (based on [17]).

Discover Stage. The purpose of the Discover stage is the discovery of the As-Is and To-Be business models to be further elaborated in the Prototype stage. The discovery of business models to be described, analyzed, experimented, innovated and evaluated across the defined phases and alternatives is done in a collaborative workshop involving the stakeholders in the value co-creation design initiative. These stakeholders include a value management analyst (that facilitates the workshop) and one or more participants of the end-user organization that are responsible for strategic planning (possibly with strategic decision-making authority) and possibly participants that are subject matter experts in the end-user organization or from other organizations in the ecosystem of the end-user organization.

Preparation: Scope + Objectives + Participants		
Session 1	Ecosystem	3 hours
Session 2	Business Model Canvas of key participants	3 hours
Session 3	Values & cause effect	3 hours
Session 4	Alternatives & phasing	3 hours
Prototyping	Sprint 1 Sprint 2 Sprint 3	2-3 weeks

Fig. 2. Organization of the strategic planning engagement (taken from [17]).

The workshop is organized in four sessions, each with their own objectives (see Fig. 2). In the first session, the business ecosystem is sketched and key participants, from whose perspective business models will be discovered, are identified. In the second session, the business models of these key participants are described and related to each other, consistent with the ecosystem. In the third session, the values that are the subject of objectives are decided upon and other values, which influence them or are influenced by them, are identified by relating them through cause-and-effect relations. In the fourth session, the strategic plan is outlined or, if already available in case of a continued engagement, further extended with one or more additional phases. The values that are the subject of objectives are defined as plan values, i.e., values that are the basis for management and measurement of plan progress and the success of plan outcomes, while the other values are related to the business models into which the strategic plan is decomposed. Phases in the plan are defined to add phase-specific objectives for the plan values. Furthermore, phase alternatives can be used to describe scenarios that analyze risks, assumptions, and strategic choices.

In the case-study, four interrelated business models owned by the LCC under study were discovered:

- Business plus: Offering cheap tickets to business passengers;
- Economy: Offering cheap tickets to economy passengers;
- Flights: Internal business model for offering of the fleet and flights to the LCC Travel business unit by the LCC Operations business unit;
- Shops: In-flight offering of customers to duty free companies.

Prototype Stage. The purpose of this stage is to develop a multi-perspective business model ecosystem by further elaborating the interrelated business models from the Discover stage, for each of the phases and alternatives in the strategic plan. This allows comparing plan values and business model values across phases and alternatives to gauge the effectiveness of the business ecosystem design and business model innovation, and to decide upon the most appropriate course of action.

The business model concept employed by the VMP is inspired by Lindgren's Business Model Cube [22]. The conceptualization of a business model as a cube implies that there are six faces:

- *Value propositions* offered and received, including *my propositions* (i.e., results of the business model as captured by the owning enterprise, also known as 'the business');
- *Customers* as business ecosystem participants that are served by the business – this determines the main purpose of the business model;
- *Partners* as business ecosystem participants that are involved in the business model to help create the values to be delivered to the customers;
- *Activities* as work performed by participants in a role (i.e., partner roles, customer roles, roles of the business) and part of value streams that pursue value propositions;
- *Competencies* as capabilities and resources that the business has and applies in order to perform the work represented by activities;
- *Values* as benefits or interests to customers and partners (in value propositions offered), as captured by the business (in my propositions), as qualifying customer rewards or partner offerings (in value propositions received), and as internal (created by activities).

In the case-study, prototyping the discovered business models for all phases and alternatives of the strategic plan involved four steps:

1. *Value network design:* Designing participant networks by defining participants, their roles, value propositions exchanged (i.e., offered and received) and their values – pertaining to value propositions, customers, partners and values faces of the business model cubes;
2. *Value stream design:* Designing value streams by defining activities that pursue value propositions, the participant roles that perform those activities, the values they create, and the values of value propositions and my propositions they contribute to – pertaining to activities and value faces of the business model cubes;
3. *Competency design:* Identifying the capabilities and resources that are needed to perform the work represented by the activities – pertaining to the competencies face of the business model cubes;
4. *Value impact design & measurement:* Designing the value aggregation structure by entering value measurements and value formulas that relate business model values (i.e., activity values, value proposition values and my proposition values) and plan values, within and if relevant across plan phases – pertaining to value propositions, activities, and values faces of the business model cubes and to plan values.

As shown in Fig. 2, the Prototype stage is performed using an agile approach in two or three weekly sprints. The prototyping is done by the value management analyst, based on the input received during the Discover stage. At the end of each sprint, feedback is obtained from the stakeholders of the involved organizations. This cycle continues until these stakeholders are satisfied with the results and are able to make decisions in the Adopt stage.

Adopt Stage. The purpose of this stage is to present the prototyping results to strategic decision-makers, allowing them to decide on adoption and initiation of the required changes. Value management professionals support the decision-making process by using the built-in dashboard, reporting, and what-if scenario analysis techniques of the VMP (see next subsection).

3.3   Techniques

The techniques used in the CBMP activities are demonstrated for the LCC case-study in the order of the process stages presented in the previous subsection.

Discover Stage. The sketching of the business ecosystem and the identification of key participants is supported by the VMP through the *business ecosystem map* (see Fig. 3), in which participants and exchanged value propositions are given pictorial representations. The business ecosystem map is based on Allee's Value Network concept [1], which is a technique used in designing value co-creation networks, allowing the analysis of value exchange, value impact and value creation.

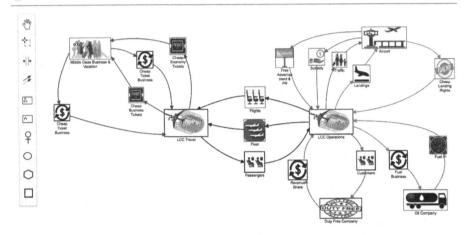

Fig. 3. VMP screenshot showing the LCC business ecosystem map.

In the LCC case-study, six business ecosystem participants were identified, from whom the LCC Travel and LCC Operations business units are key participants. These participants exchange sixteen value propositions in the ecosystem, based on which the four aforementioned business models were discovered.

The VMP supports the description of the key participants' business models through business canvasing techniques. Although different types of business canvases are supported, in the LCC case-study only use was made of the widely known *business model canvas*, which is based on Osterwalder's Business Model Ontology [23].

Figure 4 shows as an example the business model canvas for the 'flights' business model, which is owned by the LCC Operations business unit. Based on the business ecosystem map, which resulted from the first workshop session (see Fig. 3), the value management analyst had pre-filled LCC travel as the customer, airports and oil companies as partners, and fleet and flights as value propositions offered to the customer. In the second workshop session, workshop participants used the canvas to systematically think about key activities and resources required to pursue the value propositions, their costs, and the revenue streams that will be generated.

To support the identification of plan values and business model values and how values influence each other, *strategy maps* are used as a storyboard for cause-and-effect value creation. This technique is based on the homonymous technique presented by Kaplan and Norton [24].

Figure 5 shows an example strategy map for the case-study, which specifies how competencies (e.g., single model aircraft) and activities (e.g., ticketing without airport check-in) influence values for the LCC (e.g., low overhead) and for its customers (e.g., high promptness).

Finally, the last workshop session is supported through the VMP's functionality to model phases and alternatives and to enter plan values (which can for large part be derived from the strategy maps and business canvases).

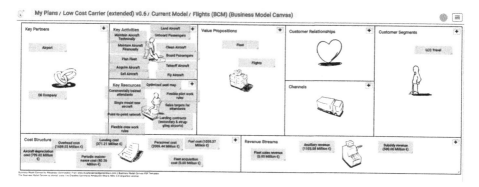

Fig. 4. VMP screenshot showing the LCC operation's 'flights' business model canvas.

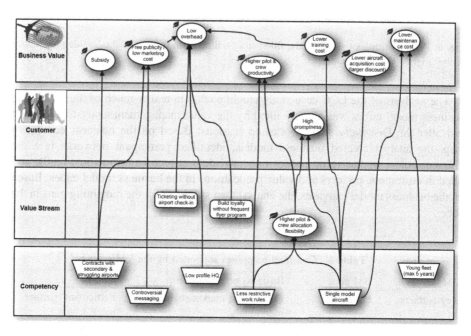

Fig. 5. VMP screenshot showing one of the strategy maps used in the LCC case-study.

Prototype Stage. The VMP provides business-friendly interfaces that employ story-telling as a technique to fill the business model cubes throughout the activities of value network design, value stream design, competencies design, and value impact design. Figure 6 shows an example of a story-telling form for the 'flights' value proposition in the 'flights' business model cube, asking for who (and in what role) offers this value proposition to whom (and in what role), delivering what values.

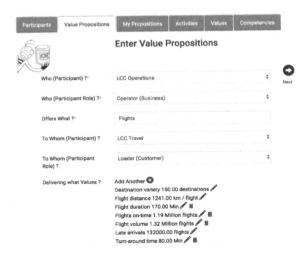

Fig. 6. VMP screenshot showing the filled story-telling form for the 'flights' value proposition in the 'flights' business model.

The analysis of the LCC case-study taught us that in reality much of the data in the business model cubes was already filled by the value management analyst in between and after the Discovery stage workshop sessions. Based on the business ecosystem map, the analyst created business models, identified participant networks (see the different colors of the value proposition provide and receive arrows in Fig. 3), and filled in customers, partners and value propositions in the business model cubes. Based on the business model canvases, the analyst then started to fill the remaining gaps in the cubes.

Table 1. Construct mappings supported by the VMP

BM canvas	BM cube	Business ecosystem map	VDML
Key partners	Partners	Enterprises, market segments, individuals	Participant (partner role)
Key activities	Activities		Activities
Key resources	Competencies		Resources
Value propositions	Value propositions	Value propositions	Value propositions
Customer segments	Customers	Enterprises, market segments, individuals	Participants (customer role)
Cost structures	Values		Value Elements
Revenue streams	Values		Value Elements

To facilitate this process, the VMP includes a mapping wizard, which provides the user an interactive mapping functionality back and forth between business ecosystem map elements and business model cube elements and between the slips on a business canvas (shown as post-it notes in Fig. 4) and business model cube elements (see Table 1 for a mapping that holds for the business model canvas). Using this mapping functionality, the structured models used in the Prototype stage can largely be generated from the graphical models used in the Discover stage. Vice-versa, a large amount of the data in the business model cubes can be visually represented in the business ecosystem map and business canvases. The mapping thus affords a great deal of flexibility in the value co-creation design process, providing two-way traceability and allowing to move back and forth between Discover and Prototype stages, and choosing between top-down, bottom-up or hybrid approaches to strategic planning.

Finally, the competencies, activities and values in the strategy map and the information on plan phases and alternatives were used by the analyst in the Prototyping sprints to further elaborate the business model cubes (e.g., entering value formulas and other measurement-related detail to complete their value aggregation structures).

Adopt Stage. The VMP offers 'dashboarding' functionality for comparing values across plan phases and alternatives. First, a report, as another type of model in the VMP, can be created to tell the story of the plan. The reporting functionality comes with an embedded rich text editor and supports the direct incorporation of Discover stage diagrams such as business ecosystem maps and business canvases. Second, interactive dashboards can be generated from the information in the strategic plan presenting comparisons of values using tables and various types of charts. Dashboards can be extended with the creation of scenarios for what-if analyses and simulation, by entering different sets of measurements for selected input values. Scenario results can then be presented in the dashboard, for comparison reasons, as well. It is also possible to promote a 'best' scenario to update the plan. Value measurements can be imported from csv-files and exported as csv-files or as xlsx-files (for further analysis in Excel). Using the import functionality, actual values can also be compared with values in the plan, typically based on a time line view in the dashboards. This functionality is essential for monitoring plan implementation, as part of the strategic planning process.

4 Discussion

The analysis of the LCC case-study taught us that the approach to value modeling with VDML using the VMP has some distinctive features. Probably the most distinctive feature is that value models are constructed through a completely transparent use of VDML. Value management professionals and other strategic planning stakeholders are offered business-friendly interfaces (e.g., business canvases, business ecosystem maps, strategy maps, business model cubes, interactive dashboards), without ever having to work directly with the modeling language VDML. In other words, no VDML knowledge is required to use the method. Nevertheless, all business model instances stored in the VMP are valid VDML value delivery models.

Another distinctive feature is that value modeling is used for strategic planning, which according to the assumed conceptual framework entails the design of value co-creation networks in different plan phases and possibly for different phase alternatives. Specifically, the discovered method regards the business model as the unit for strategic planning, employing a business model framework that is multi-perspective (resulting in an ecosystem of interrelated business models), considers structured relationships in terms of value proposition exchanges between all ecosystem participants (and not just customer value propositions and customer relationships), uses a uniform, unambiguous concept of value (for all participants), and is highly dynamic (allows for a continuous process of strategic planning). Based on the review of strategic planning theories and models in [25], we characterize the VMP-supported approach to strategic planning as assuming a *goal-based strategic model* (i.e., setting objectives aligned with mission and vision, and phased over time) and to some extent also a *scenario planning model* (i.e., exploring different alternatives per phase in the strategic plan to cope with external forces). The goal-based model is the most adopted model of strategic planning [25], which provides some justification to the design rationale of the method underlying the use of the VMP.

Of course, more research is needed on the design of the VMP-based method for value co-creation design. Our current client/context-initiated design-based research is not without limitations. First, the method as presented in this paper was the result of the analysis of a single case-study. Furthermore, this case did not result from a real-life application, but was developed for training value management professionals in the use of the VMP. To mitigate the threat that our analysis is not valid or generalizable, the two consultants from VDMbee who developed the case-study, were involved in the research and helped the researchers to interpret the case-study data. These value management experts have used the VMP in numerous real-life projects and used this experience to develop the case-study, which is not fictional but based on real data about Ryanair that is publicly available on the Internet. Nevertheless, we acknowledge that to raise the level of knowledge of the value co-creation design method to that of a design theory [19], we need to investigate more in-depth its theoretical underpinning and also evaluate the method.

Since we analyzed the LCC case-study in December 2017, the VMP tool has been further developed. The tool now includes additional techniques which can further support the strategic planning, aka value co-creation design process. The most notable additions are value stream maps and capability maps, which provide for an explicit visualization of value streams and competencies (Fig. 7). In the near future, VDMbee intends to add views for showing differences between As-Is and To-Be in a user-friendly way, such that specific differences between previous and next phases can be framed into requirements for projects to implement the plan. To support this development, we plan to conduct research on the integration of VMP-based value co-creation design with methods of capability-based planning, portfolio management, enterprise architecture management, business process management, change management, and strategic sourcing.

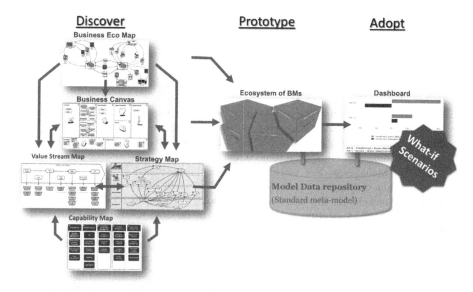

Fig. 7. Extension of the Discover stage with value stream maps and capability maps.

5 Conclusion

This paper presents an account of the practice of value modeling with VDML by means of the specific approach taken by value management analysts when using the *Value Management Platform* (VMP) tool of the company VDMbee. The contribution of this paper is to make the tacit knowledge of this approach to strategic planning, aka value co-creation design, which is currently only described at an introductory level [17], explicit by identifying and extracting the method underlying the use of the VMP. By analyzing the case-study of a low-cost carrier, we were able to describe a 'situated implementation' of the value co-creation design method artifact, in terms of purpose, process, activities, organization, participants, models, and techniques used.

Our research addresses the knowledge gap of 'how' to apply VDML as currently only anecdotic evidence is publicly available. We hope this paper helps furthering the understanding of value modeling with VDML of both value management professionals and value modeling researchers. By presenting the VMP-based method of value co-creation design, we intend to increase the maturity level of value modeling and raise the interest of practitioners and researchers in further exploring and researching this method.

References

1. Allee, V.: Value network analysis and value conversion of tangible and intangible assets. J. Intellect. Cap. **9**, 5–24 (2008)
2. Gordijn, J., Yu, E., van der Raadt, B.: E-service design using i* and e^3 value modeling. IEEE Softw. **23**, 26–33 (2006)
3. Cummins, F.A.: Building the Agile Enterprise: With Capabilities, Collaborations and Values. Morgan Kaufmann, Burlington (2016)
4. Object Management Group: Value Delivery Metamodel, Version 1.0. OMG (2015)
5. VDMbee Homepage. http://www.vdmbee.com
6. Frank, U.: Multi-perspective enterprise modeling: foundational concepts, prospects and future research challenges. Softw. Syst. Model. **13**, 941–962 (2014)
7. Blums, I., Weigand, H.: Towards a reference ontology of complex economic exchanges for accounting information systems. In: Proceedings of the 20th International Conference on Enterprise Distributed Object Computing (EDOC), pp. 1–10. IEEE (2016)
8. Gailly, F., Roelens, B., Guizzardi, G.: The design of a core value ontology using ontology patterns. In: Link, S., Trujillo, J.C. (eds.) ER 2016. LNCS, vol. 9975, pp. 183–193. Springer, Cham (2016). https://doi.org/10.1007/978-3-319-47717-6_16
9. Guarino, N., Andersson, B., Johannesson, P., Livieri, B.: Towards an ontology of value ascription. In: Proceedings of the 9th International Conference on Formal Ontology in Information Systems (FOIS), p. 331. IOS Press (2016)
10. Sales, T.P., Guarino, N., Guizzardi, G., Mylopoulos, J.: An Ontological analysis of value propositions. In: Proceedings of the 21st International Conference on Enterprise Distributed Object Computing (EDOC), pp. 184–193. IEEE (2017)
11. Roelens, B., Poels, G.: The development and experimental evaluation of a focused business model representation. Bus. Inf. Syst. Eng. **57**, 61–71 (2015)
12. Metzger, J., Kraemer, N., Terzidis, O.: A systematic approach to business modeling based on the value delivery modeling language. In: Berger, Elisabeth S.C., Kuckertz, A. (eds.) Complexity in Entrepreneurship, Innovation and Technology Research. FSSBE, pp. 245–266. Springer, Cham (2016). https://doi.org/10.1007/978-3-319-27108-8_12
13. Kiriinya, R.K.M.: Designing compliance patterns: integrating value modeling, legal interpretation and argument schemes for legal risk management. Ph.D. dissertation, University of Luxembourg, Luxembourg (2017)
14. Bettini, L., Capecchi, S.: VDML4RS: a tool for reputation systems modelling and design. In: Proceedings of the 8th International Workshop on Social Software Engineering, pp. 8–14. ACM (2016)
15. Ding, H.: Integrating value modelling into ArchiMate. Thesis, University of Twente, Enschede
16. Lankhorst, M.M., Aldea, A., Niehof, J.: Combining ArchiMate with other standards and approaches. Enterprise Architecture at Work. TEES, pp. 123–140. Springer, Heidelberg (2017). https://doi.org/10.1007/978-3-662-53933-0_6
17. Continuous business model planning with VDMbee. https://vdmbee.com/2017/12/
18. Peffers, K., Tuunanen, T., Rothenberger, M.A., Chatterjee, S.: A design science research methodology for information systems research. J. Manag. Inf. Syst. **24**(3), 45–77 (2008)
19. Gregor, S., Hevner, A.R.: Positioning and presenting design science research for maximum impact. MIS Q. **37**(2), 337–355 (2013)
20. Casadesus-Masanell, R., Ricart, J.E.: From strategy to business models and onto tactics. Long Range Plan. **43**, 195–215 (2010)

21. Zott, C., Amit, R., Massa, L.: The business model: recent developments and future research. J. Manag. **37**(4), 1019–1042 (2011)
22. Lindgren, P., Rasmussen, O.H.: The business model cube. J. Multi Bus. Model Innov. Technol. **1**(3), 135–182 (2013)
23. Osterwalder, A.: The business model ontology: a proposition in a design science approach. Ph.D. dissertation, University of Lausanne, Lausanne (2004)
24. Kaplan, R.S., Norton, D.P.: Strategy Maps: Converting Intangible Assets into Tangible Outcomes. Harvard Business Press, Boston (2004)
25. Azevedo, C.: Incorporating enterprise strategic plans into enterprise architecture. Ph.D. dissertation, University of Twente and Federal University of Espírito Santo, Enschede (2017)

Author Index

Alam, Intekhab (Ian)   72
Alt, Rainer   45
Amorim, Marlene   31, 289
Anton, Florin   358
Anton, Silvia   358
Axjonow, Alexander   261

Bader, Sebastian R.   165
Badr, Nabil Georges   316, 344
Beirão, Gabriela   303
Benz, Carina   88, 101, 112
Blöcher, Katharina   45
Borangiu, Theodor   358
Breitner, Michael H.   261

Caputo, Francesco   151
Carrubbo, Luca   151
Costa, Humberto   303

de Man, Henk   399
De Marco, Marco   316
Drăgoicea, Monica   344

Eilers, Dennis   261
Enders, Tobias   274
Engel, Christian   219

Falcão e Cunha, João   344
Feldmann, Niels   88
Floerecke, Sebastian   193

Götz, Caroline   101

Hagen, Simon   59
Herrmann, Anne   208
Hohler, Sophie   101
Holopainen, Jani   385
Hunke, Fabian   219
Husmann, Marco   177

Iacob, Iulia   358

Jannaber, Sven   59
Jussen, Philipp   177

Kampker, Achim   177
Knöll, Florian   127, 235
Kwan, Stephen K.   330

Laubis, Kevin   127
Lubarski, Aleksander   16

Mattila, Osmo   385
Megaro, Antonietta   151
Meierhofer, Jürg   208
Melão, Nuno   31

Olivotti, Daniel   261
Oltean, Virginia Ecaterina   344

Parvinen, Petri   385
Passlick, Jens   261
Patrício, Lia   373
Poels, Geert   399
Polese, Francesco   151
Pombo, Pedro   289

Raileanu, Silviu   358
Reinerth, Volkmar   138
Reis, João   31
Roelens, Ben   399
Rosa, Maria João   289

Saghezchi, Fatemeh Bashashi   289
Schmidt, Jan-Peter   165
Schwerdt, Laura   177
Seebacher, Stefan   112
Shapoval, Katerina   235
Siegel, Jörg   88
Simko, Viliam   127
Sorrentino, Maddalena   316
Stoffer, Torben   3

Teixeira, Jorge Grenha   373
Thomas, Oliver   59
Tuunanen, Tuure   373, 385

van Donge, Theodoor   399
Voessing, Michael   247
Vössing, Michael   88, 138, 165

Weissinger, Reinhard   330
Widjaja, Thomas   3
Wolff, Clemens   138, 165, 247
Wuest, Thorsten   88

Zacharias, Nicolas   3
Zeidler, Verena   127